SIMULATION APPROACHES IN TRANSPORTATION ANALYSIS

Recent Advances and Challenges

OPERATIONS RESEARCH/COMPUTER SCIENCE INTERFACES SERIES

Series Editors

Professor Ramesh Sharda
Oklahoma State University

Prof. Dr. Stefan Voß
Universität Hamburg

Other published titles in the series:

Greenberg / *A Computer-Assisted Analysis System for Mathematical Programming Models and Solutions: A User's Guide for ANALYZE*
Greenberg / *Modeling by Object-Driven Linear Elemental Relations: A Users Guide for MODLER*
Brown & Scherer / *Intelligent Scheduling Systems*
Nash & Sofer / *The Impact of Emerging Technologies on Computer Science & Operations Research*
Barth / *Logic-Based 0-1 Constraint Programming*
Jones / *Visualization and Optimization*
Barr, Helgason & Kennington / *Interfaces in Computer Science & Operations Research: Advances in Metaheuristics, Optimization, & Stochastic Modeling Technologies*
Ellacott, Mason & Anderson / *Mathematics of Neural Networks: Models, Algorithms & Applications*
Woodruff / *Advances in Computational & Stochastic Optimization, Logic Programming, and Heuristic Search*
Klein / *Scheduling of Resource-Constrained Projects*
Bierwirth / *Adaptive Search and the Management of Logistics Systems*
Laguna & González-Velarde / *Computing Tools for Modeling, Optimization and Simulation*
Stilman / *Linguistic Geometry: From Search to Construction*
Sakawa / *Genetic Algorithms and Fuzzy Multiobjective Optimization*
Ribeiro & Hansen / *Essays and Surveys in Metaheuristics*
Holsapple, Jacob & Rao / *Business Modelling: Multidisciplinary Approaches — Economics, Operational and Information Systems Perspectives*
Sleezer, Wentling & Cude/*Human Resource Development And Information Technology: Making Global Connections*
Voß & Woodruff / *Optimization Software Class Libraries*
Upadhyaya et al / *Mobile Computing: Implementing Pervasive Information and Communications Technologies*
Reeves & Rowe / *Genetic Algorithms—Principles and Perspectives: A Guide to GA Theory*
Bhargava & Ye / *Computational Modeling And Problem Solving In The Networked World: Interfaces in Computer Science & Operations Research*
Woodruff / *Network Interdiction And Stochastic Integer Programming*
Anandalingam & Raghavan / *Telecommunications Network Design And Management*
Laguna & Martí / *Scatter Search: Methodology And Implementations In C*
Gosavi/ *Simulation-Based Optimization: Parametric Optimization Techniques and Reinforcement Learning*
Koutsoukis & Mitra / *Decision Modelling And Information Systems: The Information Value Chain*
Milano / *Constraint And Integer Programming: Toward a Unified Methodology*
Wilson & Nuzzolo / *Schedule-Based Dynamic Transit Modeling: Theory and Applications*
Golden, Raghavan & Wasil / *The Next Wave In Computing, Optimization, And Decision Technologies*
Rego & Alidaee/ *Metaheuristics Optimization Via Memory and Evolution: Tabu Search and Scatter Search*

SIMULATION APPROACHES IN TRANSPORTATION ANALYSIS

Recent Advances and Challenges

edited by

Ryuichi Kitamura
Masao Kuwahara

Springer

Ryuichi Kitamura
Kyoto University
Japan

Masao Kuwahara
University of Tokyo
Japan

Library of Congress Cataloging-in-Publication Data
A C.I.P. Catalogue record for this book is available
from the Library of Congress.

ISBN 978-1-4419-3690-5
e-ISBN 978-0-387-24109-8

Printed on acid-free paper.

9 8 7 6 5 4 3 2 1

springeronline.com

CONTENTS

Part III: Representing Traffic Dynamics

Part IV: Representing User Behavior

SIMULATION APPROACHES IN TRANSPORTATION ANALYSIS: Recent Advances and Challenges

Ryuichi Kitamura and Masao Kuwahara

Preface

Achieving efficient, safe, and convenient urban automotive transportation has been the primary concern of transportation planners, traffic engineers, and operators of road networks. As the construction of new roadways becomes increasingly difficult and, at the same time, as the adverse environmental impacts of automotive traffic are more critically assessed, and as the depletion of fossil fuels and global warming loom as serious problems, it is now imperative that effective traffic control strategies, demand management schemes and safety measures be expeditiously implemented.

The advent of advanced information and telecommunications technologies and their application to transportation systems have expanded the range of options available in managing and controlling network traffic. For example, providing individualized real-time information to drivers is now almost reality. Evolving Intelligent Transport Systems (ITS) technology is making it possible to link the driver, vehicle and road system by exchanging information among them, calling for the development of new traffic control strategies. It is in this context that transport simulation is emerging as the key concept in traffic control and demand management.

Motivated by this line of thought, the International Symposium on Transport Simulation was held in Yokohama, Japan, in August 2002. It aimed at providing a forum where groups of researchers who are engaged in cutting-edge research in transport simulation would gather from all over the world, and exchange their results, discuss research issues, and identify future directions of development. Specifically, it was envisaged that the Symposium would contribute to the development and application of transport simulation methods by

- introducing state-of-the-art transport simulation models, methodologies, and examples of their applications,
- identifying short-term and long-term research issues,
- assessing promising application areas for simulation, and
- evaluating the applicability of transport simulation methods to other research areas.

It was also hoped that this Symposium would aid in establishing a worldwide network of researchers involved in transport simulation.

A total of 19 papers were presented at the Symposium. While some were concerned with specific simulation model systems and their application (most of the major simulation model systems available were represented at the Symposium), others addressed various research issues in transport simulation. This volume contains a set of papers selected from those presented at the Symposium.

This book is divided into four parts. Part I comprises four papers that represent simulation models for dynamic network assignment. The first paper by Florian *et al.*, which was the keynote speech at the Symposium, offers a concise review of existing dynamic network simulation models and an overview of issues involved in simulation-based dynamic traffic assignment. A successful application example to a medium-size network is then presented. The second paper by Liu presents the dynamic traffic network simulator, DRACULA, whose unique features include the representation of day-to-day variation in demand and drivers' learning. The following paper by Barcelo and Casas presents the microscopic traffic simulator, AIMSUN, describes car following, lane changing and other elements of the model system, and discusses its application to dynamic network simulation. The last paper of Part I by Toledo *et al.* describes the microscopic traffic simulation tool, MITSIMLab, detailing its representation of driver behavior, touching on calibration and validation issues, and demonstrating a Stockholm, Sweden, case study results.

Part II contains three papers that are concerned with the development and application of transport simulation. The first paper by Horiguchi and

Kuwahara assesses the status of simulation model application in Japan, describes the ongoing effort toward standardized model verification and validation, and report on the establishment of a forum for information exchange in Japan. The second paper by Takayama and Nakayama discusses the estimation of an origin-destination matrix in conjunction with network simulation, where paths on the network are assumed to follow an absorbing Markov chain process, whose parameters are estimated using genetic algorithms. In the third paper by Asakura *et al.*, mobile communications technology is applied to acquire space-time trajectory data from cellular phone holders, and the resultant data are applied to simulate how the participants of a sports event disperse and travel toward their respective destinations after the event is over.

Estimating and representing the dynamics of traffic flow and individual vehicle movement is the common concern of the four papers in Part III. In the first paper, Schreckenberg *et al.* attempt to combine, as "an online-tool," information from real-time traffic data from detectors and results of microscopic traffic simulation to determine the state of traffic throughout the German autobahn network. Young and Weng discuss issues involved in simulating parking in urban area, addressing the microscopic representation of vehicle movement in parking facilities, drivers' decision making including route choice, and the interaction among data collection, model accuracy and model validity. The third paper by Akahane *et al.* offers a historical summary and assessment of the development of traffic simulation models in Japan, describing the evolution of the representation of vehicle dynamics, path definition methodologies, and improvement in computational efficiency. In the last paper of Part III, Nanthawichit *et al.* propose a methodology to combine probe data and conventional detector data to estimate the state of traffic on roadway segments with improved accuracy, through the application of Kalman filters.

As its title suggests, Part IV is concerned with the representation of user behavior. The first paper by Harata addresses the issue of consistency between traffic simulation models and travel behavior choice theory, and examines specifically dynamic route choice models and time-of-day choice models. Following this, Morikawa *et al.* critically examine the traditional assumptions of shortest-path choice and user equilibrium based on perfect information, through network assignment with imperfect information. In the last paper of the volume, Kitamura *et al.* present PCATS, a micro-simulator of individuals' daily travel, which produces trip demand along a continuous time axis, and illustrate its application to the analysis of TDM measures and to long-term demand forecasting along with a dynamic network simulator, DEBNetS.

As these chapters demonstrate, transport simulation has become a powerful tool in both research and practice. It can be a practical tool for the real-time forecasting of future traffic status on road networks and evaluation of the effectiveness of alternative traffic control measures. It can be applied in the assessment of the effectiveness of alternative TDM measures, or in the selection and implementation of a variety of ITS schemes now being developed. It is hoped that this volume will aid in further development of transport simulation models and their prevalent adoption as a practical tool in traffic control and transportation planning.

A large number of individuals contributed to the organization of the Symposium and the editing of this book. In particular, we note the efforts by Drs. Akira Kikuchi, Hiroyuki Oneyama and Toshio Yoshii, who contributed tremendously throughout this project. Special thanks go to Ms. Kiyoko Morimoto; we owe the success of the Symposium to her bookkeeping and management skills. We also thank the speakers, presenters at the demo sessions, and the audience at the Symposium. Finally, we dedicate this book to those researchers and practitioners whose endeavors are contributing to better urban transportation in many significant ways.

Application of a Simulation-Based Dynamic Traffic Assignment Model

Michael Florian, Michael Mahut and Nicolas Tremblay
INRO Consultants, Inc. 5160 Décarie Blvd., Suite 610,
Montreal, Quebec, H3X 2H9 Canada
(mike, michaelm, nicolas)@inro.ca

ABSTRACT

The evaluation of on-line intelligent transportation system (ITS) measures, such as adaptive route-guidance and traffic management systems, depends heavily on the use of faster than real time traffic simulation models. Off-line applications, such as the testing of ITS strategies and planning studies, are also best served by fast-running traffic models due to the repetitive or iterative nature of such investigations. This paper describes a simulation-based, iterative dynamic-equilibrium traffic assignment model. The determination of time-dependent path flows is modeled as a master problem that is solved using the method of successive averages (MSA). The determination of path travel times for a given set of path flows is the network-loading sub-problem, which is solved using the space-time queuing approach of Mahut. This loading method has been shown to provide reasonably accurate results with very little computational effort. The model was applied to the Stockholm road network, which consists of 2100 links, 1,191 nodes, 228 zones, representing and 4,964 turns. The results show that this model is applicable to medium-size networks with a very reasonable computation time.

***Keywords** – dynamic traffic assignment, method of successive averages, traffic simulation, queuing models*

INTRODUCTION

The functional requirements of a dynamic traffic assignment (DTA) model for ITS applications may be subdivided into two major modes of use: off-line and on-line. The off-line use of DTA is for the testing and evaluation of a wide variety if ITS measures before they are implemented in practice. In particular, iterative approaches to dynamic assignment that approximate (dynamic) user equilibrium conditions are generally restricted to off-line use due to the high computation times involved. The resulting assignments can also be interpreted as imitating drivers' adaptation over time to changes in network topology or control, including the implementation of ITS measures. Due to the high number of iterations usually required, such applications are ideally suited for traffic models that have low computational requirements. On-line DTA can be used within a system that monitors and manages the network in real time. DTA and the embedded traffic models can play a key role in providing short-term forecasts of the system state that are used by adaptive traffic management, control and guidance systems. Due to the need to provide feedback in real time, on-line DTA poses rather stringent demands on the embedded models for maintaining low computation times.

The need to model the time varying network flow of vehicles for ITS applications has generated many contributions for the solution of dynamic traffic assignment methods. These contributions are varied and have been motivated by different methodological approaches. They may be classified according to the modeling paradigm underlying the temporal traffic model. In order to provide a common terminology to the various models, it is convenient to refer to two main components of any dynamic traffic model: the route-choice mechanism and the network-loading mechanism. The latter is the method used to represent the evolution of the traffic flow over the links of the network once the route choice has been determined.

Perhaps the most popular dynamic traffic models today are those based on the representation of the behavior of each driver regarding car following, gap acceptance and lane choice. These are micro-simulation models such as CORSIM (http://www.fhwa-tsis.com/corsim_page.htm), INTEGRATION

(Van Aerde, 1999), AIMSUN2 (Barceló et al, 1994) (http://www.tss-bcn.com), VISSIM (http://www.ptv.de), PARAMICS (http://www.quadstone.com) and DRACULA (http://www.its.leeds.ac.uk/software/dracula/). MITSIM (Yang, 1997) (http://web.mit.edu/its/products.html) is an academic research model that has been used in several studies in Boston, Stockholm and elsewhere.

There are many other micro-simulation models developed in universities and industrial research centers that use the same basic approach. The route choice in a micro-simulation model is either predetermined or computed while the loading of the network is being carried out. Essentially, a micro-simulation model aims to provide the traffic flows composed of individual vehicles in one network-loading step. As micro-simulation models are built by using many stochastic choice mechanisms, their proper use requires the replication of runs. The successful use of micro-simulations is commonly limited to relatively small size networks. Their application has been hindered for medium-to-large networks by the relatively high computation time and effort required for a proper model calibration. Usually, there are many parameters involved. Choosing appropriate parameter values is a relatively complex task since each computer implementation for a micro-simulation uses a large number of heuristic rules that are added to the basic car-following mechanism. A thorough understanding of how these parameter choices influence the results, when using given software package, is essential for a successful application. Nevertheless, micro-simulation models are popular and their use is enhanced by traffic animation graphics that capture the attention of non-technical staff.

The aim of handling larger networks with reasonable computational times has led to the development of so-called "mesoscopic" approaches to traffic simulation, which are less precise in the representation of traffic behavior but are less cumbersome computationally. The aim is to obtain a traffic representation that still captures the basic temporal congestion phenomena, but models the traffic dynamics with less fidelity. One of the earliest examples of such an approach is CONTRAM (Leonard et. al., 1989) (www.contram.com) which is a commercially available package that has been used in England and elsewhere in Europe.

Recently, the development of mesoscopic simulation models for off-line dynamic traffic assignment has become an area of significant research activity, as witnessed by the United States Federal Highway Administration Dynamic Traffic Assignment Project (http://www.dynamictrafficassignment.org). The development of DYNASMART (Mahmassani et al., 2001) and DYNAMIT (Ben-Akiva et al., 1998) (http://web.mit.edu/) are two significant developments. These mesoscopic models provide a path choice mechanism and a network loading method based on simplified representations of traffic dynamics. While CONTRAM represents traffic with continuous flow, as it has its roots in static traffic assignment models, DYNASMART and DYNAMIT move individual vehicles. CONTRAM and DYNAMIT provide an iterative scheme for the emulation of dynamic user equilibrium, where all cars within the same departure interval for a given origin-destination pair experience the same travel time (approximately). The approach taken in DYNAMIT is to provide an "a priori" path choice and path set by using models based on random choice utility theory. Another approach to the network loading algorithm is that based on cellular automata theory (Nagel and Schreckenberg, 1992), which has been implemented in the TRANSIMS software (http://transims.tsasa.lanl.gov), developed recently by the Los Alamos National Laboratories in the USA. In this approach, the route choice is predetermined for each traveler and the network loading method loads the vehicles on a network where each lane of a link is divided into cells of equal size. The advance of vehicles is carried out by using local rules for each vehicle that determine the next cell to be occupied and the speed of the vehicle.

Other dynamic traffic assignment models have their roots in macroscopic traffic flow theory developed during the 1950's (Lighthill and Whitham, 1955) (Richards, 1956). The work of Papageorgiou (1990) led to the development of the METACOR (Diakakis and Papageorgiou, 1996) and METANET (Messmer et al., 2000a), which has been used for the development of an iterative dynamic traffic assignment method (Messmer et al., 2000b). The route choice in this model is carried out by splitting proportions at nodes of the network, where only two arcs can originate at a given node. The network loading method is based on a second order (p.d.e.) traffic flow model.

Another line of research is that of analytical dynamic traffic assignment models, which has its roots in the mathematical programming approach to

static network equilibrium models. This area is not covered in this contribution.
The dynamic assignment model presented in this paper is based on a traffic simulation model that was designed to produce reasonably accurate results with a minimum number of parameters and a minimum of computational effort (Mahut, 2000,Astarita et al., 2001)). However, the underlying structure of the model has more in common with microscopic than with mesoscopic approaches, as it is designed to capture the effects of car following, lane changing and gap acceptance. The simulation is a discrete-event procedure and moves individual vehicles. Unlike discrete-time microscopic simulation models, where the computational effort per link is proportional to the total vehicle-seconds of travel, the computational effort per link required by this model is strictly proportional to the number of vehicles to pass through it, regardless of their travel times. As a result, the relative efficiency of this approach compared to microscopic methods increases with the level of congestion.

Another special property of this model is that the traffic dynamics are modeled without the (longitudinal) discretization of links into segments or cells. As a result the procedure only explicitly calculates the time at which each vehicle crosses each node on its path. This leads to a drastic reduction in computational effort relative to microscopic discrete time approaches, where the computational effort is a function of the total travel time experienced by the drivers.

The paper is structured as follows. The next section is dedicated to the exposition of approaches to dynamic traffic assignment. The third section is dedicated to the description of the network loading method developed; the algorithm for the dynamic traffic assignment, which combines the route-choice mechanism with the network loading method, is presented in the fourth section. Applications of the model are then given and some conclusions end the paper.

DYNAMIC TRAFFIC ASSIGNMENT

Two different approaches are commonly used to emulate the path choice behavior of drivers: dynamic assignment *en route* and dynamic *equilibrium* assignment. In this work, the approach taken is to seek an approximate solution to the dynamic equilibrium conditions.

En-Route Assignment

In the en route assignment problem, the routing mechanism is a set of behavioral rules that determine how drivers react to information received en route. Information may be available at discrete points in time (e.g. radio broadcasts), discrete points in space (e.g. variable message signs), or be continuously available in both space and time (e.g. traffic conditions visible to the driver). Some information may only be available to a certain class of vehicles; e.g., those equipped with vehicle guidance systems. Typically, the choice of what information is provided to the drivers, i.e., the information *strategy*, is an exogenous input. Moreover, how drivers respond to information is also an exogenous input and may involve one or more parameters, such as the 'penetration rate'. The output is the resulting (time-dependent) path choices given the time-dependent origin-destination demand. Another input to this problem is a suitable pre-trip assignment, i.e., path choices that represent the "do nothing" alternative and which are followed in the absence of any en route information. In many cases, an equilibrium assignment (discussed below) is used for this purpose.

En route assignment thus only requires running a single dynamic (time-dependent) loading of the demand onto the network over the time period of interest. If the information strategy or the driver response strategy is parameterized, it may be possible to design an iterative algorithm to determine the optimal values of such parameters.

Equilibrium Assignment

In the equilibrium assignment problem, only pre-trip path choices are considered. However, the path choices are modelled as a decision variable and the objective is to minimize each driver's travel time. All drivers have perfect access to information, which consists of the travel times on all paths (used and unused) experienced on the previous iterations. All drivers furthermore attempt to minimize their own travel times, and the solution algorithm takes the form of an iterative procedure designed to converge to these conditions. The solution algorithm used here consists of two main components: a method to determine a new set of time-dependent path flows given the experienced path travel times on the previous iteration, and a method to determine the actual travel times that result from a given set of path flow rates. The latter problem is referred to as the "network loading problem", and can be solved using any route-based dynamic traffic model. The algorithm furthermore requires a set of initial path flows, which are determined by assigning all vehicles to the shortest paths, based on free-flow conditions. The general structure of the algorithm is shown schematically in Figure 1.

The mathematical statement of the dynamic equilibrium problem is in the space of path flows $h_k(t)$, for all paths k belonging to the set K_i for an origin-destination $i \in I$, at time t. The time-varying demands are denoted $g_i(t)$. The path flow rates in the feasible region Ω satisfy the conservation of flow and non-negativity constraints for $t \in T_d$, where $(0, T_d)$ is the period during which the temporal demand is defined. That is

$$\Omega = h(t): \sum_{k \in K_i} h_k(t) = g_i(t), i \in I; h_k(t) \geq 0 \qquad \text{for almost all } t \in T_d \tag{1}$$

The definition of user optimal dynamic equilibrium is given by the temporal version of the static (Wardrop) user optimal equilibrium conditions, which are:

$$\begin{aligned} s_k(t) &= u_i(t) \text{ if } h_k(t) > 0 \\ s_k(t) &\geq u_i(t) \text{ otherwise} \end{aligned} \tag{2}$$

for all: $k \in K_i, i \in I$, *for almost all* $t \in T_d$
where: $h_k \in \Omega, u_i(t) = \min_{k \in K_i} \{s_k(t)\}$ for almost all $t \in T_d$ and $s_k(t)$ is the path travel time determined by the dynamic network loading. Friesz et al (1993) showed that these conditions are equivalent to a variational inequality problem, which is to find $h^* \in \Omega$ such that

$$\left(S(h^*), h - h^*\right) \geq 0, \forall h \in \Omega \tag{3}$$

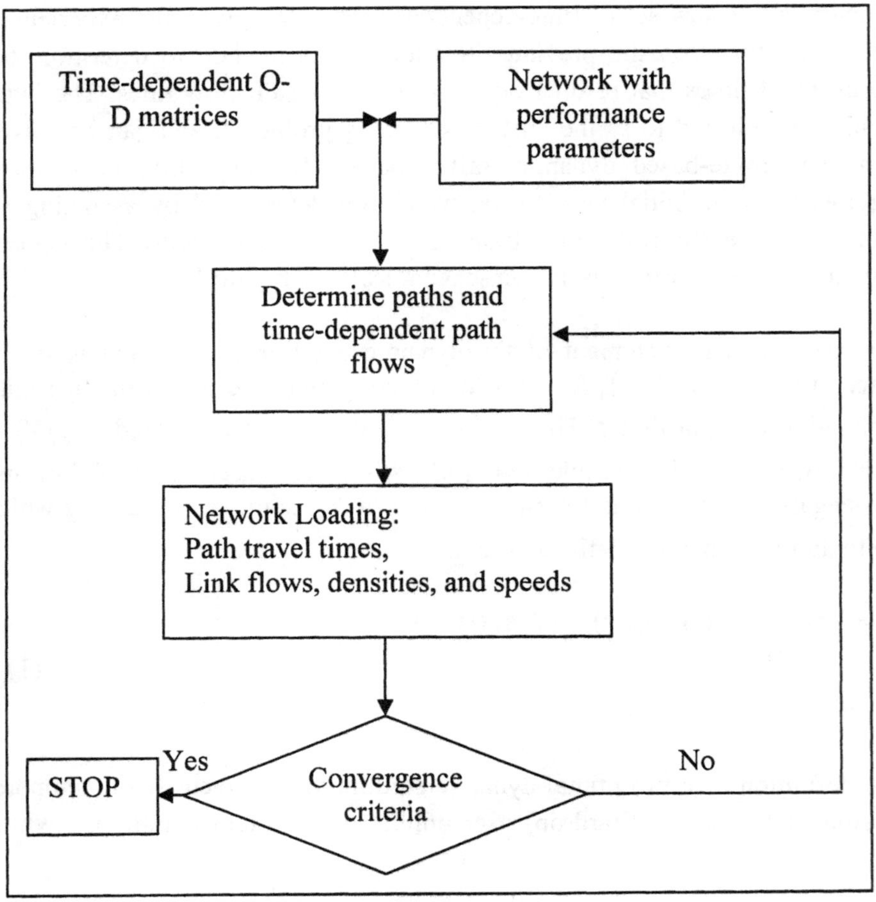

Figure 1. Structure of solution algorithm to DTA problem.

The continuous time problem (1)-(3) is usually solved by using some time discretization scheme.

THE ALGORITHM

The solution approach adopted for solving the dynamic network equilibrium model (1)-(3) is based on a time discretization into discrete time periods $\tau = 1, 2, \ldots, \left|\frac{T_d}{\Delta t}\right|$, where Δt is the chosen duration of a time interval. This results in the discretized model

$$s_k^\tau = u_i^\tau \text{ if } h_k^\tau > 0$$
$$s_k^\tau \geq u_i^\tau \text{ if } h_k^\tau = 0$$
$$\text{for all } k \in k_i,\ i \in I, \tau = 1,2,\ldots,\left|\frac{T_d}{\Delta t}\right| \quad (4)$$

where the feasible set of time dependent flows h_k^τ belong to

$$\Omega^\tau = h_k^\tau : \sum_{k \in K_i} h_k^\tau = g_i^\tau, i \in I, all\ \tau;$$
$$h_k^\tau \geq 0, k_i, i \in I, all\ \tau\} \quad (5)$$

which can be shown to be equivalent to solving the discretized variational inequality.

$$\sum_\tau \sum_{k \in K} s_k^\tau (h^\tau)(h_k^\tau - h_k^\tau) \geq 0 \quad (6)$$

where $K = \bigcup_{i \in I} k_i$ where h^τ is the vector of path flows (h_k^τ) for all k and τ.

The path input flows $h_k^\tau, k \in K$ are determined by the method of successive averages (MSA), which is applied to each O-D pair I and time interval τ.

The initialization procedure consists of an incremental loading scheme that successively assigns partial sums of the demand for each interval τ onto dynamic shortest paths. That is, the first demand increment, $g_i^1, i \in I$ is loaded onto a dynamic shortest path based on free flow travel times; there the link travel times are updated and a new dynamic shortest path is computed for interval 2; the first and second demand increments (g_i^1, g_i^2) are loaded onto the first and second computed paths and so on.

Starting at the second iteration, and up to a pre-specified maximum number of iterations, N, the time-dependent link travel times after each loading are used to determine a new set of dynamic shortest paths that are added to the current set of paths. At each iteration $n, n \leq N$, the volume assigned as input flow to each path in the set is g_i^τ/n, $i \in I, all\ \tau$. After that, for $m > N$, only the shortest among *used* paths is identified and the path input flow rates are redistributed as follows:

$$
\begin{aligned}
h_k^\tau(n) &= h_k^\tau(n-1)\left(\frac{n-1}{n}\right) + \frac{g_i^\tau}{n}\ if\ s_k^\tau(n-1) = u_i^\tau(n-1) \\
h_k^\tau(n) &= h_k^\tau(n-1)\left(\frac{n-1}{n}\right) otherwise
\end{aligned}
\tag{7}
$$

for $k \in K_i, i \in I\ and\ all\ \tau$

While no formal convergence proof can be given for this algorithm, since the network loading map does not have an analytical form, a measure of gap, inspired from that used in static network equilibrium models may be used for qualifying a given solution. It is the difference between the total travel time experienced and the total travel time that would have been experienced if all vehicles had the travel time (over each interval τ) equal to that of the current shortest path.

Hence

$$
R\ Gap^\tau(n) = \frac{\sum_{i \in I}\sum_{k \in k_i} h_k^\tau(n) s_k^\tau(n)}{\sum_{i \in I} g_i^\tau\ u_i^\tau(n)} \tag{8}
$$

Where $u_i^\tau(n)$ are the lengths of the shortest paths at iteration *n*. A relative gap of zero would indicate a perfect dynamic user equilibrium flow. Clearly this is a fleeting goal to aim for with any dynamic traffic assignment.

It is very important to note that this model, even though its general formulation is very similar to flow based models, is in fact a discrete vehicle model. The network loading procedure, as realized by the event based simulation, moves individual cars on the links of the network. It is worthwhile to note that, the nature of a dynamic traffic assignment model is determined by the choice made for the network loading mechanism.

NETWORK LOADING

As mentioned above, the input to the network-loading problem is the set of time-dependent path flows, while the output is the set of time-dependent path travel times (as a function of the departure time from the origin). Any network-loading model will simultaneously yield the time-dependent link flows, travel times and densities.

The network-loading model used here moves discrete vehicles on a network defined at the level of individual lanes. The underlying mechanism of congestion in the model is the crossing, merging and diverging – collectively referred to as *conflicts* – of vehicle trajectories. Simply stated, whenever two vehicles pass the same point in space, there must be a minimum time separation between them. How to propagate the resulting delays upstream from one vehicle to the next in a realistic way is a problem of *traffic dynamics*.

Before these delays can be propagated, the conflicts themselves must identified, which requires the drivers to make choices about which lanes they will use on the links of their pre-assigned paths. Although the paths are assigned *a priori*, the lanes are chosen as the vehicle proceeds along its path, using a set of behavioral rules. Once the conflicts are identified, they must subsequently be *resolved*: one vehicle must lead, while the other must follow, and the following vehicle incurs some amount of delay. The delay is then propagated upstream according to a simplified car following relationship. The delay experienced by the first vehicle at a traffic control device is propagated in the same way.

Simulation Approach

Most microscopic simulators (Aimsun2 (Barceló et al., 1994), VISSIM (www.ptv.de), Paramics (www.quadstone.com), MITSIM (Yang, 1997), and INTEGRATION (Van Aerde, 1999)) and some mesoscopic simulators, (DYNAMIT (Ben Akiva et al., 1998), and DYNASMART (Mahamssani et al., 2001)) use a discrete-time (fixed time step) procedure. The simulation period is discretized into small time intervals, Δt. After each Δt, all the vehicles that are present on the network are moved, which implies the computation of the new position of each vehicle. This usually implies two scans of all the vehicles: one to determine the possible movements and the other to move the vehicles. The network is updated at each clock tick $t = n(\Delta t)$, where $n \in \{0, 1, \ldots, T/\Delta t\}$, and T is the duration of the loading period.

The solution algorithm for the model used here is a discrete-event ("event-based") procedure. In a discrete-event simulation, each temporal process modeled is associated with a specific sequence of events, and each event is associated with a real-valued point in time. For instance, an event may be associated with a change of signal phase at a controlled intersection, or the arrival time of a vehicle to a link. Event based algorithms are typically used for modeling queuing systems. An event-based approach may be very efficient if one can minimize the number of events modeled and still obtain valid results.

Network Representation

The network definition required for this DTA model requires somewhat more information than that required for static network equilibrium models, yet somewhat less than is generally required for micro-simulation traffic models. Since the underlying traffic model moves individual vehicles on discrete lanes, each link must be defined by a number of lanes. Each lane furthermore is defined by an access code that determines which classes of vehicles may use the lane (e.g., taxi, bus, HOV, etc...). A length and speed limit furthermore define each link. At each node (intersection) of the network, a turn is defined for each permitted movement from an incoming link to an outgoing link. Each turn is defined by an access code and a saturation flow rate per lane. Unlike micro-simulation models, the network definition does not require geometrical information such as lane width, turning angles, and the dimensions of intersections.

Lane Choice

In contrast to continuum traffic models and static assignment models, traffic simulators model the movement of vehicles on individual lanes. How drivers utilize the available lanes of a roadway can have a significant, even drastic impact on both the total delays experienced and how these delays are distributed (spatially and temporally) in the network. Naturally, these effects will only be captured if the traffic model employed is sensitive to the effects of lane-changing activity on the effective flow capacity of a link. In this case, the pre-trip path information must be complemented by a set of lane choice rules in order to provide the necessary information to identify conflicts between vehicle trajectories. As mentioned above, such conflicts are the principal mechanism of traffic congestion in the model.

A common example of the impact of lane utilization is a congested off-ramp from a highway. Even if the ramp is only one lane wide, delays may be incurred on more than one lane of the highway. Some drivers will inevitably miss the back of the queue, intentionally or not, and then begin queuing in the neighboring lane(s) as they look for an opportunity to merge into the lane leading onto the ramp.

The degree to which the queue spills *over* onto the neighboring lanes depends to a great extent on driver behavior. Specifically, if the queue spills *back* upstream over several links on a daily basis, drivers may be able to recognize the source of congestion as they reach the end of the queue several links upstream of the ramp. Thus, those drivers who are destined for the off-ramp may decide to join the back of the queue immediately, while those remaining on the highway may choose to avoid the queue. Drivers may often make such decisions even though they are still several links upstream of the off-ramp itself, which is the critical piece of information in this decision.

By joining the back of the queue immediately, drivers destined for the off-ramp will not delay drivers remaining on the highway; i.e., the amount of queue *spill-over* is reduced. Conversely, the amount of queue spill-over could well be unrealistically high if drivers were unaware of which lane exited the highway until they were on the last link before the ramp. In the traffic simulation literature, heuristics that take into account non-local (beyond the next link or turn) information about a driver's intended path are often called "look-ahead" rules. The addition of look-ahead rules to existing heuristics based strictly on local information has been shown to significantly improve the reality of the model outputs for some specific though not uncommon network topologies (Barceló, 2000) (Ben Akiva et al., 2000).

In the model used here, vehicle trajectories along links are modelled implicitly, rather than explicitly. Specifically, each driver chooses the lanes by which he/she will enter and exit a link just before actually arriving to the link and, once on the link, the choice cannot be re-considered. The principal argument behind using such an approach is that it is sufficient to model only mandatory lane changes in order to reproduce the general congestion patterns resulting from a given set of path flows. Mandatory lane changes are those that must be made in order to exit and enter each link on the lanes permitted for the associated turns.

The permitted lanes over a sequence of downstream turns are considered here when some of the lanes immediately downstream of the driver are queuing and some are not. This logic allows a driver to join the queue if necessary, or to by-pass it if his/her path does not go through the head of queue. Preliminary tests with this look-ahead feature have indicated a significant reduction in the amount of queue spillover, as well as total delay, in the case of a congested off-ramp as discussed above.

Conflicts and Precedence

Given the network, path flow rates and lane-choice rules, conflicts may arise between vehicle trajectories at nodes and along links. A conflict between two vehicles exists when, given their positions at one moment in time, their desired arrival times to the same downstream position violates a constraint that specifies the minimum time separation between vehicles at that point (such as a specified saturation flow rate). Conflicts can arise both at nodes and on multi-lane links. In order to satisfy a minimum headway constraint, it must be decided which vehicle is to precede the other, and thus which vehicle is to be delayed. It is these delays that are the underlying mechanism of congestion in the model. The process of deciding precedence between two conflicting vehicles is referred to here as conflict *resolution*.

In reality, which vehicle precedes the other depends to some extent on human behavior. The question is typically resolved in a traffic simulation model by gap-acceptance rules (Barrel et al., 1994) (Van Aerde, 1999), which are based on one of the two vehicles having priority over the other, and the specification of a time-gap parameter. In continuum traffic models, the approach is to specify the maximum low-priority flow as a function of the prevailing high-priority flow (Leonard et al., 1999).

In the model used here, a relatively simple gap-acceptance model has been implemented to determine precedence between vehicle conflicts at nodes, while a FIFO (first-in-first-out) rule is applied on links. Simulation results for two conflicting one-lane turns at a node are shown in Figure 2.

Traffic Dynamics

Once a conflict has been identified and resolved, and the appropriate delay has been calculated, this delay (or a residual portion of it) may propagate recursively over a sequence of vehicles against the direction of the traffic flow. The propagation of delay occurs in this model much the same way as in a normal queuing model.

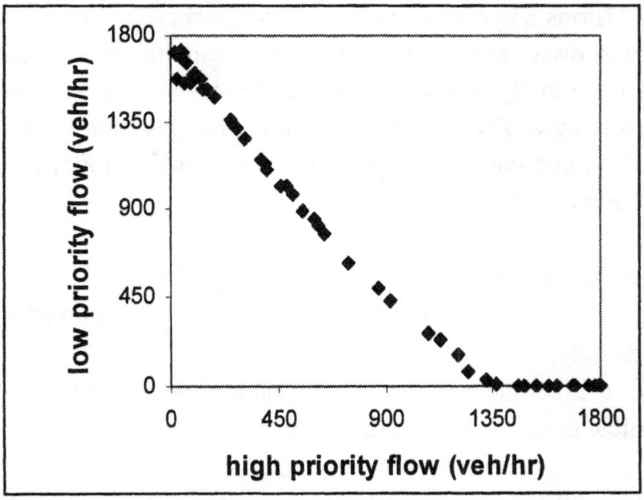

Figure 2. Maximum low-priority flow vs. high-priority flow exhibited by the gap-acceptance model.

Specifically, the *amount* of delay propagated from one vehicle to the next is exactly as would be determined by a standard queuing approach. What is different is *where* and *when* a vehicle in queue experiences each of the delays (or residuals thereof) that are propagated from downstream. This difference is due to the fact that the model employed here rigorously respects the finite speed at which delays propagate in actual traffic, sometimes called the negative wave speed. The positive (forward) wave speed is given by the speed limit. Mahut (2000) provides a detailed description of the model.

Traffic Control

The model also permits the specification of detailed traffic control information such as (pre-timed) signal timing and ramp metering plans. Traffic control specifications furthermore require the number of lanes associated with each turning movement, and the lanes (on both the upstream and downstream links) that may be used for executing a turn. These data may vary with the signal phase rather than being fixed for each turn.

A problem associated with traffic control is the issue of preventing *gridlock*, or *deadlock*, in traffic networks. This situation occurs when a sequence of stopped vehicles forms a cycle in a network, and thus each driver is ultimately waiting for his/her own vehicle to move. These vehicles can of course never achieve a positive velocity unless one of the drivers located at a node leaves the cycle by selecting a different link and thus changes paths. This problem can arise in reality, and can similarly arise in any traffic model in which the following conditions hold:

1. Vehicles (or packets) follow pre-specified paths.
2. Traffic speed (and thus flow) is equal to zero at a maximal value of traffic density.
3. There is a finite number of physical channels (lanes) on each link, and one vehicle cannot "jump over" another.

This phenomenon can occur unexpectedly in a traffic simulation model if one or more of the following conditions hold:

1. En-route path switching is not permitted.
2. The road network is under-represented in the model; i.e., relevant road sections are not coded, causing excessive congestion.
3. Information concerning roundabouts is incomplete: if signalized, control information is unknown; if not signalized, gap acceptance/priority information is unknown.
4. Path choices are naïve, causing excessive congestion.

This problem was addressed by developing an adaptive deadlock prevention algorithm that identifies when there is a high risk of a deadlocked cycle occurring. When these conditions are identified, vehicles attempting to enter the links on the cycle are forced to yield the right-of-way to those already on the cycle. The means by which the algorithm alters the traffic is not meant to represent an actual mechanism that can be implemented in reality, but rather serves as a surrogate for the cumulative effects of the missing information (e.g. signal controls) and incomplete behavioral rules (prohibiting en-route path switching) in the model. The algorithm explores the network starting at any given node using a depth-first search and continues as long as certain conditions are met. The algorithm was successfully applied to a large-scale network in which deadlocks were frequently occurring for a number of reasons, with an increase in computation time of less than 10%.

Vehicle Classes

Vehicle attributes (or parameters) can be broken down into two distinct categories: physical attributes and routing attributes. The physical attributes are the effective length (based on vehicle spacing at jam density), and the driver/vehicle response time. Together, these parameters yield the jam density and negative wave speed associated with each vehicle class. Routing attributes include the vehicle class identifier, which determines which lanes and turns of the network may be used by the class, and identifies any class-based routing strategy that may be defined. For instance, the class car uses different routing rules than the class bus, which travels along fixed itineraries and has mandatory stops. A demand matrix by class contains the flow in vehicles per hour for each origin-destination pair. The matrices are "time-sliced" in the sense that flow rates may be specified for given time intervals.

APPLICATIONS

This dynamic traffic assignment model was coded in C++ using an object-oriented approach. The original design was carried out on a SUN Workstation under Solaris 2.8. The code also runs and on an Intel PC under Linux and Windows 2000.

The Swedish Road Administration provided the authors with a road network and a time sliced origin-destination matrix for car trips in the city of Stockholm. The network consists of 1191 nodes, 2,100 links and 4,964 turns. Four 20-minute matrices provide the origin-destination demand data for 228 zones, from 6:55 am to 8:15 am. The total number of vehicles in the matrices was on the order of 108,560. The tests were run on a 2 GHz Intel PC with 768

Mb of RAM, running the Windows 2000 operating system. The RAM requirements for storing 15 trees, for each of the 8 departure intervals, was less than 100 Mb.

The dynamic traffic assignment was run for 40 iterations, each requiring roughly 1.1 minutes, for a total of about 44 minutes of computation time. The 80-minute loading interval was divided into 8 time intervals for the MSA assignment algorithm. After each iteration, the relative gap (discussed above) was calculated for the vehicles departing from the origins during each of these intervals, as shown in Figure 3. A relative gap of zero indicates a user-optimal dynamic equilibrium. Gap values ranging from 0.5 to 4.0 % were obtained by the 40th iteration. The gaps were increasing with each time interval, i.e., 0.5% was obtained for the first interval and 4.0% for the last. This increasing trend can be attributed to the fact that each driver's decision in the algorithm is based on the travel times experienced in the previous iteration. In this sense, the previous iteration serves as a prediction of the traffic conditions that will be encountered during all the time slices of the next iteration. As any given iteration (simulation) advances in time from $t = 0$ to $t = T$, the quality of this prediction degrades due to the increasing number of "unforeseen" decisions (those made for the current iteration before time t) that are affecting the actual traffic conditions on the network. The results are very promising and indicate that a reasonable level of convergence is attainable for a medium-sized network with relatively small amount of computing time.

The convergence measure is an indication of the difference between the average travel time and the best travel time *for the iteration*. This should not be interpreted as the difference between the current average travel time and that corresponding to a perfect equilibrium. A better guess of how much improvement in travel times can still be attained might be half of the gap, i.e., it might be expected that the difference between the current travel times and the true equilibrium solution is on the order of 0.25 to 2 % (depending on the time interval).

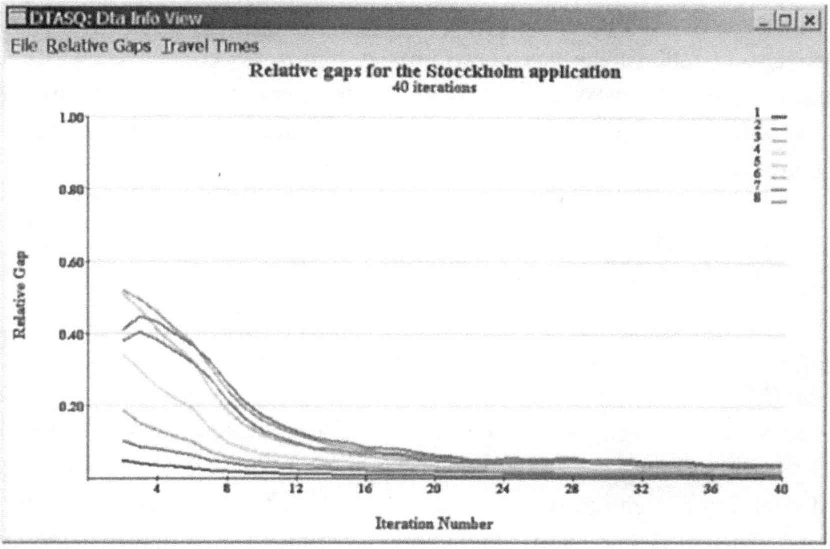

Figure 3. Relative gaps by time interval.

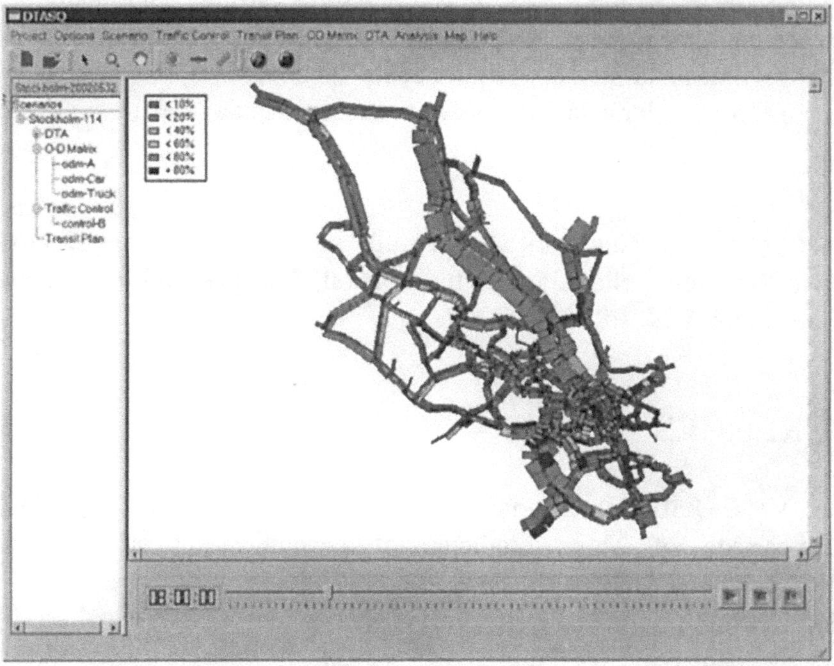

Figure 4. Link flows coloured by density at 8:00 a.m.

Network statistics were collected over 5-minute intervals. A snapshot of the network state at 8:00 of the last iteration is shown in Figure 4. The widths of the links indicate the average link outflow rates over the 10-minute interval starting at 8:00 a.m. The shades of gray colour indicates the relative density (occupancy) on each link as indicated in the legend of the plot..

CONCLUSIONS

A dynamic traffic assignment model, which uses the method of successive averages (MSA) to determine pre-trip dynamic equilibrium path choices combined with an event-based traffic simulation model, was successfully applied to a medium-sized network. The results indicate that an acceptable level of convergence can now be obtained for a medium-size network, using a realistic traffic model with a reasonable amount of computing time and memory usage.

The method has excellent potential for use in practice for a variety of applications related to the testing of ITS measures off-line. The model may have potential for further development as an on-line tool, due to the low computation times and memory requirements. Its computational efficiency is at least one order of magnitude faster than microscopic traffic simulation models.

Acknowledgement – This work was partially sponsored by a Post-Doctoral Industrial Research Fellowship of the Natural Sciences and Engineering Council of Canada (NSERC).

REFERENCES

Astarita, V., Er-Rafia, K., Florian, M., Mahut, and M., Velan, S. (2001). Comparison of Three Methods for Dynamic Network Loading, *Transportation Research Record*, **1771**, pp179-190.

Barceló, J. (2000). The Role of Traffic Simulation in Advanced Traffic Management Systems, Presented at *the Spring meeting of INFORMS*, Salt Lake City, USA. May 7-10, 2000.

Barceló, J., Ferrer, J.L., and R. Grau (1994). AIMSUN2 and the GETRAM

Simulation Environment. Internal Report, Departamento de Estadistica e Investigacion Operativa. Universitat Politecnica de Catalunya. See also http://www.tss-bcn.com.

Ben-Akiva, M., Koutsopoulis, H., Toledo, T. (2000). MITSIMLab: Recent Developments & Applications, Presented at *the Spring meeting of INFORMS*, Salt Lake City, USA. May 7-10, 2000.

Ben-Akiva, M., Koutsopoulos, H.N. Mishalani, R. (1998). DynaMIT: A Simulation-Based System for Traffic Prediction, Paper presented at *the DACCORD Short Term Forecasting Workshop*, Delft, The Netherlands. See also its.mit.edu.

Diakaki, C., and M. Papageorgiou (1996). Integrated Modelling and Control of Corridor Traffic Networks usingthe METACOR Modelling Tool, Dynamic Systems and Simulation Laboratory, Technical University of Crete. Internal Report No. 1996-8. Chania, Greece. pp. 41.

Florian, M., Mahut, M. and N. Tremblay (2001). A Hybrid Optimization-Mesoscopic Simulation Dynamic Traffic Assignment Model, *IEEE Intelligent Transportation Systems Conference Proceedings*. Oakland, California, USA. August 25-29, 2001.

Friesz, T., Bernstein, D., Smith, T., Tobin, R., and Wie, B. (1993). A variational inequality formulation of the dynamic network user equilibrium problem, *Operations Research*, **41**, pp179-191.

Lighthill, M.J. and G.B. Whitham (1955). On kinematic waves I: Flood movement in long rivers, II: A theory of traffic flow on long crowded roads, *Proceedings of the Royal Society of London*, **A229**, pp281-345.

Leonard, D.R., P. Gower and N.B. Taylor (1989). CONTRAM: Structure of the Model, Transport and Road Research Laboratory (TRRL) Research Report 178, Department of Transport, Crowthorne. See also http://www.contram.com/.

Mahmassani, H.S., A.F. Abdelghany N. Huynh, X. Zhou, Y-C. Chiu, and K.F. Abdelghany (2001). DYNASMART-P (version 0.926) User's Guide, Technical Report STO67-85-PIII, Center for Transportation Research, University of Texas at Austin.

Mahut, M. (2000). Discrete flow model for dynamic network loading, Ph.D. Thesis, Département d'informatique et de recherhe opérationelle, Université de Montréal, Published by the Center for Research on Transportation (CRT), University of Montreal.

Messmer, A (2000a). METANET A Simulation Program for Motorway Networks (Documentation), Dynamic Systems and Simulation Laboratory, Technical University of Crete. Chania, Greece.

Messmer, A (2000b). METANET-DTA An Exact Dynamic Traffic Assignment Tool Based on METANET, Dynamic Systems and Simulation Laboratory, Technical University of Crete, Chania, Greece. pp37.

Nagel, K. and M. Schreckenberg (1992). A cellular automaton model for freeway traffic, *Journal de Physique I France*, **2**, pp2221-2229.

Papageourgiou, M (1990). Dynamic Modelling, Assignment and Route Guidance in Traffic Networks,. *Transportation Research*, **24B(6)**, pp471-95.

Richards, P.I. (1956). Shock waves on the highway, *Operations Research*, **4**, pp42-51.

Van Aerde, M. (1999). INTEGRATION Release 2.20 for Windows, User's Guide. MVA and Assoicates, Kingston, Canada.

Yang, Q. (1997). A Simulation Laboratory for Evaluation of Dynamic Traffic management Systems, Ph.D. Thesis, Massachusetts Institute of Technology. See also http://web.mit.edu/its/products.html.

THE DRACULA DYNAMIC NETWORK MICROSIMULATION MODEL

Ronghui Liu
Institute for Transport Studies, University of Leeds, Leeds LS2 9JT, UK
rliu@its.leeds.ac.uk

ABSTRACT

Recent years have seen a tremendous interest world wide in the use of microsimulation techniques to model traffic in congested road networks. This interest is particularly associated with the development of real-time, high-tech based traffic management and control strategies that react to the highly dynamic and variable nature of traffic conditions and driver behaviour. This paper introduces the DRACULA dynamic traffic network model, with the main focus on its traffic microsimulation component. The paper describes the model's theoretical and behavioural foundations, and presents illustrative examples of applying the model in the evaluation of real-time management strategies. Our experience shows that this is a particularly suitable framework for the realistic modelling of real-time technological strategies.

BACKGROUND

Transport network models have played an important role in the planning and analysis of transport policies, and in evaluating their effect on road congestion and transport system designs. The analysis of traffic networks has traditionally been based on Wardrop's equilibrium principle, predicting a long-term average state of the network. They assume steady-state network supply and demand conditions from day-to-day and within different periods of a day, and have therefore had great difficulty in representing the dynamics of the transport systems and many of the contemporary transport policies that aim to respond to and influence travel demand and traffic conditions.

Recent years have seen a massive increase in "real-time" advanced technological strategies designed, for example, to reduce congestion, improve network efficiency, promote public transport, decrease pollution, increase road safety, etc.. At the network-wide level, these include: responsive, optimised traffic signal control, e.g. SCOOT (Hunt et al., 1981); congestion-based road pricing (Oldridge, 1990); dynamic route guidance/information and variable message signs (e.g. Emmerink and Nijkamp, 1999); congestion management strategies, e.g. freeway ramp metering, gating (Papageorgiou et al., 1989); and responsive priority measures for public transport (e.g. Quinn, 1992; Liu et al., 1999).

A general property of all these strategies is that they both respond to – and in turn influence - actual prevailing congestion levels, rather than being designed on the basis of long-term average conditions. That is to say, the variation in traffic conditions is just as important a consideration as the mean. Variabilities include the temporal distribution of flows both within and between days, as well as the variation in travel times and delays both within and between days. It includes not only "natural" variability associated with normal trip making decisions but also "unnatural" variability associated with incidents or accidents. In order to evaluate these systems and to determine the best strategy for implementation, it is crucial to have a reliable evaluation model that fully incorporates the effects of variability.

Recent advances in dynamic microsimulation models have produced extremely flexible frameworks whereby disaggregated, behaviour-based research can be incorporated and tested. There are generally two different approaches:

(a) "day-to-day" models have been developed to represent dynamic adjustment of driver's daily travel choice behaviour (on route, departure time and mode), based on various behavioural principles and static or dynamic traffic flow relationships (such as DYNASMART, Hu & Mahmassani 1997; TRANSIM, Nagel & Barrett, 1997; Emmerink et al 1994; Cantarella & Cascetta 1995; Jha et al 1999). Those proposed give great flexibility on the behavioural choice side, yet are more limited in their traffic flow modelling capabilities. Although, in some of these models, individual vehicles are represented, their movements are determined from a speed-flow relationship and based on the prevailing density on that segment of road. There is no representation of vehicles' lane-changing and car-following behaviour, making them difficult if not impossible to model complex traffic intersections, responsible signal control, bus priority measures, etc.

(b) "pure" traffic microsimulation models have focused on individual vehicles' detailed movements and individual system elements (eg traffic lights, intersections) to represent the within-day dynamics and variability of drivers' driving behaviour. This approach has been implemented in software packages such as CORSIM (Nsour & Santiago 1994), AIMSUN2 (Barcelo et al. 1995), VISSIM (Fellendorf et al 1997), and PARAMICS (Laird et al. 1998). These models are based on car-following and lane-changing rules and have shown themselves capable of representing real-time policies. However, they either have no concept of a route, or have routes determined exogenously by an assignment model operating at a different level of traffic flow detail.

Taking the best elements of the above two approaches and setting them within a single, consistent framework, the DRACULA (Dynamic Route Assignment Combining User Learning and microsimulAtion) model was developed as a new approach to model dynamics in transport networks. At its most detailed level, the model simulates explicitly individuals' daily travel choices and the movements of individual vehicles through the network, with a day-by-day driver learning process. Thus it provides strong interactions between the demand for travel and the network supply conditions.

The concept of the DRACULA approach and its main framework are described in Liu et al (1995), and briefly summarised in Section 2. Section 3 introduces briefly the day-to-day demand model of DRACULA. The main focus of this paper is on DRACULA-MARS (Microscopic Analysis of Road Systems), the traffic microsimulation component of the DRACULA system. Section 4 describes the theoretical and behavioural foundation of the model, namely the car-following, lane changing and gap acceptance rules which combined analytical and empirical understanding of the detailed transport operation and traffic behaviour on congested urban road networks. General properties of the traffic simulation are presented in Section 5. This is followed, in Section 6, by demonstrations of the model in a study of dynamic traffic signal controls, an evaluation of Intelligent Speed Adaptation systems and in the assessment of congestion road pricing policies. The paper concludes with a summary and current and future research activities with DRACULA.

DRACULA MODEL STRUCTURE

The dynamic network microsimulation model DRACULA has been developed at University of Leeds since 1993 (Liu et al. 1995). As with conventional models the DRACULA approach begins with the concept of demand and supply (or performance) sub-models that interact with each other. However, by contrast with conventional models, in DRACULA both the demand and supply sub-models are based on microsimulation and both evolve from day to day. In DRACULA, trip makers are individually represented and their daily route choices (demand) are made based on their past experience and their perceived knowledge of the network conditions. Individual vehicles are then moved through the network (supply) following their chosen routes according to car-following and lane-changing rules.

The demand stage predicts the level of individual demand for day k from a full population of potential drivers and the supply model for day k determines the resulting travel conditions. The costs experienced by drivers are then re-entered into their individual 'knowledge bases' which in turn affect the demand model for day $k+1$. The process continues for a pre-specified number of days. The framework combines a number of sub-models of traffic flow and drivers' choices for a given day with a day-to-day driver learning sub-model. In its most general form it has the following structure:

Day-to-day (demand) loop:

1. [Initialisation] Establish a population of potential drivers with individual characteristics and assume initial driver perceptions for each link in the network. Set day counter $k=1$.

2. [OD demand] Select the total day-k demand for each origin-destination pair according to some given probabilistic rules;
3. [Route choice] Each individual travelling on the day chooses a route based on their current perception of traffic conditions and previous experiences.
4. [Supply variability] "Global" network supply conditions are selected for day k prior to loading by some given probability laws to simulate effects such as weather and lighting conditions. For "local" variations in network conditions (such as road works, incidents occurring on the day), specify the location and duration of the incidents.

Within-day (supply) loop:

5. [Traffic loading] A microscopic simulation of traffic conditions on day k is carried out given the choices above. Drivers experience within-day variable link and turn travel times for the route they have chosen.
    a. [Initialisation] Set within-day simulation clock $t=0$.
    b. [Vehicle Generation] Vehicles enter the network at their origin following a shifted negative exponential headway distribution with the mean flow representing the average demand from the origin and a minimum headway of 1 second. Each vehicle is given a set of individual characteristics.
    c. [Vehicle Movement] Each vehicle follows the pre-specified route. Their speeds and positions are updated according to car-following rules, lane-changing rules, gap acceptance rules and traffic regulations at intersections.
    d. [Emission Calculation] Calculate emissions and fuel consumption for each individual vehicle according to their current driving mode: acceleration, deceleration, idling and cruising, and emission factors and relations to fuel consumption.
    e. [Traffic Control Update] For each signalised junction, update the stage change-over clock according to desired signal plans (fixed plans or responsive). Check if the any incident is to start or to finish.
    f. [Data Collection] Individual drivers' experience within-day are stored. Aggregated measures such as queue length, travel time, speed, flow, emissions, fuel consumption for each link, each OD pair and the whole network are recorded.

g. [End of day] If all drivers have finished their journey, terminate the day; otherwise increment the simulation clock and return to step 5b.

6. [Learning] At the end of day k, drivers update their perceptions based on their experiences on the day.
7. [Stopping test] If some stopping conditions is satisfied, terminate; otherwise increment the day counter and return to step 2.

Similar models of this structure have been considered previously by Ben-Akiva et al. (1986), Cantarella & Cascetta (1995), Vythoulkas (1990), Emmerink et al. (1994), and Mahmassani & Jayakrishnan (1991). The functionality of the day-to-day demand loop is briefly described in the next section. The main focus of this paper will be on the traffic microsimulation used in the within-day supply loop.

DAY-TO-DAY VARIABILITY OF DEMAND AND SUPPLY

Day-to-Day Demand

The DRACULA approach is based on the concept of a large "population" representing all the potential drivers in the study area. In practice a more pragmatic approach has been used in which we aim at generating a population whose trip making behaviour at the aggregated day-to-day level matches the averages and variances observed in real life. Existing conventional trip matrix T_{ij} from origin i to destination j has been used in all our applications to generate the population. On any particular day within the evolution of the model, any individual's decision as to whether to travel or not is then constrained by the predicted daily trips for their particular origin-destination pair. We assume that the day-to-day variability in demand may be described by a normal distribution whose mean is T_{ij} and whose variance is $\beta_d^2 T_{ij}^2$, where $\beta_d > 0$ is a user-set coefficient of demand variation. Hence the demand for ij trips on day k is:

$$t_{ij}^{(k)} = Nor(T_{ij}, \beta_d^2 T_{ij}^2) \tag{1}$$

Route Choice

A number of route choice mechanisms have been implemented in DRACULA. The default option is the "bounded rational choice", based on the work of Mahmassani and Jayakrishnan (1991). This model assumes that drivers will use the same (habit) route as on the last day in which they travelled, unless the cost of travel on the minimum cost route is *significantly* better than that on their habit route. The threshold is that a driver will use the same route unless:

$$C_{p1} - C_{p2} > \max(\eta \times C_{p1}, \varphi) \tag{2}$$

where C_{p1} and C_{p2} are costs along the habit and the minimum cost routes respectively, η and φ are global parameters representing the relative and absolute cost improvement required for a route switch.

The route choices are made and fixed before the trips start; drivers follow their chosen routes through the network to their destinations and will not (within the current state of model development) make en-route diversion when, e.g., encountering congestion.

Learning Model

After each journey individuals use their experienced travel times on the links used on that journey to update their perceived link travel times according to the following conditions:

(a) experiences more than M days old are forgotten; and
(b) the perceived travel cost is the average of at most the last N remembered experiences on that link.

Here M and N are global parameters set at the start of simulation, although their effect will be specific to each individual's experience. It may reasonably be argued that these parameters should be allowed to vary with the driver and/or trip type. Such options can be added to the program if future research suggests so. Generally, it is expected that N will be the main parameter affecting perceived cost; M is intended mainly as a device for drivers to ultimately forget a single bad experience of a link which may occur particularly in the atypical, initial warm-up days. Therefore, it is expected that $N < M$.

Supply Variability

The effect of day-to-day variability of network condition is represented at two levels. The global variability represents the effects of weather, daylight etc, and is represented in the model by a variable link cruise speed through a normal distribution:

$$v_a^{(k)} = Nor(V_a, \beta_s^2 V_a^2) \tag{3}$$

where $v_a^{(k)}$ is a random variable representing the cruise speed of link a on day k, V_a the average cruise speed for the link and β_s a global coefficient of variation representing the daily variation in link speed.

Locally, incidents (such as breakdowns or road closures) may occur one day but not another. This is represented before loading by specifying the location and duration of the incidents. The global and local variabilities will affect (through the traffic simulation described in the next section) the travel times of vehicles travelling on that day, but not on the routes individuals take.

THE TRAFFIC MICROSIMULATION

The traffic model in DRACULA is a microsimulation of the movements of (pre-specified) vehicles through the network. Drivers follow their pre-determined routes and en-route they encounter traffic signals, queues and interact with other vehicles on the road. A large number of such microscopic vehicle models have been developed in the past at varying levels of complexity and network size (e.g. in some the network is effectively a single intersection) - a few are mentioned in Section 1. An essential property of all such models is that the vehicles move in real-time and their space-time trajectories are determined by, e.g., car following and lane-changing models and network controls such as signals. Rather than adopting an existing model, in DRACULA we elected to develop our own microsimulation model from scratch because of the strong need to control the interaction between the supply and demand models and, in particular, the need to associate a specific route and destination with each vehicle.

The simulation is based on fixed time increments; the speeds and positions of individual vehicles are updated at an increment of one second. Spatially, the simulation is continuous in that a vehicle can be positioned at any point along a link. The simulation starts by loading the simulation parameters, network data including global and local variations and trip information (demand and routes determined by the demand model). It then runs through an interactive procedure at the pre-defined time increments, within which the following tasks are performed:

a. Update the state of traffic signal controls, and check if any incident starts or ends;
b. Generate new entry vehicles and place them on their entrance links;
c. Loop through all vehicles in the network, and for each one of them:
   i. check if the vehicle wants to change lane and, if so, whether the gaps are acceptable;
   ii. update the vehicle's speed and acceleration and advance it to its new position. At the end of the link, either remove the vehicle from the network (if it has arrived at its destination) or pass it to its next link en route;
   iii. calculate vehicle emissions and fuel consumption, and record traffic performance measures;
d. Update the graphical display if required;
e. Update the simulation clock and return to step a.

Simulation Time Periods

Three time periods were simulated in DRACULA; these are schematically depicted in Figure 1. The "demand period" is the main simulation period (time t_1-t_2 in Fig. 1); it is typically one hour representing the peak period. In addition, a warm-up period (0-t_1) is simulated with demand linearly increasing from half of the peak level to the peak level as a way of ensuring that the traffic from the demand period does not start with an empty network. At the end of the main period, the simulated demand linearly decreases to half of the peak level over a "cooling-off period" (t_2-t_3) and thereafter stays at that level until the end of the simulation (t_4). The simulation ends when all vehicles departing during the demand period have completed their journey.

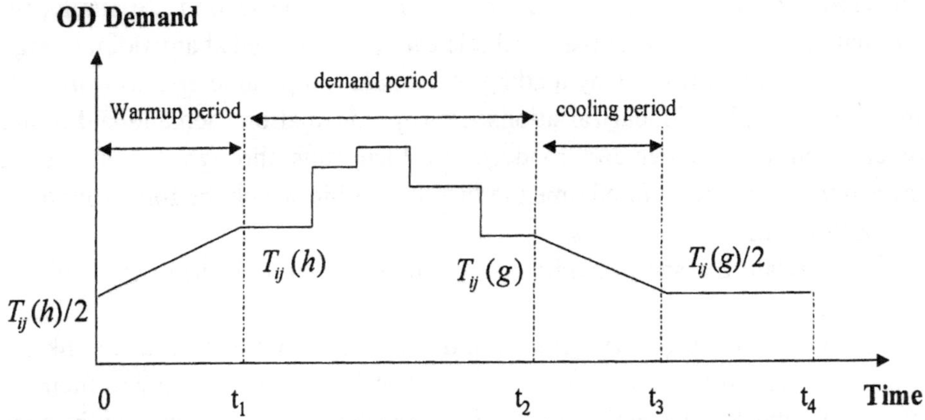

Figure 1. Time periods and the demand levels in each time period as simulated in DRACULA. ***$T_{ij}(h)$*** **and** ***$T_{ij}(g)$*** **represents the O-D demand level at the beginning and end of the main period respectively. The time t1, t2, t3 are user-defined variables, whilst the end of simulation time (t4) is variable depending on congestion levels in networks.**

Network Representation

The network is represented by nodes, links and lanes. A node is either external, where traffic enters or leaves the network, or an intersection. There is no restriction on the number of roads connected to an intersection.

A link is a directional roadway between two nodes and consists of one or more lanes. A link is specified by its upstream and downstream nodes, cruise speed, number of lanes, and turns permitted to other outbound links from the downstream node. For each permitted turn, the lane(s) in the link that can use this turn are specified and a marker describing its priority over opposing flows is given.

In the model traffic moves in lanes. A lane can be reserved for a particular type(s) of vehicles, for example, a reserved bus lane. The reserved time periods and set-backs at either end of the reserved lane can be specified. Figure 2 depicts some of the network features represented in DRACULA.

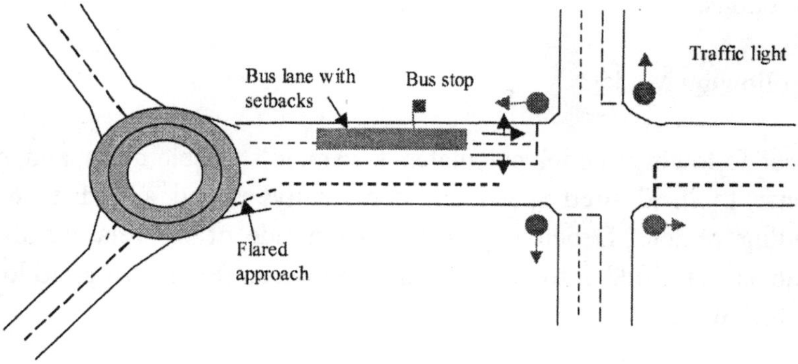

Figure 2. Network features.

Vehicle Characteristics

Vehicles are individually represented; each has a set of individual characteristics including vehicle type (e.g. car, bus, guided-bus, taxi, LGV, or HGV), vehicle length, desired minimum distance headway, normal and maximum acceleration, normal and maximum deceleration, desired speed relative to the mean speed on any individual link and acceptable gap. These characteristics are randomly sampled from normal distributions representative of that type of vehicle, subject to a lower and an upper bound:

$$p_u^n = \max\{P_u^{\min}, \min[P_u^{\max}, Nor(P_u, \beta_{pu}^2 P_u^2)]\} \quad (4)$$

where p_u^n is the value of parameter *p* for vehicle *n* of type *u*. P_u, $P_u^{\min}$ and $P_u^{\max}$ are the average, lower and upper bounds respectively of parameter *p* and vehicle type *u*, and β_{pu} the coefficient of variation for variable *p* and type *u*. The default values are based on a number of sources including May (1990) and ITE (1982).

Vehicle Movement Simulation

Vehicle movements in a network are determined by the vehicle's desired movement, its response to traffic regulations and interactions with neighbouring vehicles. The simulation maintains a linked list of vehicles in each lane and moves individual vehicles according to a car-following model and a lane-changing model, and their response to traffic controls at intersections.

Car-Following Model

The car-following model calculates a vehicle's acceleration and speed in response to its desired speed and the relative speed and distance of the preceding vehicle. Depending on the magnitude of the relative distance, a vehicle is classified into one of three regimes: free-moving, following or close-following.

Free-moving: when a vehicle is the leading vehicle in its lane and its position relative to the stop-line of the link is larger than a pre-defined threshold d^h, or if its preceding vehicle is more than d^h further ahead, the vehicle will accelerate or decelerate freely in order to maintain its desired speed.

Following: when the space headway becomes shorter than d^h but longer than a lower threshold d^l, the vehicle will take a controlled speed which is derived from the relative speed and distance of the preceding vehicle in a manner similar to that used in NEMIS (Mauro, 1991):

$$v_n^{following}(t+\tau_n) = c_1 v_n(t) + c_2 v_{n-1}(t) + c_3[x_{n-1}(t) - x_n(t) - L_{n-1} - s_n^{\min}] \quad (5)$$

where n and n-1 denote the subject and its preceding vehicle, v and x the speed and position of the vehicle. τ_n is the reaction time, s_n^{min} the minimum safety distance of vehicle n, and L_n the length of vehicle n-1. Parameters c_1, c_2 and c_3 are constants.

Close-following: when the space headway is below d^l, vehicle n will prepare to stop in case the preceding vehicle brakes suddenly. The Gipps' (1981) safety speed is used here:

$$v_n^{close}(t+\tau_n) \le d_n\tau_n + \sqrt{d_n^2\tau_n^2 - d_n\{2[x_{n-1}(t) - x_n(t) - L_{n-1} - s_n^{\min}] - v_n(t)\tau_n - v_{n-1}^2(t)/d'_{n-1}\}} \quad (6)$$

where d_n is the deceleration of vehicle n and d'_{n-1} the deceleration of vehicle n-1 perceived by vehicle n.

The actual speed of the following vehicle n is chosen as the minimum of the two speeds derived from equations (5) and (6). In all cases, drivers will not want to move at a speed exceeding their desired one, accelerate at a rate exceeding their maximum acceleration, or decelerate above their maximum deceleration rate. When a vehicle moves at a speed below a minimum speed, the vehicle is regarded as stationary.

Lane-Changing Model

The lane-changing model contains three steps: (1) obtain the lane-changing desires and define the type of changing, (2) select the target lane, and (3) change lane if gaps are acceptable.

The model divides drivers' lane-changing desires into one of five types when drivers have to or want to change lane in order to:

(a) reach a bus stop on the link;
(b) avoid a restricted-use lane or incident;
(c) make their turn from the next junction;
(d) move into a lane reserved for their type; or
(e) gain speed by overtaking a slower moving vehicle.

The first three types are "mandatory", i.e. the lane-changing has to be carried out by a certain position on the current link; the other two types are "discretionary". Whether a discretionary lane-change can be carried out depends on the actual traffic conditions. For example, a vehicle would only change lane to gain speed if the speed offered by the adjacent lane is higher by a threshold.

When a vehicle wishes to change lane, it looks for a target lane. The target lane is generally determined by the lane-changing requirement, except in the case of overtaking which is only permitted from the nearside to the offside. Once it has chosen a target lane, it examines the "lead" and "lag" gaps in its target lane and makes the lane-changing movement immediately if both gaps are acceptable. For discretionary lane-changing, a gap is acceptable if it is greater than a minimum safety distance G_n^{min} which vehicle n wants to keep in case the preceding vehicle breaks suddenly:

$$G_n^{\min}(t) = v_n(t)\tau_n + v_n^2(t)/2d_n - v_{n-1}^2(t)/2d'_{n-1} + s_n^{\min} \quad (7)$$

The acceptable gap for mandatory lane-changing decreases as the vehicle gets closer to its "target point". The target point can be a bus-stop, the position of an incident, or the end of the queue from the stopline (in the case of lane-changing for next junction turning). If a vehicle gets nearer to its target point but has not been able to change to the target lane, the vehicle may slow down and eventually stop and wait for an opportunity to change lanes. When the speed on the target lane is below a pre-defined threshold, some drivers on the target lane may deliberately slow down in order to create gaps for the subject vehicle to join. These drivers are randomly selected from a pre-defined proportion which is related to the type of subject vehicle (for example, there might be a higher proportion of people willing to giveway to buses than to cars). Vehicles can only change one lane at a time. After one such manoeuvre, the vehicle has to wait for some time before making another lane-changing attempt.

Simulation outputs

The traffic simulation records the link travel times for each demand trip (those depart during the "demand period") and passes this information to their individual knowledge base which in turn update that individual's perception of the network. To measure the performance of a network, the simulation also provides summary statistics on link-, route- and network-wide average travel time, speed, queue length, fuel consumption and pollutant emission over regular time periods. A distinction is made in DRACULA between the "supply" costs for a given demand and the "performance" measures over a specified space-time area. This distinction is described in detail in the next sub-section.

The most detailed records are the second-by-second individual vehicles' locations and speeds. The model also provide point- or loop-based detector measures on headway distribution, flow, occupancy and speed. For each bus service, the model summarizes the mean and standard deviation of total journey time and journey time between stops, a measure which can help distinguish service delay due to traffic congestion from that due to poor management. A graphical animation of the vehicles' movements can also be

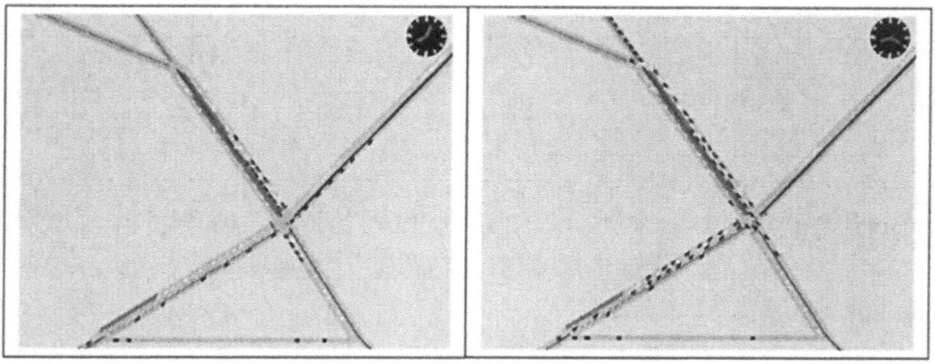

Figure 3. Simulated traffic conditions at Clifton Green intersection in the City of York. Two snapshots were taken at time 08:04 (a) and 08:20 (b) as shown by the clock on each snapshot. Vehicles are shown as coloured rectangles. Parts of the network were blocked due to roadwork, which is shown in dark grey.

shown in parallel with the simulation, giving the user a direct view of the traffic condition on the network. Fig. 3 displays two snapshots from a simulation of a signalised intersection in York; they show clearly the changes in traffic conditions between the beginning and the middle of a morning peak period.

Network performance and supply measures

DRACULA makes clear distinction between the performance of a network and costs associated with a given demand (the supply costs). The performance of a network or a single link can be measured in terms of flow performed and time performed in a defined period. They are engineering description of the performance of the link or network at a given point in time or over a given time period, and can be used to estimate the link or network equivalent of speed-flow relationships (the "performance curves").

The performance measures are based upon time-sliced approach (see Figure 4), whereby the simulation period is divided into a number of equal performance periods. The traffic flow (q), traffic density (k) and average speed (v) for link a of length L_a over time period $(h, h+\omega)$ can be calculated as:

$$q_a(h) = \frac{\sum_{n=1}^{N_a(h)} x_n(h)}{L_a \omega}, \; k_a(h) = \frac{\sum_{n=1}^{N_a(h)} s_n(h)}{L_a \omega}, \; v_a(h) = \sum_{n=1}^{N_a(h)} x_n(h) / \sum_{n=1}^{N_a(h)} s_n(h) \qquad (8)$$

where $x_n(h)$ is the distance and $s_n(h)$ the time travelled by vehicle n in the space-time area $L_a \times \omega$ at the start of time h, and $N_a(h)$ the number of vehicles on link a during the period $(h, h+\omega)$.

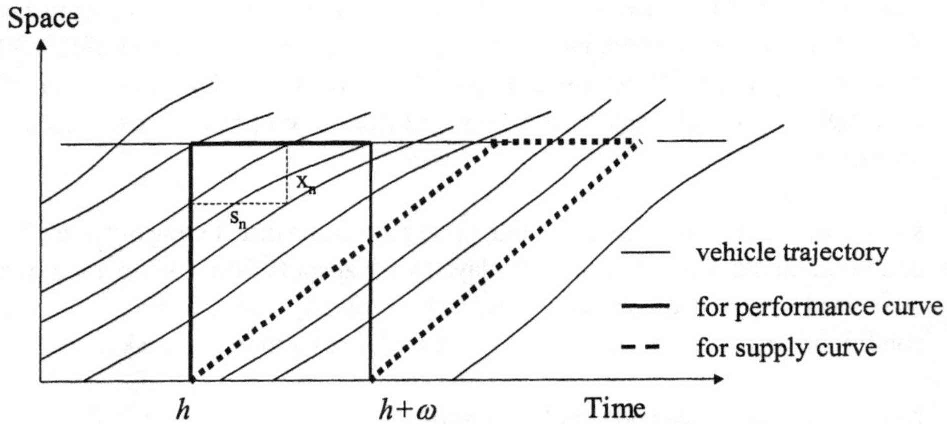

Figure 4. Space-time domains used to measure performance and supply costs.

The supply costs reflect the costs experienced by a driver using the network at a given level of demand; and they can be used to describe the way in which costs of using a network rise as demand-levels increases (the "supply curve"). Since any journey through a network will pass through a number of different traffic states and the costs incurred will be affected by both the journey length and the route taken, as well as by the impacts of other demands on the network both at that time and in earlier time periods. In order to measure these costs, individual vehicles need to be "tracked" through the network. Thus the space-time domains used to measure performance curves and supply curves are different, as shown in Fig. 4, and supply curves cannot be readily observed in the way that performance curves can.

Supply measures track individual vehicles through the network and summarise their trajectories over a given time period. The summation can be done with over a "departure time period" or an "arrival time period". In DRACULA, the departure-time aggregated supply measures are recorded.

In tracking them through the network, DRACULA collects individual vehicles' link journey time. Let us denote $y_a^n(t)$ as the journey time traversing link *a* by vehicle *n* who entered the link at time $t=(h, h+\omega)$. Then the supply cost for vehicle *n* travelling on link *a* will be:

$$C_a^n(t) = VOT^n \times y_a^n(t) + VOD^n \times L_a \tag{9}$$

where VOT^n and VOD^n are values of travel time and distance (operating costs) for vehicle *n*. The average generalised cost for trips departed in time period $(h, h+\omega)$ along path *p* from OD-pair *ij* will be:

$$C_{pij}(h) = \sum_{t=h}^{h+\omega} \sum_{n \in N_{pij}(t)} C_p^n(t) / \sum_{t=h}^{h+\omega} N_{pij}(t) \tag{10}$$

where $N_{pij}(t)$ is the number of individuals entering the network at time *t* using path *pij*. The supply costs for individual OD pair *ij* can be obtained by a trip-weighted average for all paths:

$$C_{ij}(h) = \sum_{p \in \Pi_{ij}(h)} N_{pij}(h) C_{pij}(h) / T_{ij}(h) \tag{11}$$

where $N_{pij}(h)$ is the number of vehicles using path *p* between origin *i* and destination *j* departing in period $(h, h+\omega)$, $\Pi_{ij}(h)$ the set of paths used and $T_{ij}(h)$ the number of vehicles travelling between origin *i* and destination *j* by vehicles departing in period $(h, h+\omega)$. The supply curve for the whole network is then calculated as a trip-weighted average over all O-D pairs.

MODEL PROPERTIES

Model Implementation and Performance

The program is originally written in C and later incorporated C++

object-oriented programming. The program operates under the PC Windows environment. The implementation imposes no limitation on the size of the network, nor the demand level. The processing speed does not appear to be affected significantly by the size of the network, either. It does, however, decreases as the number of vehicles travelling on the network at the any one time increases.

Figure 5 shows the simulation processing speed (measured as the ratio of the time simulated to CPU time) as a function of traffic density in a network using a Pentium II-300 PC. The network is based on the city of Leeds which covers a triangular area of the city centre and north part of the city, with some 200 intersection and 23,000 trips/hr in the morning peak period. It can be seen from the figure that the processing speed decreases exponentially as flow density increases. Even at the full demand (23,000 vehicles/hour) the simulation ran 20 times faster than real time.

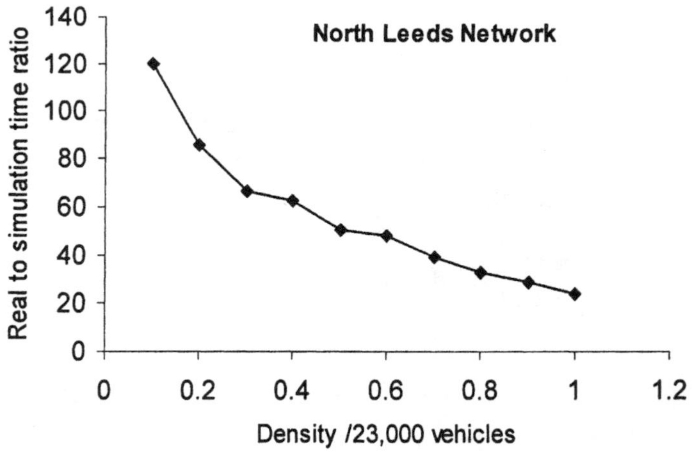

Figure 5. Simulation processing speed versus traffic density.

General Properties

Fig. 6 shows the relationship between traffic flow and density as simulated by DRACULA on a single-lane circular test track. Measurements were taken from virtual detectors located on the test track. The figure shows several distinct traffic flow regimes typically exhibit in uninterrupted traffic flow operations (see May, 1990). The part of the curve pointed by A in Fig. 6 represents the free-flow condition, whilst the area pointed by C shows the typical characteristics of discharge traffic after a jam. The complete flow-density plot resembles the mirror image of a reversed λ, with the maximum flow of free-flow traffic (point B in Fig. 6) being considerably higher than that of the congested traffic (point B'). This is another prominent feature of traffic flow, known as "capacity drop" as first discussed in Edie (1961) and evident in other experimental data (e.g. Koshi et al, 1983).

Saturation flow is an important measure of traffic performance at signal-controlled intersections. Fig. 7 shows the discharge rates of vehicles crossing a signalised intersection versus the green time as simulated by DRACULA. The number of vehicles crossing the stop-line was recorded and averaged over one-hour simulation to produce the average discharge flow rates. Three levels of arrival flows are modelled. It can been seen that as the demand flows increase the peak of discharge rates gets more and more stable; the stable (flat) discharge rate gives the saturation flow for that approach. The example demonstrates that the traffic simulation of DRACULA has a fair representation of the travel behaviour and traffic operation at signalised intersection.

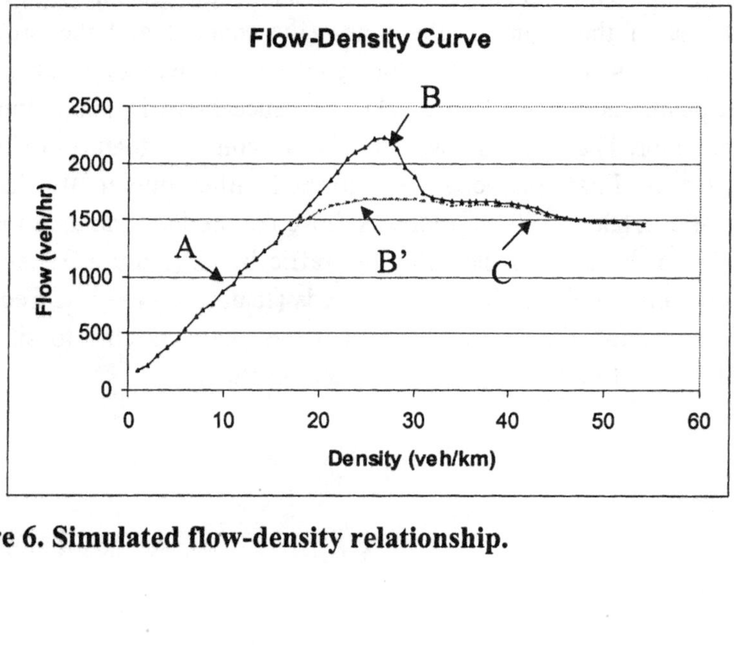

Figure 6. Simulated flow-density relationship.

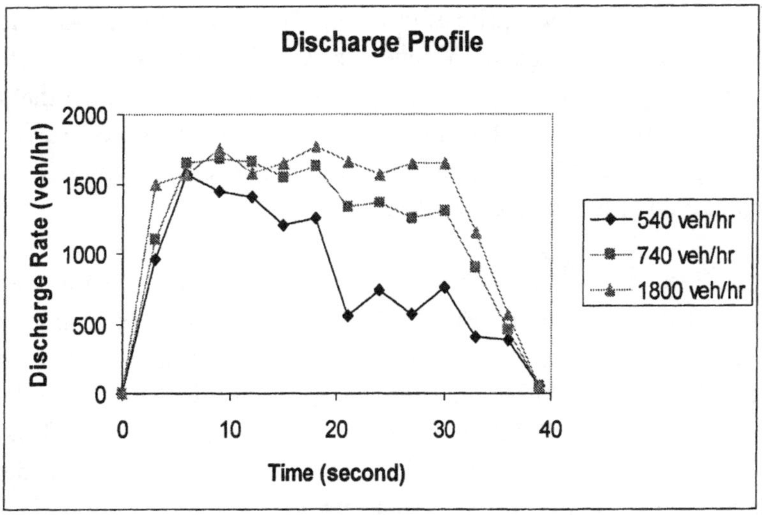

Figure 7. Discharge profiles at a signalised intersection. Time is measured from the start of green.

The trajectories of vehicles travelling along a single-lane corridor (without lane-changing) with two signalised intersections en-route are shown in Figure 8. The abscissa of the figure is the times (in seconds) and the ordinates represent the spaces travelled along the lane: each trajectory indicates the sequence of positions occupied by a vehicle in successive instants along the axis of the corridor. The figure shows a number of common features exhibited in traffic streams. First, platoons were formed either due to the lack of opportunity of overtaking (for example, as those featured around area marked as A in Fig. 8) or because of the delay by traffic lights (feature B). Second, there is significant variability in vehicle speeds (feature C) even at free-flow conditions. In general, the figure shows that the model is able to simulate responses of drivers to traffic signal control successfully.

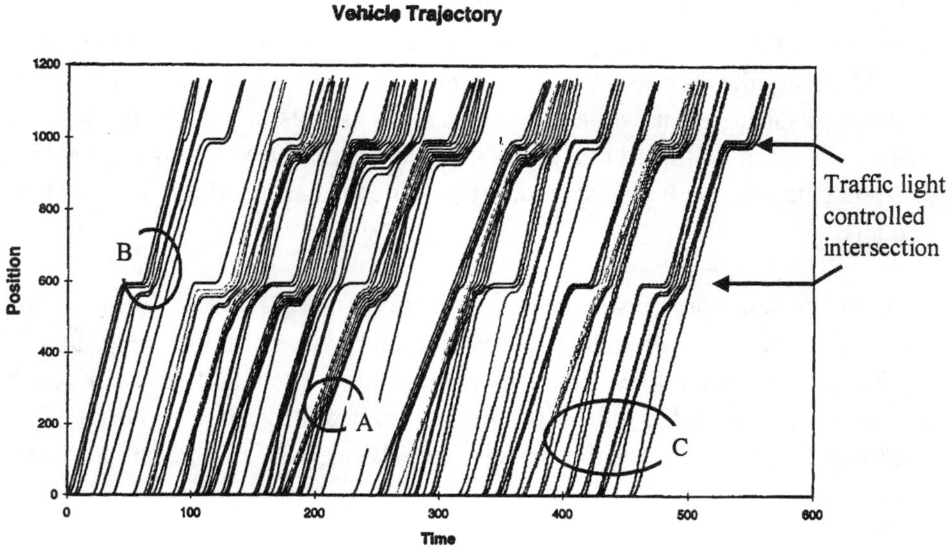

Figure 8. Trajectories of vehicles simulated along a single-lane corridor.

NOVEL APPLICATIONS

DRACULA has been developed as a flexible framework through modular implementation of its sub-models. At its most detailed level, DRACULA represents individual drivers' day-to-day choice making processes and individual vehicles' movements through a network. In practice, however, it may be desirable to run the model with a number of simplifications. Thus, the traffic supply model may be based on a more conventional static network model with macroscopic flow-delay functions but with variable parameters such as link capacity, while the demand model is based on the full evolution of driver choices from day to day.

Similarly the demand route choice can be derived from a static equilibrium assignment, but applied to the vehicle-by-vehicle simulation. DRACULA is compatible with the equilibrium model SATURN (van Vliet, 1982) such that it can use the network and route assignment from SATURN and combine them with its detailed microsimulation to model the supply-side effect of real-time strategies. The microsimulation model requires essentially the same basic data as a macroscopic model such as SATURN - nodes, links, number of lanes per link, lane markings, signal operations, giveway rules, etc., with some extra data related to the geometry and size of intersections for example. Applications of the DRACULA raffic microsimulation model with route choice generated from SATURN equilibrium assignment are presented later. The flexibility of the framework ensures that, while keeping its novel aspects in one way or the other, DRACULA can be linked to a greater or a lesser extent with existing models. Current data bases will almost certainly provide the best starting points for new models.

Next we present some results from applications of DRACULA in modelling dynamic systems on drivers' route choice and system performance, and in evaluating new technology and traffic management. The results and discussion are primarily intended to illustrate the applicability of the DRACULA approach and to show that the model responds logically to changes in model parameters.

Responsive Traffic Signals

In this example, we apply DRACULA to a study of the effect of responsive signals on network performance and drivers' route choice. The full DRACULA model was tested on a small artificial network with 2 O-D pairs, 4 possible routes and 4 signals (see Figure 9).

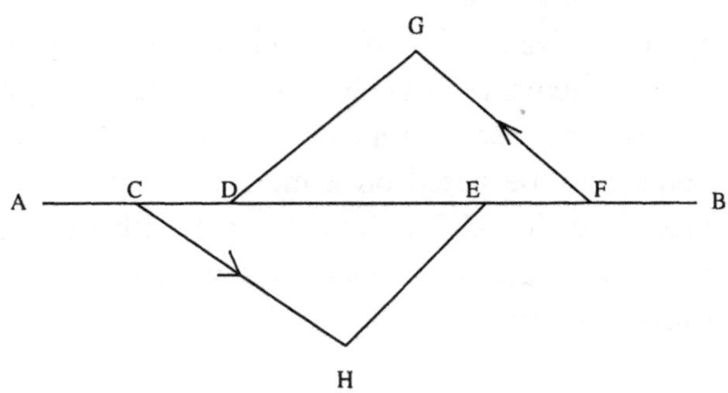

Fig. 9. The network for testing the signal control policies. Intersections C, D, E and F are signalized and the two OD pairs are A to B and B to A. One-way streets are indicated by arrows.

The signals may be set by a simple responsive "equi-saturation" policy where the green proportions allocated to each stage are determined based on the number of vehicles discharged in the previous cycle. Here, signal cycles were kept constant and a minimum green period of 8 seconds was maintained. In addition, a fixed plan optimised to the average traffic condition is used for comparision. A total of 100 days and two levels of variability in daily demand (β_d = 0.05 and 0.2) were simulated. The averages and standard deviations in network total travel times (in vehicle-hours) are summarized in Table 1. Day-to-day total vehicle-hours are shown in Figures 10 for the low and high levels of variability.

Table 1. Network total travel times (in vehicles-hours) under the two signal control policies.

Demand Variability	Signal Policy	Mean	Std. Dev.
	Fixed	101.1	15.6
β_d = 0.05	Responsive	79.7	12.2
	Difference	21.4	
	Fixed	111.5	44.0
β_d = 0.2	Responsive	84.6	36.0
	Difference	26.9	

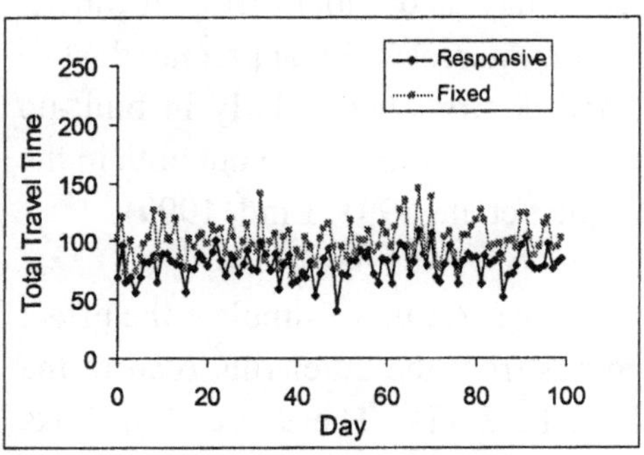

(a)

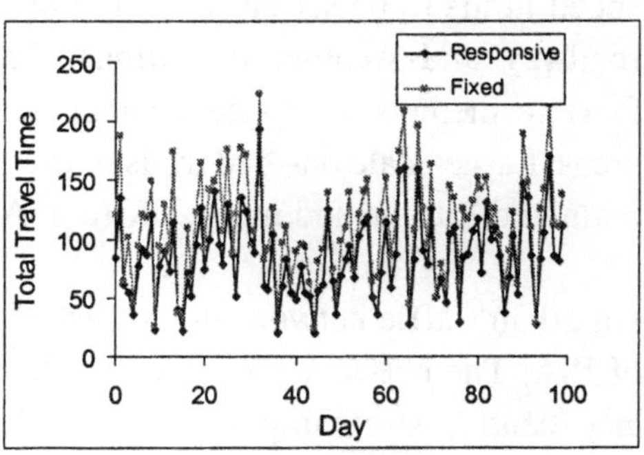

(b)

Fig. 10. Network total travel time under the fixed and the responsive signals for demand variability of 5% (a) and 20% (b).

It can be seen that:

(a) Under both signal control policies, both the average and variance in vehicle-hours are higher at higher demand variability;

(b) Average travel times are lower under the responsive policy than under the fixed plan; and

(c) The responsive policy performanced even better over the fixed signals under higher demand variability; the average difference in travel times between the responsive signals and the fixed plans is 26.7sec with β_d=0.2 compared with 20.4sec when β_d=0.05 (Fig. 10).

The better travel performances produced by the responsive signals have also played an important role in drivers' route choice. Changes in signals were seen to attract drivers to the more direct routes. With the responsive plans all drivers were assigned to the two minimum distance routes by the end of 100 days, whereas with the fixed signal all four routes were used.

Intelligent Speed Adaptation (ISA)

ISA systems use in-vehicle electronic devices to control the maximum speeds of vehicles to the prevailing speed limits through, for example, communication with a Global Positioning System. The main advantage of ISA systems relative to other forms of urban speed control measures (such as 20 mph zones or traffic calming measures) is its flexibility. The system allows for different control speed limits to be set up for different time of day, and under different traffic, roadway and weather conditions. They are increasingly appreciated as a flexible method for speed management and control, particularly in built-up areas. Large-scale on-road trials of the systems are being carried out both in the mainland Europe and in the UK (e.g. Almquist et al., 1991; Lind, 1999)

An urban traffic network in the east of Leeds was set up to simulate the effect of ISA. The network covers two radial routes from the outer ring road to the city centre, stretching over some 15km (Fig. 11). There are 240 links connecting 120 intersections. Two levels of speed limits were set according to road type: a 40mph speed limit is set for the two radial routes and a 30 mph limit for all residential streets. In addition, on one of the entry links on the ring road, a national speed limit of 70mph was defined. Figure 11 shows the speed limit distribution over the network. A morning peak network (with some 18,000 car-trips per hour) and an off-peak network (with some 12,000 trips/hour) were set up in order to examine the performance of ISA under different traffic congestion levels. The base networks were calibrated against observed link counts and floating-car journey times (Liu & Tate 2000).

The ISA penetration rate was introduced as a control variable in DRACULA. Simulations for both the morning and off-peak periods were carried out with 10 ISA penetration rates: 10, 20,…,100% of the total fleet were equipped with ISA. The results were compared with the base case where there was no ISA speed control. For each scenario, the model was run 10 times with different random number seeds to establish a distribution of the results.

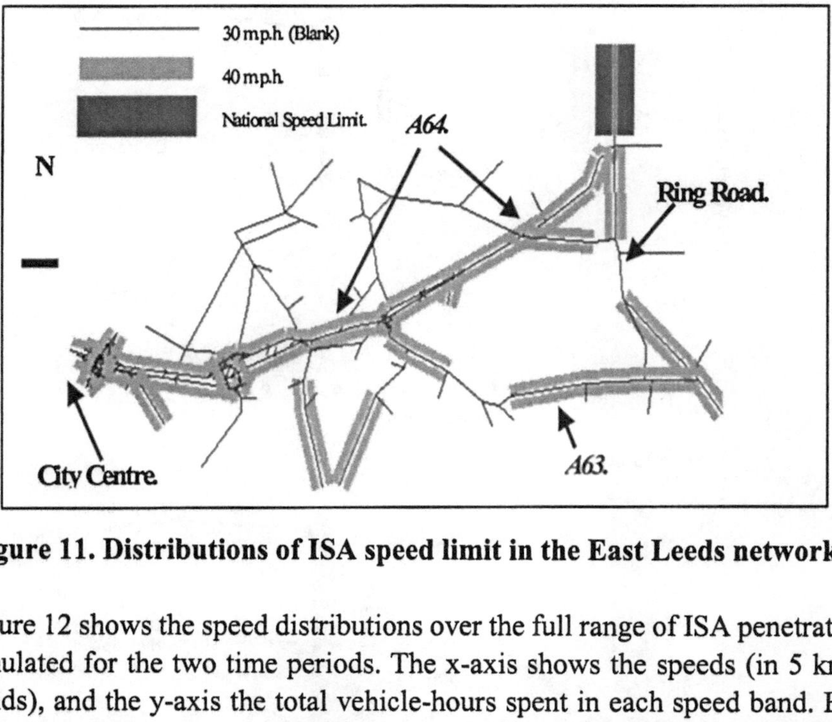

Figure 11. Distributions of ISA speed limit in the East Leeds network.

Figure 12 shows the speed distributions over the full range of ISA penetrations simulated for the two time periods. The x-axis shows the speeds (in 5 km/hr bands), and the y-axis the total vehicle-hours spent in each speed band. Each row of data represents one ISA penetration rate as a percentage. The results clearly demonstrate that the proportion of vehicles exceeding the speed limits (30 and 40 mph) decreases with increasing levels of ISA penetration. In the off-peak scenario, calculations established that in the no-control case there were 34% of vehicles exceeding the 30 mph speed limit. This was zero under full ISA control. Due to more congestion this figure is reduced to 20% during the peak period. Similarly, the level of vehicle-hours exceeding the 40 mph speed limit was 16% and 12% for the off-peak and peak periods respectively. Another feature worth noting, is that, in the morning peak period, there was a substantially high proportion of vehicle-hours spent at speed below 10 km/hr. ISA implementation did not alter the proportion of such congested traffic by any significant amount. This suggests that, whilst ISA is effective in reducing excessive speeds, it does not induce further congestion to the network, a feature which may make such system more acceptable to the general public and the network managers.

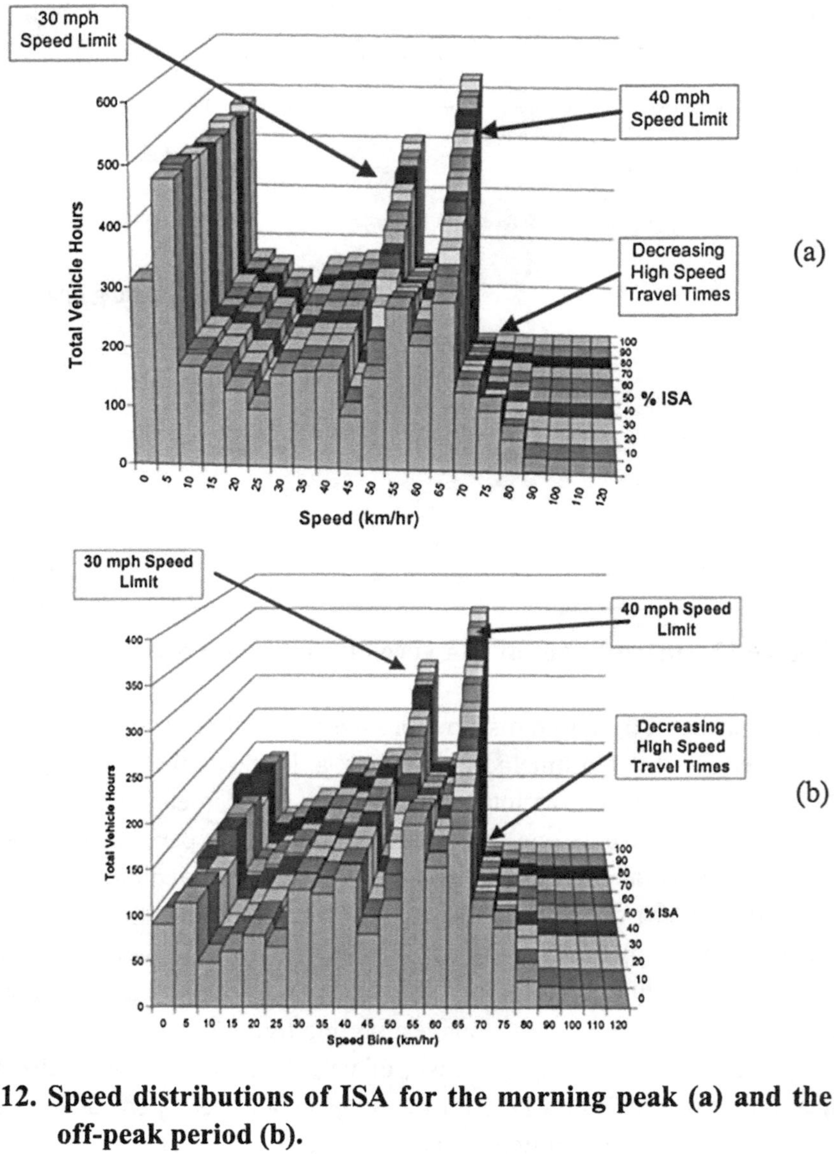

Figure 12. Speed distributions of ISA for the morning peak (a) and the off-peak period (b).

Congestion Road Pricing

This section presents an application of DRACULA in investigating the variability of *congestion metering (CM)* charges in continuous time and space, as would be experienced by individual drivers. *CM* is a congestion-based road pricing system first proposed by Oldridge (1990). The underlying concept of the *CM* system is that charges would only be levied when delays occur. This can be achieved with advanced technology on-board a vehicle through an electronic in-vehicle unit (IVU) which debits value from an inserted smartcard only when a certain, pre-specified level of delay is encountered. Linking a clock and an odometer, in similar fashion to a taxi meter, allows the IVU to perform continuous calculations of the time taken to travel a certain unit of distance. By specifying a critical time value, it is then possible to set a *congestion threshold*, above which a charge would be levied. This threshold approach has many interesting features, in particular:

(a) charges would be expected to vary by route rather than by link;
(b) charges may vary significantly over a relatively short timescale, related to quite small variations in network conditions close to the threshold level;
(c) the definition of the threshold would need to include a margin of error, charging for less than 100% of delayed time, to ensure that vehicles would not be penalised for being courteous and obeying traffic regulations in un-congested conditions; and
(d) there may be significant variations in charges on a day-to-day basis.

Previous modelling work to investigate the route and demand choice aspects of a range of charging technologies in a static modelling context has been forced to rely on a much coarser specification (May and Milne, 2000). In addition, for *CM*, often the point at which a charge occurs may not be the same point at which the delay that triggers the charge happens (as demonstrated in Fig. 14). This raises the question of how to model drivers' reactions to the road charge in terms of their perception of the network. For this reason, only a route-based, individual vehicle second-by-second simulation model can fully represent the technology envisaged for time-dependent road charging systems. DRACULA simulation is novel in that it tracks vehicles along pre-specified routes to their destination, rather than using junction-by-junction turning percentages, and is thus able to monitor the experience of individual drivers following fixed routes to assess variations in charges between drivers and under different level of congestion and charging scenarios.

Implementing *CM* in DRACULA has required modification of the model to include four additional user specified parameters:

(i) a charging band, representing the unit of distance over which charges are levied;
(ii) a charging segment, representing the frequency at which charging information is assessed in distance terms;
(iii) a charging threshold, representing the travel time allowed for the coverage of the charging segment before any charge is levied; and
(iv) a charging rate, representing the level of charge levied once the threshold has been exceeded.

The initial values of these parameters were assumed to be 500 metres, 10 metres, 3 minutes and 30 pence per minutes respectively. Thus, for each individual vehicle on the network, it is assumed that an IVU is fitted which assesses the state of the threshold every 10 metres travelled. At that point, it reviews the travel time taken for the previous 500 metres. If time is less than or equal to 3 minutes, the vehicle is uncharged. If time exceeds 3 minutes, the vehicle is charged at a rate of 30 pence per minute for the excess time. Once a charge has been levied the state of the threshold is not assessed again until a further 500 metres has been covered. Values for the first three parameters were taken directly from those suggested in Cambridge. The unit rate of charge was chosen based on optimum levels identified in previous work (Milne 1997).

The DRACULA simulation of *CM* system has been applied to a real-world urban network of Leeds. The network has 175 nodes, 70 zones and 1748 routes used by a peak-hour demand of 23,000 trips, which is about 25% of the total morning peak demand for the city of Leeds.

Figure 13 shows the variability of *CM* charges and travel time along one of the three major radial route inbound to the city centre. It plots, for each individual vehicles travelled along the route, the *CM* charges and the effective time-based charges incurred by individual vehicles against their departure time. It can be seen that the charges levied by both charging systems, and more significantly by the *CM*, can be extremely volatile over very short periods of time and that, therefore, drivers would have extreme difficulty in predicting the costs of their journey both within and between days, even if they were aware of the likely overall traffic levels on any given day.

Figure 14 shows link-based simulation summary statistics. It plots, in bandwidth, the delays to each link and the amount of charges levied on each link. It can be seen that the location of charges may not necessarily be the same as where delay occurred. In fact, distribution of charges is more spread spatially than delays. This difference is in part due to the way these two measures are estimated. Link delay is simply the difference between free-flow travel time and actual travel time on the link. Whilst the *CM* charges are levied by looking back a specific length (500 metres in this case) and checking the time threshold. If, for example, a congestion occurred 150m upstream of an intersection, a charge could be levied on the downstream link.

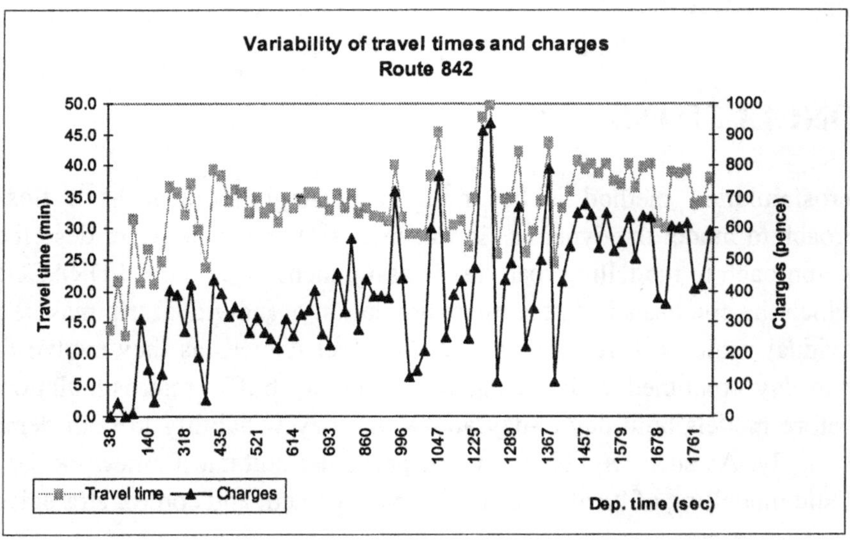

Figure 13. Variability of travel time and CM charges along a major corridor in North Leeds.

The results suggest that a practical threshold-based *CM* system may levy charges which are extremely variable over time and, thus, very difficulty to predict with any accuracy by both the drivers and system managers and that the location where a charge is levied differs from the location where actual delay occurred. The results obtained suggest some potentially important implications for the suitability of congestion-dependent charging mechanism, both for approximating the marginal costs of road travel and for providing useful incentives to drivers towards more efficient travel behaviour.

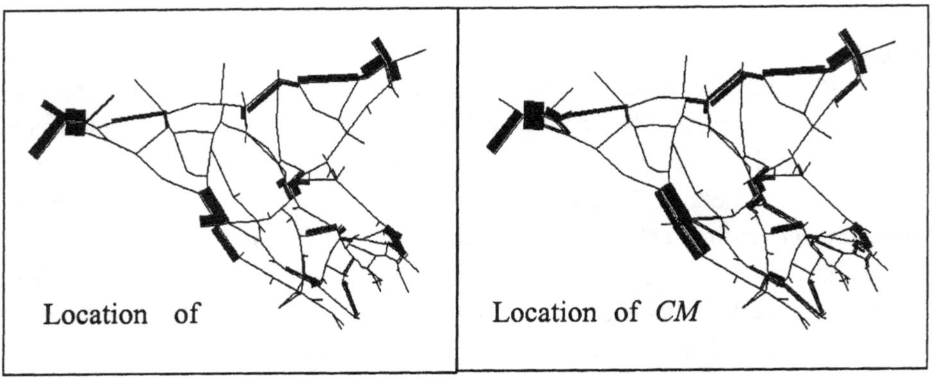

Figure 14. Spatial distribution of link delays and congestion charges levied.

CONCLUSIONS

Microsimulation method is becoming increasingly popular as a flexible approach to model the dynamics in transport systems. This paper describes a new approach to modelling road traffic assignment, code-named DRACULA, in which the emphasis is on the microsimulation of individual trip makers and individual vehicles. It represents directly driver choices as they evolve from day to day combined with a detailed within-day traffic microsimulation. It therefore models both day-to-day and within-day variability in both demand and supply. As such we believe it is a particular suitable framework for the realistic modelling of "real-time" traffic management and control strategies.

The traffic microsimulation component of the DRACULA suite is similar to many other traffic microsimulation models in that it models individual vehicles' movements based on car-following, lane-changing and gap-acceptance rules. However, it differs significantly in that it tracks vehicles along pre-specified routes from origin to their destination, rather than junction-by-junction turning percentages. The latter can lead to implausible, cyclic routes. Most of the other microscopic traffic simulators – some of them are mentioned in Section 1 – perform over a fixed time period and hence can only provide measures on network performance. By tracking all vehicles departing from a particularly time period to their destination, DRACULA is able to monitor the experiences of individual drivers to assess variations in the costs of travel between drivers, and through its day-to-day demand model, to model individuals' choices of travel based on their individual experience rather some aggregated system measures.

The framework is undergoing further research and development, reflecting on-going research in developing real-time strategies and in the dynamic evolution of traffic networks. A model of public transport operations has been developed which allows public transport priority measures such as guided bus to be evaluated (Liu et al. 1999). Further research is underway to combine a public transport assignment model with microsimulation of bus passengers and to study the effect of passenger demand and bus scheduling on service reliability. A pedestrian microsimulation model has been development and embedded under the general framework of DRACULA. The model simulates explicitly the complex interactions between vehicular traffic and pedestrians at signalised intersections and real-time traffic signal control strategies aimed to maximize the capacity while giving pedestrian priority (Liu et al. 2002). Research is under way to explore artificial intelligent methods, in particular multi-agent approach, to model the complex decision-making processes of drivers (Rossetti et al. 2002) and information provision and processing. It is also planned to connect the simulation model to a real-time signal control system and to incorporate dynamic en route diversion into the framework.

REFERENCES

Almquist, S., Hyden, C. & Risser, R. (1991). Use of speed limiters in cars for increased safety and a better environment, *Transport Research Record*, **1318**, pp34-39.

Barcelo, J., Ferrer, J., Grau, R., Florian, M. and Chabini, E. (1995). A route based version of the AIMSUN2 microsimulation model, *2nd World Congress on ITS,* Yokohama.

Ben-Akiva, M., De Palma, A. and Kanaroglou, P. (1986). Dynamic model of peak period traffic congestion with elastic arrival rates. *Transportation Science,* **20(2)**, pp164-181.

Cantarella, G.E. and Cascetta, E. (1995). Dynamic process and equilibrium in transportation networks: towards a unifying theory, *Transportation Science* **29(4)**, pp305-329.

Edie, L.C. (1961). Following and steady-state theory for non-congested traffic, *Oper. Res.*, **9**, pp66-76.

Emmerink, R.H.M., Axhausen, K.W., Nijkamp, P. and Rietveld, P. (1994). Effects of information in road transport networks with recurrent congestion, *Transportation,* **22**, pp21-53.

Emmerink, R.H.M. and Nijkamp, P. (ed.) (1999). *Behavioural and Network Impacts of Driver Information Systems*, Avebury, Aldershot, UK.

Fellendorf, M., MacAongusa, C. and Pierre, M. (1997). LRT priority within the SCATS environment in Dublin - a traffic flow simulation study, *Proc. Urban Transport and the Environment*, Computational Mechanics Publications, Oct. 1997.

Gipps, P.G. (1981). A behavioural car-following model for computer simulation,. *Transportation Research* **15B**, pp105-111.

Hu., T.-Y. and Mahmassani, H.S. (1997). Day-to-day evolution of network flows under real-time information and reactive signal control, *Transportation Research,* **5C**, pp51-69.

Hunt, P.B., Robertson, D.I., Bretherton, R.D. and Winton, R.I. (1981). SCOOT – a traffic responsive method of coordinating signals, *TRRL Laboratory Report 1014.*

ITE (1982). *Transportation and Traffic Engineering Handbook*, 2nd edition. Institute of Transportation Engineers, Prentice-Hall Inc., Englewood Cliffs, N.J.

Jha M, Madanat S and Peeta S (1999). Perception updating and day-to-day travel choice dynamics in traffic networks with information provision, *Transportation Research* **6C**(3), pp189-212.

Koshi, M., Iwasaki, M. and Ohkura, I. (1983). Some findings and an overview on vehicular flow characteristics, Proc. *8th International Symposium on Transportation and Traffic Flow Theory* (eds. V. Hurdle, E. Hauer, & G. Stuart), pp403-426.

Laird J, Druitt S & Fraser D (1999). Edinburgh city centre: a microsimulation case-study, *Traffic Eng. & Contr*, **40**(2), pp72-76.

Lind, G. (1999). Large-scale testing of intelligent speed adaptation – important evaluation issues, *Proc. AET European Transport Conference,* Seminar D, Vol P432, pp91-103, Cambridge.

Liu R, Van Vliet D and Watling DP (1995). DRACULA: Dynamic Route Assignment Combining User Learning and microsimulAtion, Proc. *PTRC*

Annual Conference, Seminar E, pp143-152.

Liu R, Clark SD, Montgomery FO and Watling DP (1999). Microscopic modelling of traffic management measures for guided bus operation, *Selected Proceedings of 8th World Conference on Transport Research* (eds. Oum, TH, Meersman, H & Winkelmans W), Vol 2, 367-380, Elsevier.

Liu R. and Tate J. (2000). Microsimulation modelling of intelligent speed adaptation systems, *Proc.European Transport Conference,* Cambridge 2000.

Liu, R., Silva, J. and Seco, A. (2002). A bi-modal microsimulation model for the assessment of pedestrian delays and traffic management, Paper submitted to *Transportation Research* **C.**

Mahmassani, H.S. and Jayakrishnan, R. (1991). System performance and user response under real-time information in a congested traffic corridor, *Transportation Research,* **25A**, pp293-308.

Mauro, V. (1991). Evaluation of dynamic network control: simulation results using NEMIS urban microsimulator, *Transportation Research Board 70th annual meeting*, Washington DC.

May, A.D. (1990). Traffic Flow Fundamentals, Prentice Hall, Englewood Cliffis, New Jersey.

May, A.D. and Milne, D.S. (2000). Effects of alternative road pricing systems on network performance,*Transportation Research,* **34A,** pp407-436.

Milne, D.S. (1997). Modelling the Effects of Urban Road User Charging, PhD, University of Leeds.

Nagel, K. and barrett, C.L. (1997). Using microsimulation feed back for trip adaptation for realistic traffic in Dallas, *International Journal of Modern Physics C,* **8(3)**, pp505-526.

Nsour S and Santiago A (1994). Comprehensive plane development for testing, calibration and validation of CORSIM, In *Proc. of 64th ITE Annual Transportation Engineers*, Dallas, pp486-490.

Oldridge, B. (1990). Electronic road pricing - an answer to traffic congestion?, *Proc. Information Technology and Traffic Management,* HMSO, London.

Papageorgiou, M., Hadi-Salem, H., Blosseville, J.M. and Bhouri, N. (1989). Modelling and real-time control of traffic flow on the Boulevard Peripherique in Paris, *IFAC Control, Computers, Communications in Transportation*, Paris, Pergamon, Oxford, pp205-211.

Rossetti, RJF, Bordini, R.H., Bazzan, A.L.C., Bampi, S., Liu, R. and Van Vliet, D. (2002). Using BDI agents to improve driver modelling in a commuter scenario, *Trans. Res.* **10C**, pp47-72.

Quinn, D.J. (1992). A review of queue management strategies, *Traffic Eng. & Contr.*, **33(11)**, pp600-605.

Van Vliet, D. (1982). SATURN - a modern assignment model, *Traffic Eng.& Contr.*, **23(12)**, pp578-581.

Vythoulkas, P.C. (1990). A dynamic stochastic assignment model for the analysis of general networks, *Transportation Research* **24B**, pp453-469.

DYNAMIC NETWORK SIMULATION WITH AIMSUN

Jaime Barceló
Dept.of Statistics and Operations Research, Universitat Politècnica de Catalunya, Pau Gargallo 5, 08028 Barcelona, Spain
barcelo@aimsun.com

Jordi Casas
TSS-Traffic Simulation Systems, Paris 101, 08029, Barcelona, Spain
casas@aimsun.com

ABSTRACT

The deployment of ITS must be assisted by suitable tools to conduct the feasibility studies required for testing the designs and evaluating the expected impacts. Microscopic traffic simulation has proven to be the suitable methodological approach to achieve these goals. This paper discuses some of the most critical aspects of the dynamic simulation of road networks, namely the heuristic dynamic assignment, the implied route choice models, and the validation methodology, a key issue to determine the degree of validity and significance of the simulation results. The paper is structured in two parts, the first provides an overview on how the main features of microscopic simulation have been implemented in AIMSUN, and the second is devoted to discus in detail the heuristic dynamic assignment.

AN OVERVIEW OF MICROSCOPIC TRAFFIC MODELING IN AIMSUN

The large investments generally required by transportation projects have to be justified in a robust way. Therefore feasibility studies need to be carried out that validate the proposals, assess their expected impacts and provide the basis for sound cost benefit analysis. This is especially so when studying the possible deployment of Intelligent Traffic Systems (ITS) to discover the benefits that should be expected from their operation. Microscopic traffic simulators are simulation tools that realistically emulate the flow of individual vehicles through a road network. They are proven tools for aiding transportation feasibility studies. This is not only due to their ability to capture the full dynamics of time dependent traffic phenomena, but also because they are capable of using behavioral models that can account for drivers' reactions when exposed to ITS. GETRAM, presented in this paper, is a simulation environment with a microscopic traffic simulator (AIMSUN) at its heart. It provides:

- The ability to accurately represent any road network geometry: An easy to use Graphic User Interface (TEDI) that can use existing digital maps of the road network allows the user to model any type of traffic facility.
- Detailed modeling of the behavior of individual vehicles. This is achieved by employing sophisticated car following and lane changing models that take into account both global and local phenomena that can influence each vehicle's behavior. The high quality of these models has been proven in numerous field tests.
- An explicit reproduction of traffic control plans: pre-timed as well as those defined by TRANSYT, SYNCHRO or Nema's standards. Auxiliary interfacing tools that allow the simulator to work with almost any type of real-time or adaptive signal control systems, as C-Regelaar, Balance, SCATS, SCOOT, and UTOPIA are also provided.
- Animated 2D and 3D output of the simulation runs. This is not only a highly desirable feature but can also aid the analysis and understanding of the operation of the system being studied and can be a powerful way to gain widespread acceptance of complex strategies.

Vehicles are assigned to routes according to a route choice model. Additionally AIMSUN allows vehicles to change their chosen route from origin to destination according to variations in traffic conditions as they travel through the road network. This provides the basis for heuristic traffic assignment.

The recent evolution of the AIMSUN microscopic simulator has taken advantages of the state-of-the-art in the development of object-oriented simulators, and graphical user interfaces, as well as the new trends in software design and the available tools that support it adapted to traffic modeling requirements (Banks, 1998). A proper achievement of the basic requirements of a microscopic simulator implies building models as close to the reality as possible. The closer the model is to reality the more data demanding it become. This has been traditionally the main barrier preventing a wider use of microscopic simulation. Manual coding of geometric data, turnings movements at intersections, timings and so on, is not only cumbersome and time consuming but also a potential source of errors. It is also hard to debug if the appropriate tools are not available. AIMSUN (**A**dvanced **I**nteractive **M**icroscopic **S**imulator for **U**rban and Non-Urban **N**etworks; http://www.aimsun.com) is imbedded in GETRAM (**G**eneric **E**nvironment for **TR**affic **A**nalysis and **M**odeling), a simulation environment inspired by modern trends in the design of graphical user interfaces adapted to traffic modeling requirements. A way of overcoming these drawbacks has been to provide GETRAM/AIMSUN with the proper user friendliness based on the versatility of the TEDI traffic network graphical editors, which can import the geometric background of the road network to draw the network model on top, as shown on the left part of Figure 1. The background can be imported as a .dxf file from a CAD or GIS system, or any other graphic format as .jpg, bit map and so on. All objects comprising the road model can be built with the graphic editor. Their attributes and parameters are defined and assigned values by means of window dialogues such as the one in the right part of figure 1, which shows the definition of the shared movements in a phase of a pre-timed signal control, and the allocation of the timings.

In summary, this software environment for traffic modeling make an easy task of the model building process, ensure accurate geometry, prevent errors, provide powerful debugging tools and can model any type of traffic related facility.

All transportation modes can be virtually modeled with the use of vehicle types, provided there is available the corresponding behavioral model. GETRAM includes a library or default vehicle types and a vehicle type editor to edit or create new ones. Figure 2 depicts an example of the model of a complex intersection in Amsterdam, courtesy of DHV Environment and Infrastructure that combines six different transportation modes. Figure 3 shows a window dialogue to define the parameters of a vehicle type. Once the simulation model of a basic scenario has been built, before being used for sophisticated applications the microscopic traffic simulator has to prove that it can reproduce to an acceptable degree of significance the observed traffic conditions or, in other words, that it is capable of emulating the reality with enough accuracy. The calibration and validation of the simulator are the required proofing exercises. A very good example of a data collection process for the calibration and validation of a traffic simulation can be found in (Hughes, 1998).

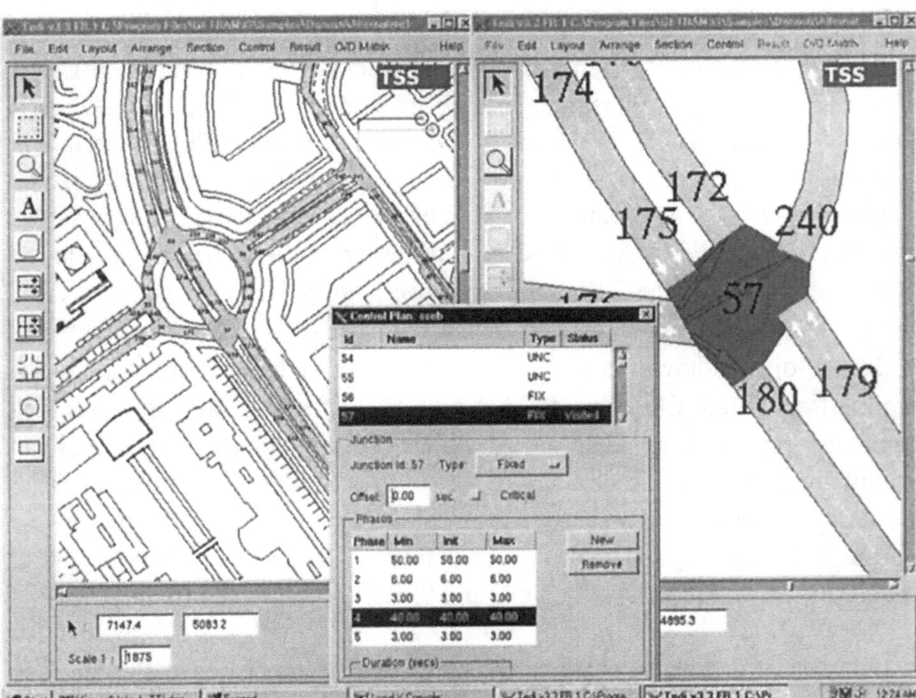

Figure 1: Example of GETRAM graphic user interface for building microscopic simulation models.

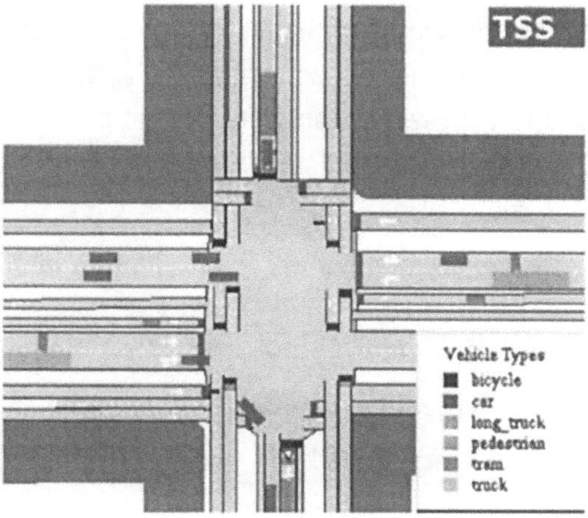

Figure 2

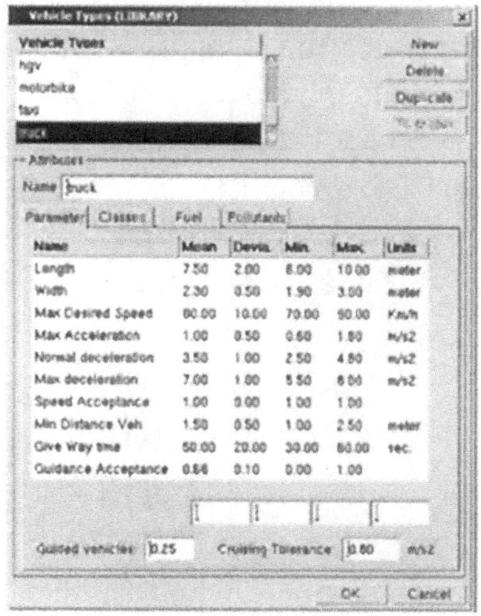

Figure 3

Car-following in microscopic simulation

Most of the currently existing microscopic traffic simulators are based on the family of car-following, lane changing and gap acceptance models to model the vehicle's behavior. Some of the most used car-following models derive from the extensive research undertaken in the late fifties by the General Motors Group and are based on comprehensive field experiments and the development of the mathematical theory bridging micro and macro theories of traffic flows. This research led to the formulation of the car-following models as a form of stimulus-response equation, (Gerlough and Huber, 1975), where the response is the reaction of a driver to the motion of the vehicle immediately preceding him in the traffic stream. The response is always to accelerate or decelerate in proportion to the magnitude of the stimulus at time t and is begun after a time lag T, the reaction time of the follower. The General Motors Group developed a series of models whose basis equation is of the form:

Response (t+T) = Sensitivity * Stimulus (t)

Models vary according to the various answers to the key questions: What is the nature of the driver's response? To what stimulus does he react and how do we measure his sensitivity?. For an overview of car-following models see (Gabbard, 1991). The first, and simplest model corresponds to the case when the response is represented by the acceleration or deceleration of the follower driver and the stimulus is represented by the variation in the relative speeds. This simple model considers that the sensitivity is constant. If $\mathbf{X_n(t)}$ and $\mathbf{X_{n+1}(t)}$ are the positions of the leader and follower respectively at time t then the basic model is:

$$\ddot{\mathbf{X}}_{n+1}(\mathbf{t}+\mathbf{T}) = \lambda\left[\dot{\mathbf{X}}_n(\mathbf{t}) - \dot{\mathbf{X}}_{n+1}(\mathbf{t})\right] \quad \text{And}$$

$$\begin{array}{lll} \text{If } \dot{\mathbf{X}}_n(\mathbf{t}) > \dot{\mathbf{X}}_{n+1}(\mathbf{t}) & \text{then} & \ddot{\mathbf{X}}_{n+1}(\mathbf{t}+\mathbf{T}) > \mathbf{0} \\ \text{If } \dot{\mathbf{X}}_n(\mathbf{t}) < \dot{\mathbf{X}}_{n+1}(\mathbf{t}) & \text{then} & \ddot{\mathbf{X}}_{n+1}(\mathbf{t}+\mathbf{T}) < \mathbf{0} \\ \text{If } \dot{\mathbf{X}}_n(\mathbf{t}) = \dot{\mathbf{X}}_{n+1}(\mathbf{t}) & \text{then} & \ddot{\mathbf{X}}_{n+1}(\mathbf{t}+\mathbf{T}) = \mathbf{0} \end{array} \qquad (1)$$

The underlying hypothesis behind these models, is that a driver will place himself at a distance from the lead vehicle such that in the event of an emergency stop by the leader the follower will come to rest without striking the lead vehicle. These models assume that the follower driver will adjust his reactions to a change in velocity of the leader accelerating and decelerating at the same rate for a given perception of the stimulus. However, it is obvious that the deceleration capabilities are usually greater that the acceleration capabilities. This was already observed by (Herman and Rothery, 1959), who proposed to modify the basis model (1) so that:

$$\begin{array}{ll} \ddot{\mathbf{X}}_{n+1}(\mathbf{t}+\mathbf{T}) = \lambda_{+}\left[\dot{\mathbf{X}}_n(\mathbf{t}) - \dot{\mathbf{X}}_{n+1}(\mathbf{t})\right] & \text{for relative velocity positive} \\ \ddot{\mathbf{X}}_{n+1}(\mathbf{t}+\mathbf{T}) = \lambda_{-}\left[\dot{\mathbf{X}}_n(\mathbf{t}) - \dot{\mathbf{X}}_{n+1}(\mathbf{t})\right] & \text{for relative velocity negative} \end{array}$$

Taking into account these different behaviors, and the fact that from the behavioral point of view other factors, as for example the target or desired speed of a driver, should also be taken into account, (Gipps, 1981), (Mahut, 2000), develops an empirical (behavioral instead of "response to an stimulus") model consisting of two components: acceleration and deceleration, defined as a function of variables that can be measured. The first represents the intention of a vehicle to achieve certain desired speed, while the second reproduces the limitations imposed by the preceding vehicle when trying to drive at the desired speed. This model states that, the maximum speed at which a vehicle (n) can accelerate during a time period (t, t+T) is given by:

$$V_a(n,t+T) = V(n,t) + 2.5a(n)T\left(1 - \frac{V(n,t)}{V^*(n)}\right)\sqrt{0.025 + \frac{V(n,t)}{V^*(n)}} \quad (2)$$

Where: V (n, t) is the speed of vehicle n at time t; V*(n) is the desired speed of the vehicle (n); a(n) is the maximum acceleration for vehicle n; T is the reaction time. On the other hand, the maximum speed that the same vehicle (n) can reach during the same time interval (t, t+T), according to its own characteristics and the limitations imposed by the presence of the leader vehicle is:

$$V_b(n,t+T)$$
$$= d(n)T + \sqrt{d(n)^2T^2 - d(n)\left[2\{x(n-1,t) - s(n-1) - x(n,t)\} - V(n,t)T - \frac{V(n-1,t)^2}{d'(n-1)}\right]} \quad (3)$$

Where: d (n) (< 0) is the maximum deceleration desired by vehicle n; x (n, t) is position of vehicle n at time t; x (n-1, t) is position of preceding vehicle (n-1) at time t; s (n-1) is the effective length of vehicle (n-1); d' (n-1) is an estimation of vehicle (n-1) desired deceleration. The final speed for vehicle n during time interval (t, t+T) is the minimum of those previously defined speeds:

$$V(n,t+T) = \min\{\, V_a(n,t+T), V_b(n,t+T) \,\} \quad (4)$$

The position of vehicle n inside the current lane is updated by taking the speed into the movement equation:

$$x(n,t+T) = x(n,t) + V(n,t+T)T \quad (5)$$

Rewriting equation (4) as **v_f(t+T)=min {v_a(t+T), v_b(t+T)}** at each instant in time the follower vehicle can be in one of the two states:

1. **v_f(t+T) = v_a(t+T)** maximum speed of the follower at time t+T allowed by the acceleration constraint, when the safe deceleration to stop constraint is not active and the vehicle is moving "freely".
2. **v_f(t+T) = v_b(t+T)** maximum speed of the follower at time t+T allowed by the safe deceleration to stop constraint, when the safe deceleration to stop constraint becomes active.

A common drawback of the way in which most of these models are implemented in simulation packages is that the model parameters are global i.e. constant for the entire network, whereas it is well known that driver's behavior is affected by traffic conditions. Therefore a more realistic way of implementing car-following modeling for microscopic simulation should account for local behavior. That implies that some of the model parameters must be local depending on local geometric and traffic conditions.

The AIMSUN car following model evolved after the seminal Gipps model, which was improved to meet the requirements, described earlier. The first improvement is related to the vehicle speed $V^*(n)$ used in the Gipps model. In AIMSUN implementation $V^*(n)$ is the desired speed of vehicle n for the current section, and is therefore a local parameter. Which is calculated according to the procedure described next.

Calculating the speed of a vehicle on a section

The car-following model is such that a leading vehicle, i.e. a vehicle driving freely, would try to drive to its maximum desired speed. Three parameters are used to calculate the maximum desired speed of a vehicle while driving on a particular section or turning; two are related to the vehicle and one to the section or turning:

1. Maximum desired speed of the vehicle *n*: $v_{\max}(n)$
2. Speed acceptance of vehicle *n*: $\theta(n)$ (A parameter measuring the driver's degree of accomplishment of the speed limits on the section)
3. Speed limit of the section or turning s: $S_{\lim nt}(s)$

The speed limit for a vehicle *n* on a section or turning *s*, $S_{\lim nt}(n,s)$, is calculated as:

$$s_{\lim nt}(n,s) = S_{\lim nt}(s) \cdot \theta(n)$$

Then, the maximum desired speed of vehicle *n* on a section or turning *s*, $v_{\max}(n,s)$ is calculated as:

$$v_{\max}(n,s) = MIN\left[s_{\lim nt}(n,s), v_{\max}(n)\right]$$

This maximum desired speed $v_{max}(n,s)$ is the one referred above, in the Gipps car following model, as V*(n).

Modeling the influence of adjacent lanes in the car following model and effects of grades

When the vehicle is driving along a section, the modified car-following model considers the influence that certain number of vehicles (*Nvehicles*) driving slower in the adjacent right-lane –or left-lane, when driving on the left–, may have on the vehicle. The model calculates first the mean speed for *Nvehicles* driving downstream of the vehicle in the adjacent slower lane (*MeanSpeedVehiclesDown)*. Only vehicles within a certain distance (*MaximumDistance*) from the current vehicle are taken into account. We distinguish two cases: 1) the adjacent lane is an on-ramp, and 2) the adjacent lane is any other type of lane. Apart from *Nvehicles* and *MaximumDistance* parameters, the user can define two additional parameters, *MaximumSpeedDifference and MaximumSpeedDifferenceOnRamp*. Then, the final desired speed of a vehicle on a section is calculated as follows:

if (the adjacent slower lane is an On-ramp)
{MaximumSpeed = MeanSpeedVehiclesDown + MaximumSpeedDifferenceOnRamp}
else {MaximumSpeed = MeanSpeedVehiclesDown + MaximumSpeedDifference}
DesiredSpeed = Minimum ($v_{max}(n,s), \theta(n)$ * MaximumSpeed)

This procedure ensures that the differences of speeds between two adjacent lanes will approximately be always lower than *MaximumSpeedDifference* or *MaximumSpeedDifferenceOnRamp*, depending on the case.

The influence of the section grade in the vehicle movement is modeled by means of an increase or reduction of the acceleration and braking capability. The maximum acceleration for a vehicle on a section that will be used in the car-following model is a function of the grade and the maximum desired acceleration for the vehicle given by:

accel = Maximum(vehicle_acc - grade*9.81/100.0, vehicle_acc*0.1)

In order to avoid zero or negative acceleration values, a minimum value of 10% of the maximum desired acceleration for the vehicle is used.

Model calibration and testing

In addition to numerous tests performed by the research team, the car-following model in AIMSUN has been tested and calibrated in various real life projects; due to space limitations we only present the benchmark test performed by a research group from Robert Bosch GmbH, (Manstetten *et al.*, 1998a) (Manstetten *et al.*, 1998b), (Bleile *et al.* 1996). This test employed a set of field data and most of the micro-simulator developers in Europe and North America were invited to participate and asked to use the same error metric to measure the accuracy between measured and simulated values, in order to get comparable results. To avoid overrating discrepancies for large distances the following relative metric was chosen weighted by the logarithm and squared: $Em = \sum \left[\log\left(\frac{d\_sim}{d\_meas} \right) \right]^2$ where *d_sim* is the distance of the simulated vehicle, *d_meas* is the distance measured with the test vehicle, and *log* denotes the logarithm base 10. The results show that the AIMSUN car-following model is able of a fairly good reproduction of the observed values. The numerical value of the error metric outperforms those provided for most of the currently used models (see Manstetten et al (1998b) for details), as the following table shows.

Table 1. Comparative results of the Bosch Car-Following Test

Model	MITSIM	AIMSUN	Wied/Pel	Wied/Vis	NSM	OVM	T^3M
Deviation	3.75	3.36	14.01	10.67	24.51	9.37	2.40

MiITSIM is the simulator developed at MIT, (Qi and Kotsopoulos, 1996), Wied/Pel, and Wied/Vis stand for Wiedemans/Pelops and VISSIM models respectively, (Wiedemann, 1974) and (Fellendorf, 1994). NSM is the cellular automaton model of Nagel-Schreckenberg (Schreckenberg *et al.*, 1995), and the remaining two are particular models referenced in (Manstetten *et al.*, 1998b).

An additional test to analyze the quality of the microscopic simulator is to check the ability to reproduce macroscopic behavior. Also the research team at Bosch proposed in Manstetten et al. (1998a) a test to compare various microscopic simulators: "*The macroscopic behavior of a microscopic model can be most easily tested by simulating the traffic on cyclic one lane roads. This excludes any effects of lane changes and node passing and concentrates on the car-following task. For this study, a cyclic road of 1000 m length was used. A fixed number of vehicles have been initially set with speed value 0 km/h at randomly determined positions. All vehicles had the same length of 4.5 m and the drivers had the same free flow speed of 54 km/h. Starting with this initial situation a 10 minute time period was simulated without any measurements to reach traffic conditions which are achievable by the model's behavior itself. After the starting phase the traffic behavior has been recorded at one local measurement point during a simulation time of 2 hours (exact passing time and speed value of each vehicle). The fixed number of vehicles for the simulation run was varied in discrete steps to realize different traffic densities. To visualize the results the traffic flow has been drawn versus the density (given as the number of initially set vehicles on the 1 km ring). The maximal mean traffic flow value of about 1800 veh/h is known as a quite realistic value for longer periods of measurement time. Under urban traffic conditions this maximal flow is typically reached at higher density values than for freeway traffic*". The results of AIMSUN for the simulated flow density curve versus the empirical one for the second test are displayed in figure 4, and they appear to be fairly reasonable. This subjective perception is confirmed by the values of the error metric to measure the fitting between the measured and simulated values as before, that in this case is *Em*=0,011411. The graphics in figure 4 also shows the sensitivity of the AIMSUN Car-following model to variations in the values of the model parameters. In absence of microscopic measurements the adjustment of model parameters to fitting macroscopic empirical curves for the relationships between the fundamental traffic variables could also be used as an alternative procedure for model calibration. A subset of the simulation experiments to determine the values of the model parameters best fitting the observed values is summarized in Table 2.

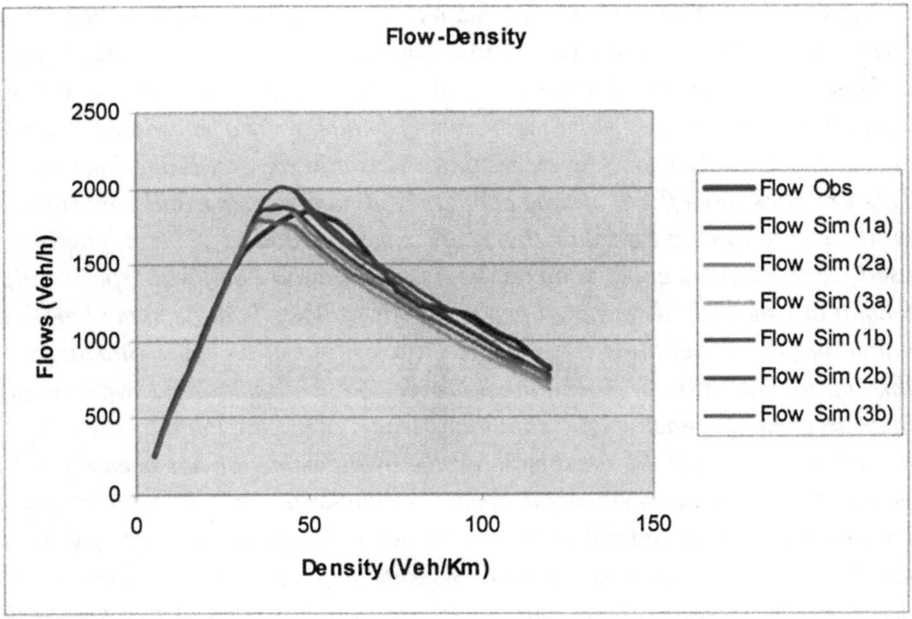

Figure 4: Empirical versus simulated flow-density curves

Table 2: Model quality as a function of reaction time and effective vehicle length

Model Parameters: Reaction Time (RT, seconds) and Effective Vehicle Length (Vehicle Length+DM, meters)						
	Simulation 1a	Simulation 2a	Simulation 3a	Simulation 1b	Simulation 2b	Simulation 3b
	RT0.9 /DM1.0	RT0.95 /DM1.0	RT1.0 /DM1.0	RT0.90 /DM0.75	RT0.95 /DM0.75	RT1.0 /DM0.75
Em	0,01917175	0,04267196	0,07880027	0,011411	0,01865812	0,04366021

As Table 2 shows, the best fitting is achieved in the simulation experiment 1b with a reaction time of 0.9 seconds, and an effective length equal to the vehicle length plus 0.75 meters.

Lane changing model

The lane change model in AIMSUN can also be considered as a further evolution of the Gipps lane change model (Gipps, 1986). Lane change is modeled as a decision process analyzing the necessity of the lane change (as in the case of turning maneuvers determined by the route), the desirability of the lane change (as for example to reach the desired speed when the leader vehicle is slower), and the feasibility conditions for the lane change that are also local, depending on the location of the vehicle on the road network. The lane-changing model is a decision model that approximates the driver's behavior as follows:

Each time a vehicle has to be updated the model draws up the question: *Is it necessary to change lanes?* The answer to this question depends on several factors: the turning feasibility at current lane, the distance to next turning and the traffic conditions in the current lane. The traffic conditions are measured in terms of speed and queue lengths. When a vehicle is driving slower than he wishes, he tries to overtake the preceding vehicle. On the other hand, when he is traveling fast enough, he tends to go back to the slower lane.

If the answer to the previous question is affirmative, to succeed in the lane changing two more questions have to be answered:

a) *Is it desirable to change lanes?* This requires checking if there will be any improvement in the traffic conditions for the driver as a result of the lane changing. This improvement is measured in terms of speed and distance. If the speed in the future lane is faster (i.e. a user specified threshold is exceeded) than the current lane or if the queue is shorter than a threshold, then it is desirable to change lanes.

b) *Is it possible to change lanes?* This requires verifying if there is a sufficient gap to do the lane change with complete safety. For this purpose, we calculate both the braking imposed by the next downstream vehicle to the changing vehicle and the braking applied by the changing vehicle to the future upstream vehicle. If both braking ratios are acceptable then lane changing is possible.

In order to achieve a more accurate representation of the driver's behavior in the lane changing decision process, three different zones inside a section are considered, each one corresponding to a different lane changing motivation. The distance up to the end of the section characterizes these zones and which is the next turning point. The figure 5 depicts the structure of these zones that are defined as follows:

- **Zone 1**: This is the farthest from the next turning point. The lane changing decisions are governed by the traffic conditions of the lanes involved; the feasibility of the next desired turning movement is not yet taken into account. To measure the improvement that the driver will get on changing lanes several parameters are considered: the desired speed of the driver, speed and distance of the current preceding vehicle and speed and distance of the future preceding vehicle.

- **Zone 2**: This is the intermediate zone. Mainly it is the desired turning lane that affects the lane changing decision. Vehicles who are not driving on a valid lane (i.e. a lane where the desired turning movement can be done) tend to get closer to the correct side of the road where the turn is allowed. In this zone vehicles look for a gap and may try to accept it without affecting the behavior of vehicles in the adjacent lanes.

- **Zone 3**: This is the nearest to the next turning point. Vehicles are forced to reach their desired turning lanes, reducing the speed if necessary and even coming to a complete stop in order to make the lane change possible. Also, vehicles in the adjacent lane can modify their behavior in order to allow a gap big enough for the lane-changing vehicle

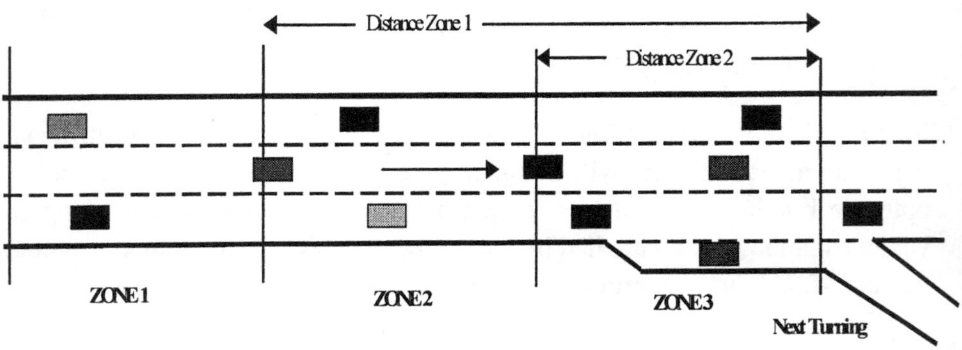

Figure 5: Lane Changing Zones

Lane changing zones are defined by two parameters, Distance to Zone 1 and Distance to Zone 2. These parameters are defined in time (seconds) and they are converted into distance whenever it is required for each vehicle at each section using the Vehicle Desired Speed at a Section. This means that these distances are then local parameters their value depending on the current traffic conditions on the section. When a vehicle crosses from zone 1 to zone 2 there is a change in the vehicle's behavior, as now it becomes relevant the next turn. Also the crossing from zone 2 to zone 3 produces a change in the behavioral rules of the vehicles, as now reaching the turning lane becomes urgent. In order to distribute these changes of behavior along a longer distance a greater variability is given to the Lane Changing Zones. These zones are calculated particularly for each vehicle according to the following equation:

Distance Zone n for vehicle ***v*** in section ***s*** (in meters)
= Distance Zone n (in seconds) * Speed Limit of Section ***s***
* Vehicle ***v*** Coefficient

Vehicle ***v*** Coefficient
= Speed Limit of Section ***s*** / Desired speed of Vehicle ***v*** in section ***s***

This algorithm ensures that for vehicles whose desired speed is slower than the speed limit the lane changing zones will be longer than for vehicles whose desired speed is greater than the speed limit. It means for instance that a heavy truck will try to reach the appropriate turning lane earlier than a speed car.

Look Ahead

When traffic conditions are very congested it may happen that some vehicles cannot reach the appropriated turning lane and consequently miss the next turn. This situation could appear either in urban networks where there are short sections or in freeways where weaving sections may be relatively short. It gets worst as the sections get more congested. Tuning some modeling parameters such as lane changing zone distances, simulation step, acceleration rates etc., could improve the behavior in order to minimize the number of lost vehicles. Also using polysections in modeling the geometry instead of sections, when feasible, to model streets or weaving areas might help to improve the situation, but it was not enough. To override these drawbacks a major improvement has been done in the lane change model consisting in modeling a Look Ahead process. The objective is to provide vehicles with the knowledge of various next turning movements and not only one, so they will be able to make decisions not only based on the immediate next turning movement, but on a set of next turning movements. The Look Ahead consists of four steps:

1. At any time, each vehicle knows the next two turning movements, so the lane changing decisions are influenced by two consecutive turns.
2. The lane changing zones 2 and 3 of any section is extended back beyond the limits of the section, therefore affecting the upstream sections.
3. The next turning movement also influences the turning maneuvers so the selection of destination lane is done based also on the next turn.
4. A greater variability is given to the Lane Changing Zones in order to distribute the lane changing maneuvers along a longer distance.

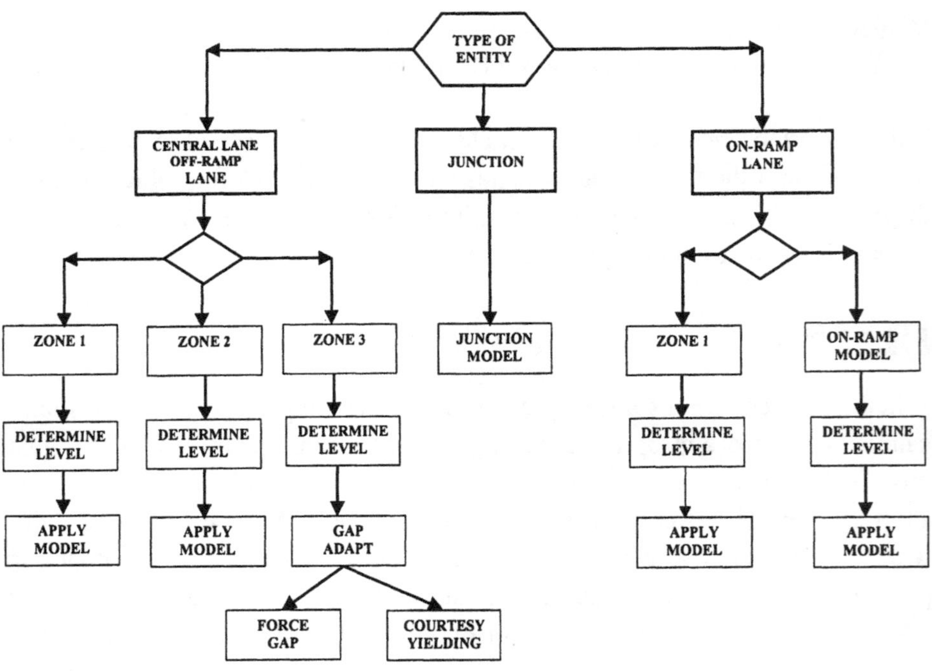

Figure 6: Lane changing decision tree

Lane Changing Modeling at ON-Ramps

A special Lane Changing Modeling is applied at entrance ramps. An additional zone parameter may be defined, TimeDistanceOnRamp. This is the Distance (in seconds, converted into distance as before) from those lateral lanes considered to be on-ramp lanes, in order to distinguish between a common lateral lane, that is a long lane used for overtaking which drops down, from the proper on-ramp lanes, which are never used for overtaking. Vehicles driving on a lateral lane that are farther than *TimeDistanceOnRamp* from the end behave as if they were in the Zone 1 of a normal lane. When they are closer than *TimeDistanceOnRamp* to the end of the lane, they behave as having to merge from an on-ramp. Merge from on-ramp model takes into account whether a vehicle is stopped or not, if it is stopped whether it is at the beginning of the on-ramp queue or not and how long has been waiting. There is another vehicle parameter, *Maximum Waiting Time*, that determines how long a vehicle is willing to wait before getting impatient.

The Lane-Changing decision process

The whole lane change process is modeled formally as a decision tree model whose logic structure is depicted in the diagram of figure 6. The process identifies the type of entity (central lane, off-ramp lane, junction, on-ramp etc.) in which the maneuver is going to be done, next determines how the zone modeling should be applied. The current traffic conditions are analyzed and the level at which the lane change can be performed is determined, and then the corresponding model is applied, this also includes the specific modeling of lane change prohibitions and overtaking maneuvers, as well as specific Gap Acceptance models for each situation. For further details see AIMSUN User's Manual (2002).

Gap acceptance models in AIMSUN

Two typical situations in which gap acceptance should be applied are:

- In lane changing, and
- To model give way behaviour

Gap Acceptance Model in Lane changing

To answer the question *"Is it possible to change lanes?"* AIMSUN applies the following algorithm to check whether the gap is acceptable or not.

```
Get downstream and upstream vehicles in target lane
Calculate gap between downstream and upstream vehicles: TargetGap
if ((TargetGap > VehicleLengh) & (it is aligned)) then
   Calculate the distance between vehicle and downstream vehicle in target
     lane: DistanceDown
   Calculate the speed imposed by downstream vehicle to vehicle, according
     to Gipps Car-following Model: ImposedDownSpeed
   if (ImposedDownSpeed is acceptable by vehicle, according to the
       deceleration rate) then
      Calculate the distance between upstream vehicle in target lane and
       vehicle: DistanceUp
      Calculate the speed imposed by vehicle to upstream vehicle, according
       to Gipps Car-following Model: ImposedUpSpeed
      if (ImposedUpSpeed is acceptable by upstream vehicle, according to
          the deceleration rate) then
         Lane Change is Feasible
         CarryOutLaneChange
      else
         The gap is not acceptable because of the upstream vehicle
      endif
   else
        The gap is not acceptable because of the downstream vehicle
   endif
else
   There is no gap aligned with the vehicle
endif
```

Gap Acceptance Model in give way

The Gap-Acceptance model used to model give way behaviour determines whether a lower priority vehicle approaching a junction can or cannot cross depending on the circumstances of higher priority vehicles (position and speed). This model takes into account the distance of vehicles from the hypothetical collision point, their speeds and their acceleration rates. It then determines the time needed by the vehicles to clear the junction and produces a decision to cross or not which is also a function of the level of risk for each driver. Several vehicle parameters may influence the behaviour of the gap-acceptance model: acceleration rate, desired speed, speed acceptance and maximum give-way time. Other parameters, such as visibility distance at the junction and turning speed, which are related to the section, may also have an effect. Among these, the acceleration rate, the maximum give-way time and the visibility distance at junctions are the most important. The acceleration rate gives the acceleration capability of the vehicle and therefore has a direct influence on the required safety gap. The maximum give-way time is used to determine when a driver starts to get impatient if he/she cannot find a gap. When the driver has been waiting for more than this time, it reduces the safety margin (normally two simulation steps) by half (only one step). The following algorithm is applied in order to determine whether a vehicle approaching a give-way sign can cross or not (see Figure 7):

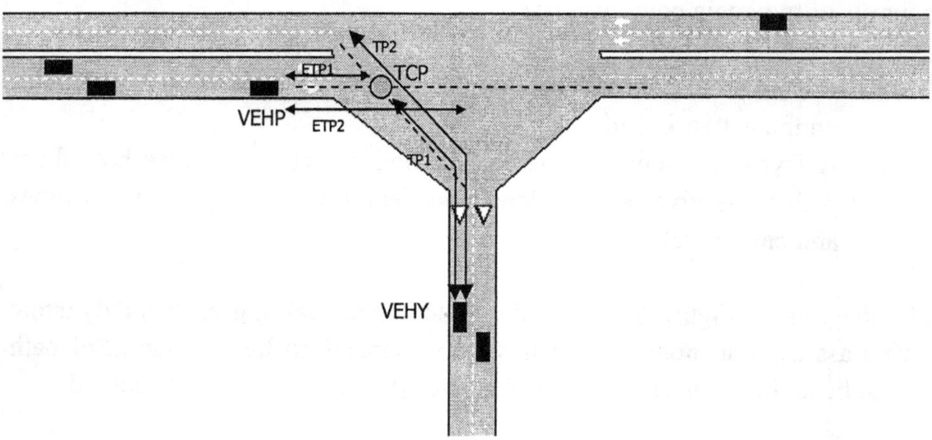

Figure 7

Given a vehicle (VEHY) approaching a Yield (Give Way) junction,

Obtain the closest higher priority vehicle (VEHP),
Determine the Theoretical Collision Point (TCP),
Calculate the time (TP1) needed by VEHY to reach TCP,
Calculate the estimated time (ETP1) needed by VEHP to reach TCP,
Calculate the time (TP2) needed by VEHY to cross TCP,
Calculate the estimated time (ETP2) needed by VEHP to clear the junction,

If TP2 (plus a safety margin) is less than ETP1, vehicle VEHY has enough time to cross, therefore it will accelerate and cross,

Else, if ETP2 (plus a safety margin) is less than TP1, vehicle VEHP will have already crossed TCP when VEHY reaches it, then search for the next closest vehicle with a higher approach, Next VEHP and go to step 2.

Else, vehicle VEHY must give way, decelerating and stopping if necessary.

HEURISTIC DYNAMIC ASSIGNMENT BASED ON AIMSUN MICROSCOPIC TRAFFIC SIMULATION

According to (Florian *et al.*, 2001), a dynamic traffic assignment model consists of two main components:

1. A method to determining the path dependent flow rates on the paths on the network, and
2. A Dynamic Network Loading method, which determines how these path flows give raise to time-dependent arc volumes, arc travel times and path travel times

The diagram in Figure 8 depicts the logic conceptual approach for dynamic traffic assignment models. Path flow rates depend on the emulation of path choice behavior of drivers. Two alternative approaches can be considered:

- Dynamic assignment *en route*: at each time period the corresponding fraction of the demand is assigned to the currently available paths for each Origin-Destination pair according to the probabilities estimated by a route choice model. Driver can be allowed to dynamically change the route *en route* if a better path from their current position to their destination is available.

- Dynamic Equilibrium Assignment: path flows are determined by an approximate solution to the mathematical model for the dynamic equilibrium conditions. (Florian *et al.*, 2001), (Barceló *et al.*, 2002).

The Dynamic Network Loading, also known as Dynamic network Flow Propagation, (Cascetta, 2001), "models simulate how the time-varying continuous path flows propagate through the network inducing time-varying in-flows, out-flows and link occupancies". A wide variety of approaches, from analytical, (Wu, 1991; Wu *et al.*, 1998a; Wu *et al.*, 1998b; Xu *et al.*, 1998; Xu *et al.*, 1999), to simulation based, (Florian *et al.*, 2001), have been proposed. This paper explores the feasibility of a Dynamic Network Loading mechanism based on the AIMSUN microscopic simulation model, (Barceló *et al.*, 1995, 1998, 2002), (Codina and Barceló, 1995). This traffic simulation approach is proposed not only due to its ability to capture the full dynamics of time dependent traffic phenomena, but also for being capable of dealing with behavioral models accounting for drivers' reactions in the way required by the Dynamic Network Loading.

However, it should be noticed that the simulation of such systems requires a substantial change in the traditional paradigms of microscopic simulation, in which vehicles are generated at the input sections in the model, and perform turnings at intersections according to probability distributions. In such model vehicles have neither origins nor destinations and move randomly on the network. The required simulation approach should be based on a new macroscopic simulation paradigm: a route based microscopic simulation. In this approach, vehicles are input into the network according to the demand data defined as an O/D matrix (preferably time dependent) and they drive along the network following specific paths in order to reach their destination. In the main Route Based simulation new routes are to be calculated periodically during the simulation, and a Route Choice model is needed, when alternative routes are available.

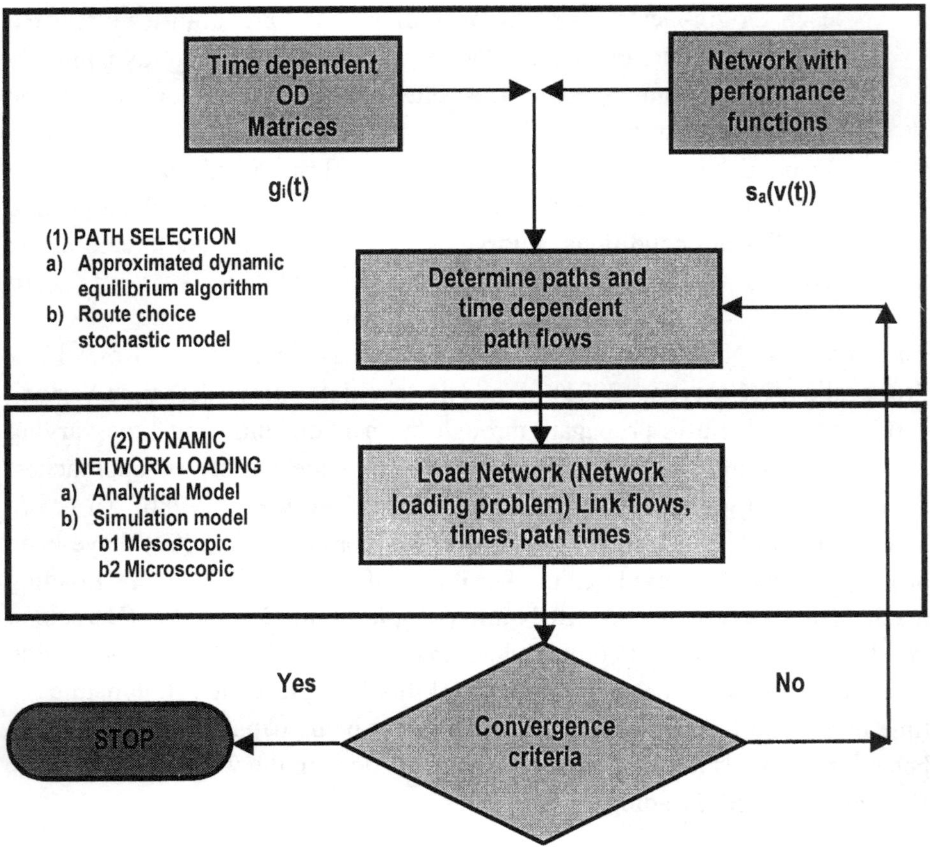

Figure 8. Conceptual approach to dynamic traffic assignment

The Heuristic Network Loading

The simulation process based on time dependent routes consists of the following steps:

1. Calculate initial shortest routes for each O/D pair using the defined initial costs.
2. Simulate for a predefined time interval (e.g. 5 minutes) assigning to the available routes the fraction of the trips between each O/D pair for that time interval according to the selected route choice model and obtain new average link travel times as a result of the simulation.

3. Recalculate shortest routes, taking into account the current average link travel times.
4. If there are guided vehicles, or variable message signs suggesting rerouting, provide the information calculated in 3 to the drivers that are dynamically allowed to reroute.
5. Go to step 2.

Which repeat until all the demand has been assigned instead of the convergence criteria in Figure 8 used by the analytical model to control the length of the simulation.

Two types of section cost functions are provided by default for calculating the shortest path trees, depending on whether or not there are simulated data available to be used for. These are the Initial Cost Function and the Current Cost Function. In both cases, the cost function represents section travel time in seconds, including the penalty of the turning movement, if it exists. The *Initial Cost Function* is applied at the beginning of simulation when there is no simulated data gathered to calculate the travel times. In this case, the cost of each section is calculated as a function of the travel time in free flow conditions and the capacity of the section. The initial cost of section s, $IniCost_{s,i}$, taking into account the penalty for the i-th turning at the end of the section, and the capacity $CT_{s,i}$ of the i-th turning on section s is calculated as follows:

$$IniCost_{s,i} = TravelTFF_{s,i} + TravelTFF_{s,i} * \varphi * (1 - CT_{s,i} / MaxCapacity)$$

Where $TFF_{s,i}$ is the estimated travel time in free flow conditions, φ is a user-defined capacity weight parameter that allows the user to control the influence that the section capacity has in the cost in relation with the travel time, and *MaxCapacity* is the theoretically estimated maximum section capacity in the network. The *Current Cost Function* is applied when there is simulated travel time data available. The default current cost for each section is the mean travel time, in seconds, for all simulated vehicles that have crossed the section during the last time period.

The heuristic approach proposed in this paper may be considered as *preventive*, see (Wu, 1991; Wu *et al.* 1998a; Wu *et al.*, 1998b; Xu *et al.*, 1998;

Xu *et al.*, 1999), since it tries to describe the evolution of the flows when users make route choice decisions based on actual experienced travel times, represented here in terms of the path costs for the previous time interval.

Estimation of path flow rates: assignment based on discrete route choice models

The vehicles follow paths from their origins in the network to their destinations. So the first step in the simulation process is to assign a path to each vehicle when it enters the network. In the implementation done in AIMSUN the candidate paths can be of different types:

- User-defined Paths (UdP): Predefined using the network editor or as an output from other traffic simulators or transportation models, either macroscopic (i.e. a transport planning) or microscopic.
- Calculated Shortest Path Trees (CSPT): Shortest path trees calculated using the default or user defined cost functions. There are two types of CSPT:

    - Initial Shortest Path Tree (ISPT): For each destination centroid it provides a shortest path tree, using an initial default or user defined cost function for each turning movement.

    - Statistical Shortest Path Tree (SSPT): For each destination centroid and every user defined time interval it provides a shortest path tree, using the values of the cost function for each turning movement for that time interval. The value of the cost function is updated for each time interval with the statistical data previously gathered during the simulation.

A vehicle of vehicle type *vt* traveling from origin O_i to destination D_j, can choose one path according to a discrete choice model from the set of alternative paths:

- *N* User-defined Paths: $UdP_k(O_i,D_j)$ $k=1..N$
- *1* Initial Shortest Path Tree: $ISPT(D_j)$,
- *P* Statistical Shortest Path Trees: $SSPT_k(D_j)$ $k=1..P$

With probabilities

$P(UdP_k(O_i,D_j), vt)$: Probability of use $UdP_k(O_i,D_j)$ by a vehicle type vt
$P(ISPT(D_j), vt)$: Probability of use $ISPT(D_j)$ by a vehicle type vt

Satisfying the condition:

$$\sum_{k=1}^{N} \mathrm{P}(\mathrm{UdP_k}(\mathrm{O_i},\mathrm{D_j}), \mathrm{vt}) + \mathrm{P}(\mathrm{ISPT}(\mathrm{D_j}), \mathrm{vt}) \leq 1$$

At the beginning of the simulation, shortest path trees are calculated from every section to each destination centroid, taking as arc costs the specified initial costs. During the simulation, new routes are recalculated at every time interval, taking the specified arc costs updated for each arc after the statistics gathered during a number of statistics gathering interval defined by the analyst. The user may define the time interval for recalculation of paths, that is the frequency at which paths are refreshed according to the prevailing traffic conditions, and the maximum number of path trees to be maintained during the simulation. When the maximum number of path trees (K) is reached, the oldest paths will be removed as soon as no vehicle is following them. It is assumed that vehicles only choose among the most recent K path trees. Therefore, the oldest ones will become obsolete and disused. From the point of view of the modeling approach there are two alternatives determining how the dynamic network loading will be performed:

- The concept of cost used in updating the routes
- The route choice model used in assigning vehicles to available routes

Assuming that route cost is the sum of the costs of the arcs composing the route, a wide variety of arc costs can be proposed: travel times at each simulation interval, toll pricing, historical travel times representing driver's experience from previous days, combinations of various arc attributes as for instance travel times, length and capacity, etc. The version 4.1 of AIMSUN, GETRAM/AIMSUN (2002), provides the user with two alternatives: use default arc costs or use the Function Editor included in the TEDI set of graphical editors of the GETRAM traffic modeling environment to define his/her own arc cost function using as arguments any of the numerical attributes of the road sections, statistical values or vehicle characteristics.

Calculation of shortest paths is carried out per vehicle type, taking into account the reserved lanes. Therefore, the set of paths from which a vehicle may select, either when entering the network or when being re-routed, may be different for different vehicle types even though they travel to the same destination, depending on the presence of reserved lanes. Also the travel time used in the cost function for recalculation of shortest paths is taken as the travel time per vehicle type. The default cost assigned to each arc is a function of the travel time of the section and the turning movement, the arc capacity can also be taken into account and then the default cost of arc *a* for vehicle type *vt* is calculated as:

$$Cost(a,vt) = CurrentCost(a,vt) + CurrentCost(a,vt) \times \varphi \times \left(1 - \frac{Capacity(a)}{MaxCapacity}\right)$$

As an alternative to the Default Initial and Cost Functions users can define their own Cost Functions. This is done via the Function Editor in Tedi. If no User-Defined Function is assigned to an arc, the Default Cost Function is applied. To define an Initial or Cost Function, the user can use any of the most common mathematical functions and operators (+, -, *, /, ln, log, exp, etc.). The function parameters can be constants or variables that are data related to the description of the network, sections, turning movements and vehicle types, in other words the function arguments could be any of the numerical attributes available in the simulator. For Cost Functions, as there is also simulated output data available, variables corresponding to simulated statistical data can also be used in the cost function. Figure 9, illustrates an example of a simulation model for which alternative arc cost functions have been defined by the user. The open window shows part of the algebraic expression for a cost function per vehicle type.

In a similar way when using these dynamic assignment abilities it could be raised the question of which is the most suitable route choice function. Route choice functions represent implicitly a model of user behavior, representing the most likely criteria employed by the user to decide between alternative routes: perceived travel times, route length, expected traffic conditions along the route, etc.

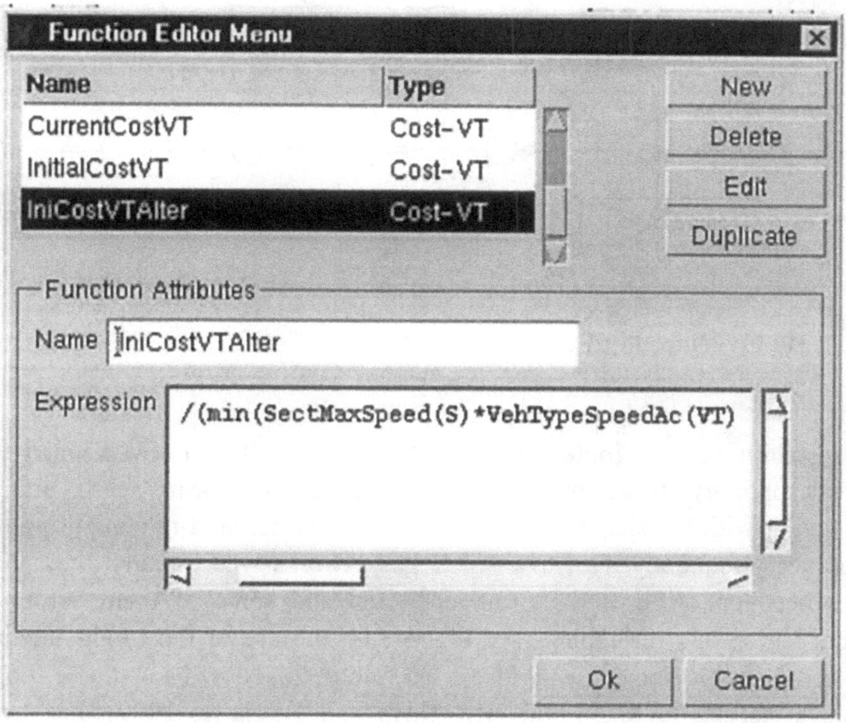

Figure 9: Example of user defined link cost

The solution implemented in the version 4.1 of AIMSUN also provides the user two alternatives: use the default functions or define his/her own route choice function by means of the Function Editor. The most used route choice functions in transportation analysis are those based on the discrete choice theory, i.e. Logit functions assigning a probability to each alternative route between each origin-destination pair depending on the difference of the perceived utilities. A drawback reported in using the Logit function is the observed tendency towards route oscillations in the routes used, with the corresponding instability creating a kind of flip-flop process. According to our experience there are two main reasons for this behavior. The properties of the Logit function and the inability of the Logit function to distinguish between two alternative routes when there is a high degree of overlapping.

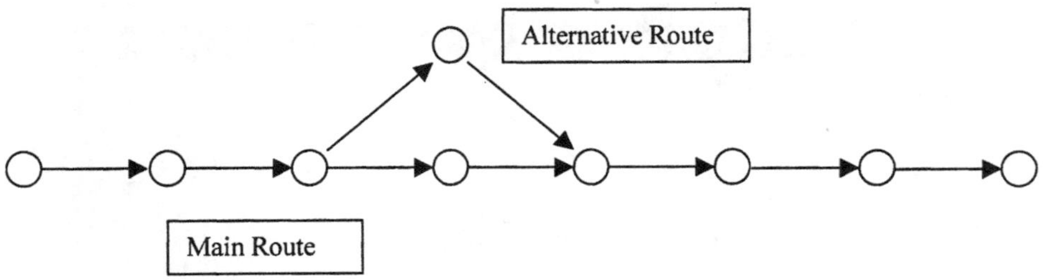

Figure 10: Overlapping Routes

The instability of the routes used can be substantially improved when the network topology allows for alternative routes with little or no overlapping at all, playing with the shape factor of the Logit function and re-computing the routes very frequently. However, in large networks where many alternative routes between origin and destinations exist and some of them exhibit a certain degree of overlapping (see Figure 10), the use of the Logit function may still exhibit some weaknesses. To avoid this drawback the C-Logit model, (Cascetta *et al.*, 1996; Ben-Akiva and Bierlaire, 1999), has been implemented. In this model, the choice probability P_k, of each alternative path k belonging to the set I_{rs} of available paths connecting an O/D pair, is expressed as:

$$P_k = \frac{e^{\theta(V_k - CF_k)}}{\sum_{l \in I_{rs}} e^{\theta(V_l - CF_l)}}$$

where V_i is the perceived utility for alternative path i, and θ is the scale factor, as in the case of the Logit model. The term CF_k, denoted as 'commonality factor' of path k, is directly proportional to the degree of overlapping of path k with other alternative paths. Thus, highly overlapped paths have a larger CF factor and therefore smaller utility with respect to similar paths. CF_k is calculated as follows:

$$CF_k = \beta \cdot \ln \sum_{l \in I_{rs}} \left(\frac{L_{lk}}{L_l^{1/2} L_k^{1/2}} \right)^{\gamma}$$

where L_{lk} is the length of arcs common to paths l and k, while L_l and L_k are the length of paths l and k respectively. Depending on the two factor parameters β and γ, a greater or lesser weighting is given to the 'commonality factor'. Larger values of β means that the overlapping factor has greater importance with respect to the utility V_i; γ is a positive parameter, whose influence is smaller than β and which has the opposite effect. The utility V_i used in this model for path i is the opposite of the path travel time tt_i, (or path cost depending on how has been defined by the user).

A heuristic adaptive algorithm to estimate an origin-destination dependent scale factor for the logit route choice function

Taking into account the role of the scale factor in smoothing out the shape of the logit function, the right value of parameter θ should be different for each O-D pair depending on the travel time variability. From a practical point of view the easy way is to define θ as a global simulation parameter with the same value for all O-D pairs, and this works for road networks in which path travel times are of the same order of magnitude, but it could be unrealistic for large networks where long path travel times coexist with shorter ones. But, how to input the θ_i values for each O-D pair $i \in I$, the set of all O-D pairs when $|I|$ is a large number. This raises the question on how to define the right value for each O-D pair. To cope with this problem the following adaptive-heuristics is proposed. Let $tt_1^i, \ldots, tt_n^i$ be the ordered set of travel times for the n paths of the r-th O-D pair at time interval i, that is $tt_1^i < \ldots < tt_n^i$. When the variability of the travel times for the r-th O-D pair between successive time intervals changes beyond some predefined thresholds the estimation by a maximum likelihood procedure of the scale factor θ_r for the route choice function of the r-th O-D pair is as follows:

1. **<u>Estimation</u> of target probabilities:** Let $s = \sum_{j=1}^{n} tt_j^i$ be the sum of the current travel times for the n paths between the r-th O-D pair.

   <u>Compute</u> the set of weights: $w_j = \frac{tt_n^i}{tt_j^i}$, j=1,…,n;

   <u>Define</u>: $\overline{tt}_j^i = (s - tt_j^i)(w_j)^\beta \quad j=1,\ldots,n$;

Calculate: $\bar{s} = \sum_{j=1}^{n} \bar{tt}_j^i$; **and Compute the estimation of the target probabilities as:** $\hat{P}_j = \frac{\bar{tt}_j^i}{\bar{s}}$, **j=1,...,n**

2. **Estimation of the scale factor** θ_r: **Select** $\hat{P}_j$ for some j and

Define
$$f(\theta_r) = \ln \hat{P}_j + \ln\left(\sum_{k=1}^{n} e^{-\theta_r tt_k^i}\right) + \theta_r tt_j^i$$
And

$$f'(\theta_r) = \frac{\sum_{k=1}^{n} \left(-tt_k^i\right) e^{\theta_r tt_k^i}}{\sum_{k=1}^{n} e^{-\theta_r tt_k^i}} + tt_j^i$$

Initialize θ_r^0

Iterative step k+1 Let $\theta_r^{k+1} = \theta_r^k - \frac{f(\theta_r^k)}{f'(\theta_r^k)}$

If $\left|\theta_r^{k+1} - \theta_r^k\right| \leq \varepsilon$ then stop

Otherwise let k←k+1 and repeat

On basis to the computational experiments conducted with network models of various sizes the recommended value for β, is 2,5 ≤ β ≤ 3, although the determination of the right value in each case will be the outcome of the validation process.

Path Analysis

To get the insight on what is happening in a heuristic dynamic assignment for the proper calibration and validation of the simulation model the user should have access to the analysis of the used routes. To support the user in this analysis process AIMSUN includes a path analysis tool. Figure 11 depicts the path dialogue window.

The path list box contains the list of section identifiers composing the path and the following information is displayed for each section:

- The cost in time (seconds) from each of the sections in the path to the destination centroid. This can be calculated as either the sum of *IniCost(a)* or *Cost(a, vt)* of all the arcs composing the path.
- The travel time in seconds from each of the sections in the path to the destination centroid. This is equal to the cost only if the capacity weight parameter is set to zero.
- The distance (meters) from each of the sections in the path to the destination centroid.

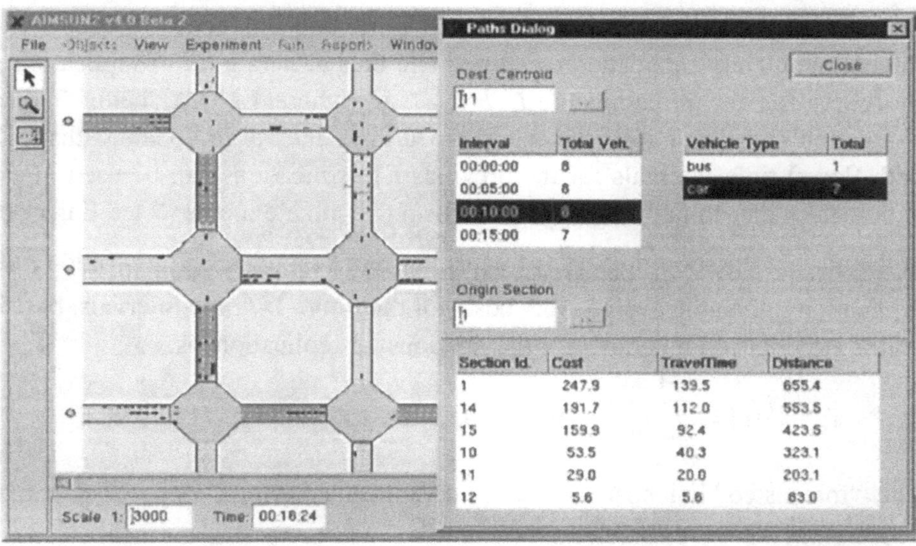

Figure 11: Path Dialog Window

In the example shown in Figure 11, the shortest path from section 1 to centroid 11 goes through sections 14, 15, 10, 11 and 12. The cost of the whole path is 247.9 seconds, the travel time is 139.5 seconds and the distance is 655.4 meters. In this case Cost and Travel Time are different, as the Capacity Weight has been set to 1.25. Vehicles are initially assigned to a route from a set of available routes on a probabilistic way determined by the selected route choice model.

Apart from the initial assignment of route, which is made at the vehicle's departing time, there is the possibility of making a route reassignment during the trip. This is the Dynamic route choice model in which a vehicle can make a new decision about what route to follow at any time along its trip, whenever there are new shortest routes available, whereas in the Static model, a vehicle will always follow its initially selected route until reaching the destination, although new shortest route could be available during the trip. In the Dynamic model all vehicles or only a special class, that of the guided vehicles can take the decision of changing to a new shortest route during the trip, assuming that this information is available to them.

Route choice based on a day-to-day learning mechanism

In this case the simulation is replicated N times and link costs for each time interval and every replication are stored and thus at iteration l of replication j the costs for the remaining $l+1$, $l+2,\dots$, L (where L=T/Δt, being T the simulation horizon and Δt the user defined time interval to update paths and path flows) time intervals for the previous j-1 replications can be used in an anticipatory day-to-day learning mechanism to estimate the expected link cost at the current iteration. Let $s_a^{jl}(v)$ be the current cost of link a at iteration l of replication j, then the average link costs for the future L-l time intervals, based on the experienced link costs for the previous j-1 replications is:

$$\bar{s}_a^{j,l+i}(v) = \sum_{m=1}^{j-1} s_a^{m,l+i}(v); \quad i = 1,\dots, L-l$$

The "forecasted" link cost can then be computed as:

$$\tilde{s}_a^{j,l+1}(v) = \sum_{i=0}^{L-l} \alpha_i \bar{s}_a^{j,l+i}(v);$$

$$where \quad \sum_{i=0}^{L-l} \alpha_i = 1, \alpha_i \geq 0, \forall i; \; are \;\; weighting\, factors$$

The resulting cost of path k for the i-th OD pair is $\tilde{S}_k(h^{l+1}) = \sum_{a \in A} \tilde{s}_a^{j,l+1}(v)\delta_{ak}$ where, as usually δ_{ak} is 1 if link a belongs to path k and 0 otherwise. The path costs $\tilde{S}_k\left(h^{l+1}\right)$ are the arguments of the route choice function (logit, C-logit, user defined, etc.) used at iteration $l+1$ to split the demand g_i^{l+1} among the available paths for OD pair i.

MODEL VALIDATION

The reliability of any simulator depends on its ability to produce results close to reality. The process of determining whether the simulation model is close enough to the actual network being simulated is generally achieved through the validation of the model; this is generally an iterative process involving calibration of the model parameters, comparing the model to the actual system behavior and using the discrepancies between the two, improve the model until the accuracy is judged to be acceptable. In the case of traffic the behavior of the actual system is usually defined in terms of measurable traffic variables such as flows, speeds, occupancies, queue lengths, and so on, which for practical purpose are measured by detectors at specific locations in the road network. To validate the simulation model the simulator should be able of emulating the traffic detection process and produce a series of simulated measurements, which are close to the actual. The statistical methods and techniques for validating simulation models are presented in most textbooks and specialized papers Balci (1998), Kleijnen (1992, 1995), Law and Kelton (1991). When several days of real measurements are available it is recommended that the calibration and the final validation of the model be based on different datasets. This will generate the trust that the model is not only accurate for the day(s) were it was calibrated but that the calibration also holds in general. Calibration means to determine the right values for the parameters governing the basic models (i.e. car following, lane changing, etc.). An example of calibration exercise has been discussed in section 2.2 for the car-following model in AIMSUN. This procedure can be considered as an extension to the methodology proposed by Yoshii (1999). At each step in the iterative validation process a simulation experiment is conducted. Each of these simulation experiments is defined by the data input to the simulation model and the set of values of the model parameters that identify the experiment. The output of the experiment is a set of simulated values of the variables of interest, such as the flows measured at each detector station in the road network per counting interval. For example, if the counting interval is five minutes, the model statistics are gathered every five minutes. In the case where the validation variable is the simulated flow **w**, the output of the simulation model will be characterized by the set of values $\mathbf{w_{ij}}$, of the simulated flow at detector station i at time j, where index i identifies the detector station (i=1,2,...,n, being n the number of detectors), and index j the sampling interval (j=1,2,...,m, being m the number of sampling intervals in the simulation horizon T). If $\mathbf{v_{ij}}$ are the corresponding actual measurements for detector i at sampling interval j, a typical statistical technique to validate the model would be compare both sets of simulated and actual counts to determine if they are close enough. For detector i the comparison could be based on testing whether the error of the j-th "prediction" $\mathbf{d_i = w_{ij} - v_{ij}}$, **j=1,...,m,** is small enough. A typical way of estimating the error of the predictions for the detector i is "Root Mean Square Error", $\mathbf{rms_i}$, or **the** "Root Mean Square Percentage Error", $\mathbf{rmsp_i}$, defined respectively by:

$$\mathbf{rms_i} = \sqrt{\frac{1}{m}\sum_{j=1}^{m}\left(w_{ij} - v_{ij}\right)^2} \quad \text{and}$$

$$\mathbf{rmsp_i} = \sqrt{\frac{1}{m}\sum_{j=1}^{m}\left(\frac{w_{ij} - v_{ij}}{v_{ij}}\right)^2}$$

This error estimate has been perhaps the most used in traffic simulation, and although obviously the smaller $\mathbf{rms_i}$ is the better the model is, it has a quite important drawback, as far as it squares the error it emphasizes large errors. Therefore it would be helpful to have a measure that considers both, the disproportionate weight of large errors and provides a basis for comparison with other methods. On the other hand it is quite frequent in traffic simulation that neither the observed values nor the simulated ones are independent, namely when only single sets of traffic observations are available (i.e. flows, speeds and occupancies for one day of the week during the rush hour). A family of statistical tests for the validation of simulation model achieving these objectives is rooted in the observation that the measured and the simulated series, $\mathbf{v_{ij}}$ and $\mathbf{w_{ij}}$ respectively, are time series. In this case the measured series could be interpreted as the original one and the simulated series the "prediction" of the observed series, and thus the quality of the simulation model can be established in terms of the quality of the prediction, and that means to resort to time series forecasting techniques for that purpose.

If one considers that what is observed as output of the system as well as output of the model representing the system is dependent on two type of components: the functional relationships governing the system (the pattern) and the randomness (the error), and that the measured as well as the observed data are related to these components by the relationship: **Data = pattern + error.** Then the critical task in forecasting can be interpreted in terms of separating the pattern from the error component so that the former can be used for forecasting. The general procedure for estimating the pattern of a relationship is through fitting some functional form in such a way as to minimize the error component. A way of achieving that could be through the regression analysis.

Theil's U-Statistic, Theil (1966) is the measure achieving the above-mentioned objectives of overcoming the drawbacks of the rmse index and taking into account explicitly the fact that we are comparing two autocorrelated time series, and therefore the objective of the comparison is to determine how close both time series are. In general, if X_j is the observed and Y_j the forecasted series, j = 1,...,m, then, if $\mathbf{FRC}_{j+1} = \frac{Y_{j+1} - X_j}{X_j}$ is the forecasted relative change, and $\mathbf{ARC}_{j+1} = \frac{X_{j+1} - X_j}{X_j}$ is the actual relative change, Theil's U-Statistic is defined as:

$$U = \sqrt{\frac{\sum_{j=1}^{m-1}\left(FRC_{j+1} - ARC_{j+1}\right)^2 \Big/ (m-1)}{\sum_{j=1}^{m-1}\left(ARC_{j+1}\right)^2 \Big/ (m-1)}} = \sqrt{\frac{\sum_{j=1}^{m-1}\left(\frac{Y_{j+1} - X_{j+1}}{X_j}\right)^2}{\sum_{j=1}^{m-1}\left(\frac{X_{j+1} - X_j}{X_j}\right)^2}}$$

An immediate interpretation of Theil's U-Statistic, is the following:

$\mathbf{U = 0 \Leftrightarrow FRC_{j+1} = ARC_{j+1}}$, and then the forecast is perfect
$\mathbf{U = 1 \Leftrightarrow FRC_{j+1} = 0}$, and the forecast is as bad as possible

In this last case the forecast is the same as that that would be obtained forecasting no change in the actual values. When forecasts Y_{j+1} are in the opposite direction of X_{j+1} then the U statistic will be greater that unity. Therefore the closer to zero the Theil's U-Statistics is the better the forecasted series is or, in other words, the better the simulation model. When Theil's U-statistic is close to or greater than 1 the forecasted series, and therefore the simulation model, should be rejected. When the forecast efficiency is based on the regression model $\mathbf{E(v|w)} = \beta_0 + \beta_1\mathbf{w} + \varepsilon$ (ε random error term) the most efficient forecast would correspond to $\beta_0 = \mathbf{0}$ and $\beta_1 = \mathbf{1}$, that can be tested by the application of variance analysis to the regression model as indicated earlier. But taking into account that the average squared forecast error:

$$\mathbf{D_m^2} = \frac{1}{m}\sum_{j=1}^{m}\left(Y_j - X_j\right)^2$$

can be decomposed (Theil) in the following way:

$$\mathbf{D_m^2} = \frac{1}{m}\sum_{j=1}^{m}\left(Y_j - X_j\right)^2 = \left(\overline{Y} - \overline{X}\right)^2 + \left(S_Y - S_X\right)^2 + 2(1-\rho)S_Y S_X$$

where $\overline{\mathbf{Y}}$ and $\overline{\mathbf{X}}$ are the sample means of the forecasted and the observed series respectively, $\mathbf{S_Y}$ and $\mathbf{S_X}$ are the sample standard deviations and ρ is the sample correlation coefficient between the two series, the following indices can be defined:

$$\left.\mathbf{U_M} = \frac{\left(\overline{Y} - \overline{X}\right)^2}{D_m^2};\ \mathbf{U_S} = \frac{\left(S_Y - S_X\right)^2}{D_m^2};\ \mathbf{U_C} = \frac{2(1-\rho)S_Y S_X}{D_m^2}\right\} \Rightarrow \mathbf{U_M + U_S + U_C = 1}$$

U_M is the "Bias proportion" index and can be interpreted in terms of a measure of systematic error, $\mathbf{U_S}$ is the "variance proportion" index and provides an indication of the forecasted series ability to replicate the degree of variability of the original series or, in other words, the simulation model's ability to replicate the variable of interest of the actual system. Finally $\mathbf{U_C}$ or "Covariance Proportion" index is a measure of the unsystematic error. The best forecasts, and hence the best simulation model, are those for which $\mathbf{U_M}$ and $\mathbf{U_S}$ do not differ significantly from zero and $\mathbf{U_C}$ is close to unity. It can be shown that this happens when β_0 and β_1 in the regression do not differ significantly from zero and unity respectively.

A Case Study

To illustrate in practice the proposed methodology, a set of simulation experiments has been conducted with the model depicted in Figure 12. This is the AIMSUN model of the Borough of Amara of the city of San Sebastian. The urban road network comprises 365 road sections, 100 junctions and intersections, 24 of which are signalized, 13 centroids defining 135 O-D pairs, and 15 traffic detectors measuring traffic flows. The simulation horizon runs from 18:00 to 20:00 corresponding to the evening rush hour. The dialogue window shown in figure 12 illustrates the definition of a simulation experiment in which, according to the descriptions in the above sections, the route choice function selected is the logit, routes are updated every 5 minutes, statistics for the 3 previous time intervals are used to estimate the new route costs, a maximum of 5 shortest paths trees is kept for each O-D pair, and the option of estimating adaptively the values for the logit scale factor is used. The values of these simulation parameters have been determined empirically during the model validation process. The validation criterion has been the comparison between the measured flow for the traffic detectors and their emulation. The Figure 13 depicts the comparison between the observed and the measured flow values for one of the detectors, a similar analysis has been conducted for all other detectors. The comparison is based on the analysis of the U Theil's statistics.

The corresponding values are:

$$\mathbf{U = 0.023364 \quad U_M = 0.280401 \quad U_S = 0.29644}$$
$$\mathbf{R^2 = 0{,}974429 \quad rmsp = 0{,}049314}$$

The U, R2 and rmsp values show a fairly good agreement between the real system and the simulation model, although the bias and the variance proportions are at the limited of the acceptability.

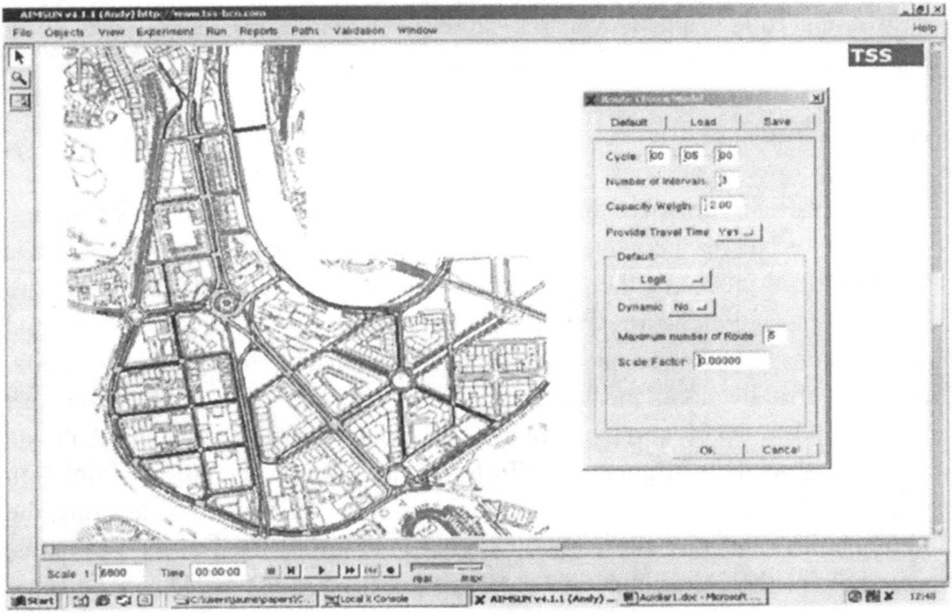

Figure 12. Simulation model of the road network of the Borough of Amara in San Sebastian

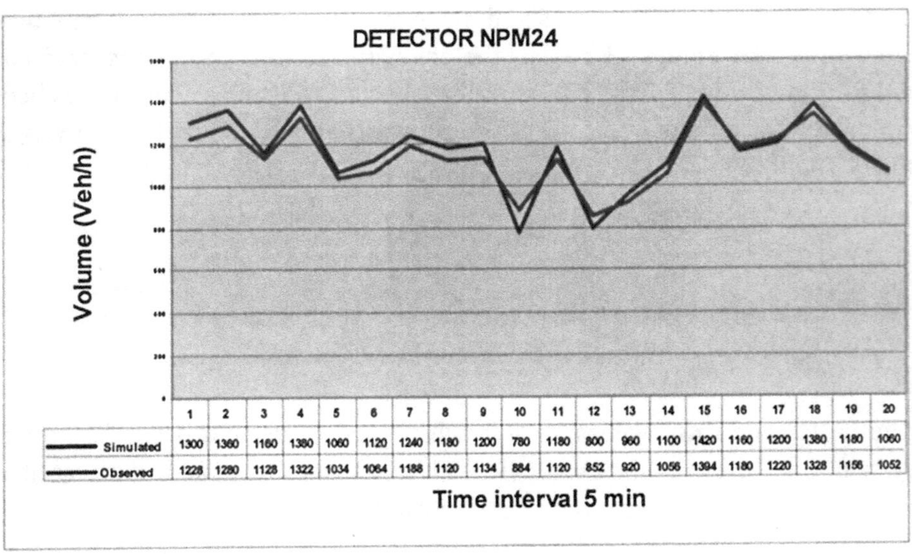

	1	2	3	4	5	6	7	8	9	10	11	12	13	14	15	16	17	18	19	20
Simulated	1300	1360	1160	1380	1060	1120	1240	1180	1200	780	1180	800	960	1100	1420	1160	1200	1380	1180	1060
Observed	1228	1280	1128	1322	1034	1064	1188	1120	1134	884	1120	852	920	1056	1394	1180	1220	1328	1156	1052

Figure 13. Comparative analysis of measured versus simulated values for detector NPM24 in Amara

CONCLUSIONS

This paper has been structured along three main components. In the first part some of the main basic modeling principles on which microscopic simulation is based, car-following, lane changing, etc, have been discussed in some detail and the improved way in which they have been included in AIMSUN has been described. Next a main contribution on how the dynamic simulation of complex networks works, from the point of view of the heuristic dynamic traffic assignment, has been analyzed in detail, and its implementation in AIMSUN has been reported. The last part of the paper has discussed a relevant issue for microscopic simulation, that of the model validation. A methodology has been proposed and its application has been demonstrated by a real life case study.

ACKNOWLEDGEMENTS

The developments and part of the research reported in this paper have possible by the collaboration of Pablo Barceló, Jaime Ferrer and David García, former members of the research group LIOS of the Department of Statistics and Operations Research of the Technical University of Catalonia (Barcelona, Spain), and current staff of TSS-Transport Simulation Systems.

REFERENCES

AIMSUN Version 4.1 User's Manual, TSS-Transport Simulation Systems, 2002.

Balci O. (1998). Verification, Validation and Testing, *in Handbook of Simulation: Principles,*
Methodology, Advances, Applications and Practice, Ed. by J. Banks, John Wiley, 1998.

Banks J. Editor, (1998). *Handbook of Simulation: Principles, Methodology, Advances, Applications and Practice*, John Wiley and Sons.

Barceló J., J.L. Ferrer, R. Grau, M. Florian, I. Chabini and E. Le Saux, (1995). A Route Based variant of the AIMSUN Microsimulation Model, *Proceedings of the 2nd World Congress on Intelligent Transport Systems,* Yokohama.

Barceló J., J.L. Ferrer, D. García, M. Florian and E. Le Saux (1998). Parallelization of Microscopic Traffic simulation for ATT Systems Analysis. In *Equilibrium and Advanced Transportation Modeling,* (P. Marcotte and S. Nguyen), Kluwer Academic Publishers.

Barceló J. and J. Casas (2002). Heuristic Dynamic Assignment based on Microscopic Traffic Simulation, *Proceedings of the 9th Meeting of the Euro Working Group on Transportation*, Bari, Italy.

Ben-Akiva M. and M. Bierlaire (1999). Discrete Choice Methods and Their Applications to Short Term Travel Decisions, in *Transportation Science Handbook*, Kluwer.

Bleile T., W. Krautter, D. Manstetten and T. Schwab (1996). Traffic Simulation at Robert Bosch GmbH, *Proc. Euromotor Seminar Telematic / Vehicle and Environment*, Aachen, Germany, Nov. 11-12,

Cascetta E., A. Nuzzolo, F. Russo and A. Vitetta (1996). A Modified Logit Route Choice Model Overcoming Path Overlapping Problems, in *Proceedings of the 13th International Symposium on Transportation and Traffic Flow Theory*, Pergamon Press.

Codina E. and J. Barceló (1995). Dynamic Traffic Assignment: Considerations on Some Deterministic Modeling Approaches, *Annals of Operations Research*, **60**, pp1-58.

Fellendorf M. (1994). VISSIM: A microscopic simulation tool to evaluate actuated signal control including bus priority, Technical paper, Session 32, *64th ITE Annual Meeting*, Dallas.

Florian M., M. Mahut and N. Tremblay (2001). A Hybrid Optimization-Mesoscopic Simulation Dynamic Traffic Assignment Model, *Proceedings of the 2001 IEEE Intelligent Transport Systems Conference*, Oakland, pp120-123.

Gabbard J.F. (1991). Car-Following Models. In: M. Papageorgiou (Ed.), *Concise Encyclopedia of Traffic and Transportation Systems*, Pergamon Press, Oxford.

Gerlough D.L. and M.J. Huber (1975). Traffic Flow Theory – A Monograph, *Transportation Research Board*, Special Report 165, Chapter 6, TRB, 1975.

Gipps P.G. (1981). A behavioral car-following model for computer simulation. *Transp. Res. B*, **Vol. 15**, pp105-111.

Gipps P.G. (1986). A Model for the Structure of Lane-Changing Decisions, *Transpn. Res.*, **20B**, pp403-414.

Herman R., E. W. Montroll, R. Potts and R.W. Rothery (1959). Traffic Dynamics: Analysis of Stability in Car-following. *Operations research*, **1(7)**, pp86-106.

Hughes J. (1998). Intensive Traffic Data collection for Simulation of congested Auckland Motorway, *Proceedings 19th ARRB Transport Research Conference*, Sydney.

Kleijnen J.P.C. and W. Van Groenendaal (1992). *Simulation: A Statistical Perspective*, John Wiley.

Kleijnen J.P.C. (1995). Theory and Methodology: Verification and Validation of Simulation Models, *European Journal of Operational Research*, **Vol. 82**, pp145-162.

Law A.M. and W.D. Kelton (1991). *Simulation Modeling and Analysis*, McGraw-Hill.

Mahut M. (2000). Behavioral Car Following Models, Ph. D. Thesis, Chapter 3, Centre de Recherchesur les Transports, CRT, Université de Montréal.

Manstetten D., W. Krautter and T. Schwab (1998a).Traffic Simulation Supporting Urban Control System Development, Robert Bosch GmbH, Corporate Research and Development, Information and Systems Technology, P.O. Box 10 60 50, 70049 Stuttgart, Germany.

Manstetten D., W. Krautter and T. Schwab (1998b). Traffic Simulation Supporting Urban Control System Development, *Proceedings of the 4th World Conference on ITS,* Seoul.

Schreckenberg M., A. Schadschneider, K. Nagel and N. Ito (1995). Discrete stochastic models for traffic flow, *Physical Review E*, **Vol. 51, No. 4**.

TEDI (2002). Version 4.1 User's Manual, TSS-Transport Simulation Systems.

Theil H. (1966). Applied Economic Forecasting, North-Holland.

Wiedemann R. (1974). Simulation des Verkehrsflusses. Schriftenreihe des Instituts für Verkehrwessen, Heft 8, Universität (TH) Karlsruhe.

Wu J.H. (1991). A Study of Monotone Variational Inequalities and their Application to Nertwork Equilibrium Problems, Ph. D. Thesis, Centre de Recherche sur les Transports, Université de Montréal, Publication #801.

Wu J.H., Y. Chen and M. Florian (1998a). The Continuous Dynamic Network Loading Problem: A Mathematical Formulation and Solution Method, *Trans. Res.-B*, **Vol. 32, No. 3**, pp173-187.

Wu J.H., M. Florian, Y.W. Xu and J.M. Rubio-Ardanaz (1998b). A projection algorithm for the dynamic network equilibrium problem, Traffic and Transportation Studies, *Proceedings of the ICTTS'98*, pp379-390, Ed. By Zhaoxia Yang, Kelvin C.P. Wang and Baohua Mao, ASCE,

Xu Y.W., J.H. Wu and M. Florian (1998). An Efficient Algorithm for the Continuous Network Loading Problem: a DYNALOAD Implementation, in *Transportation Networks: Recent Methodological Advances*, Ed. By M.G.H. Bell, Pergamon Press.

Xu Y.W., J.H. Wu, M. Florian, P. Marcotte and D.L. Zhu (1999). Advances in the Continuous Dynamic Network Problem, *Transportation Science*, **Vol. 33, No. 4**, pp341-353.

Yang Qi and H.N. Koutsopoulos (1996). A Microscopic Traffic Simulator for Evaluation of Dynamic Traffic Management Systems, *Transp. Res. C.*, **Vol.4, No. 3,** pp113-129.

Yoshii T. (1999). Standard Verification Process for Traffic Simulation Model – Verification Manual, Kochi University of Technology, Kochi, Japan,.

MICROSCOPIC TRAFFIC SIMULATION: MODELS AND APPLICATION

Tomer Toledo
Massachusetts Institute of Technology, Center for Transportation and Logistics, 77 Massachusetts Ave., NE20-208, Cambridge MA, 02139, USA
toledo@mit.edu

Haris Koutsopoulos
Northeastern University, Department of Civil and Environmental Engineering, 437 Snell Engineering Center, Boston MA 02115,USA
haris@coe.neu.edu

Moshe Ben-Akiva
Massachusetts Institute of Technology, Department of Civil and Environmental Engineering, 77 Massachusetts Ave., 1-181, Cambridge MA, 02139, USA
mba@mit.edu

Mithilesh Jha
Jacobs Civil Inc., 222 South Riverside Plaza, Chicago IL, 60606, USA
mithilesh.jha@jacobs.com

ABSTRACT

Microscopic traffic simulation is an important tool for traffic analysis, particularly in the presence of intelligent transportation systems (ITS). In this paper we describe the main components of a microscopic traffic simulation model and illustrate the discussion with examples drawn from our experience with a microscopic traffic simulation tool, MITSIMLab. With respect to the use of simulation models in practical applications we discuss issues related to their calibration and validation. We demonstrate this part with a case study of applying MITSIMLab in Stockholm, Sweden.

INTRODUCTION

Traffic congestion is a major problem in urban areas. It has a significant adverse economic impact through deterioration of mobility, safety and air quality. A recent study (FHWA 2001) estimated that 32% of the daily travel in major U.S. urban areas in 1997 occurred under congested traffic conditions. The annual cost of lost time and excess fuel consumption during congestion was estimated at $72 billion, which represents a 300% increase from 1982. Schrank and Lomax (2001) estimated that 1,800 new freeway lane-miles and 2,500 new urban street lane-miles would have been required in the U.S. in order to keep congestion from increasing from 1998 to 1999. The budgets required for such infrastructure investments far exceed available resources. Moreover, in many urban areas, land scarcity and environmental constraints would limit construction of new roads or expansion of existing ones even if funds were available.

As a result, the importance of better management of the road network to efficiently utilize existing capacity is increasing. In recent years, a large array of traffic management schemes have been proposed and implemented. Methods and algorithms proposed for traffic management need to be calibrated and tested. In most cases, only limited, if any, field tests are feasible because of prohibitively high costs and lack of public acceptance. Furthermore, the usefulness of such field studies is deterred by the inability to fully control the conditions under which they are performed. Hence, tools to perform such evaluations in a laboratory environment are needed.

Intelligent Transportation Systems (ITS) applications, such as dynamic traffic management and route guidance, have emerged as important tools for traffic management. These applications involve information dissemination from a traffic management center to drivers and deployment of management and control schemes. The impact of information and control strategies on traffic flow can be realistically modeled only through the response of individual drivers to the information. For example, evaluation of different incident response strategies that utilize lane use signs requires modeling of drivers' response to the signs and a plausible model of their lane changing behavior. Microscopic simulation models, which analyze traffic phenomena through explicit and detailed representation of the behavior of individual drivers, have been widely used to that end by both researchers and practitioners. The detailed level of behavior modeling in microscopic simulation models is particularly critical when disaggregate relations between vehicles are more important than aggregate traffic flow characteristics. An example is the study of safety impacts, for which headway distributions, frequency of emergency braking and the number and locations of lane changes may provide better indication of the impact on safety of different geometric design plans than aggregate measures such as average speed, flow and density.

The purpose of this paper is to describe the main components of a microscopic traffic simulation model and discuss calibration and validation of these tools to practical applications. We illustrate the discussion with examples drawn from our experience using the microscopic traffic simulator MITSIMLab (Yang and Koutsopoulos 1996, Yang et al 2000).

OVERALL STRUCTURE

While the software implementation may vary significantly, three main components are common to most existing microscopic traffic simulation tools:

1. Traffic flow model (vehicle-mover)
2. Traffic management system representation
3. Output and graphical interfaces

In MITSIMLab these components are implemented as separate modules, which exchange information with each other according to the paradigm outlined in Figure 1.

Within the microscopic traffic simulator (MITSIM) module, the movements of individual vehicles are represented via detailed travel and driving behavior models. Traffic flow characteristics emerge from the individual behaviors. Vehicles traveling in the network activate surveillance devices (e.g. loop detectors, video-sensors). The information gathered by the surveillance system is transferred to the traffic management simulator (TMS), which represents the traffic control and routing logic under evaluation. This information is used to determine traffic control settings and route guidance. The control and routing strategies generated by the traffic management module determine the states of traffic control and route guidance devices. In turn, drivers respond to the various traffic controls and guidance while interacting with each other. Output from the simulation can be obtained both in the form of numerical data and via the graphical user interface (GUI), which is used for both debugging purposes and demonstration of traffic impacts through vehicle animation.

In the next sections we describe some of the important elements and models within the simulation framework.

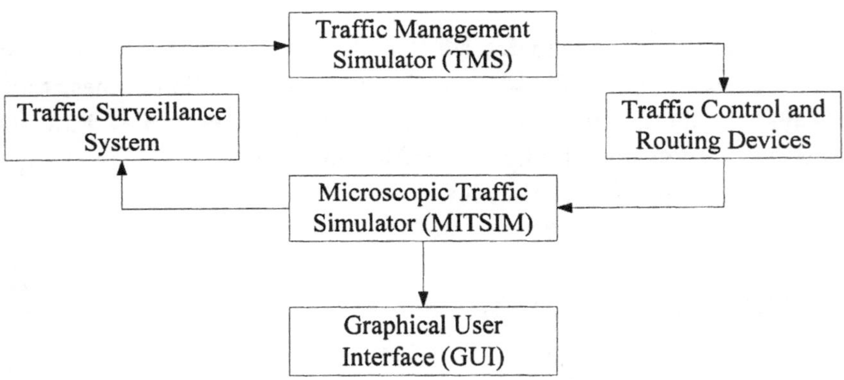

Figure 1. Elements of MITSIMLab and their Interactions.

TRAFFIC FLOW MODELING

In microscopic simulation, traffic flow is captured by detailed modeling of the behavior of individual vehicles, including representation of travel demand and behavior (e.g. route choice models) and driving behavior (e.g. speed/acceleration, lane changing and gap acceptance models). Simulated traffic patterns emerge from the collective travel and driving decisions made by the simulated drivers. Hence, the realism of the simulator depends on the richness and fidelity of these models and therefore development of sound behavior models and rigorous calibration and validation of these models are critically important to any application of the simulation tool.

Travel demand

Travel demand is most commonly represented in microscopic traffic simulation tools by time-dependent origin to destination (OD) tables. During a simulation run, individual vehicles are generated based on the demand rates specified in these tables.

An alternative specification of the demand utilizes generation rates at the origins and turning fractions at intersections. The advantage of this approach is that it uses data that is directly measurable in the field. This approach is suitable for simulation of traffic corridors, which do not allow for route choices. However, in more complex networks, it may lead to significant over-estimation of traffic flows caused by vehicles circulating around the network. Moreover, since simulated vehicles do not have defined destinations, route choice behavior may not be modeled realistically and therefore, for example, the impact of advanced traveler information systems (ATIS) may not be captured.

To illustrate this problem, consider the network shown in Figure 2. Suppose that the true demand is for 50 trips for each of the following OD pairs: 1→3, 1→6, 4→3, 4→6 and that only the simple paths (1-2-3, 1-2-5-6, 4-5-2-3 and 4-5-6, respectively) are used. Based on field observations, the following generation rates and turning fractions would be defined:

- 100 vehicles generated at nodes 1 and 4.
- At node 2: 50% turning to node 3, 50% turning to node 5.
- At node 5: 50% turning to node 2, 50% turning to node 6.

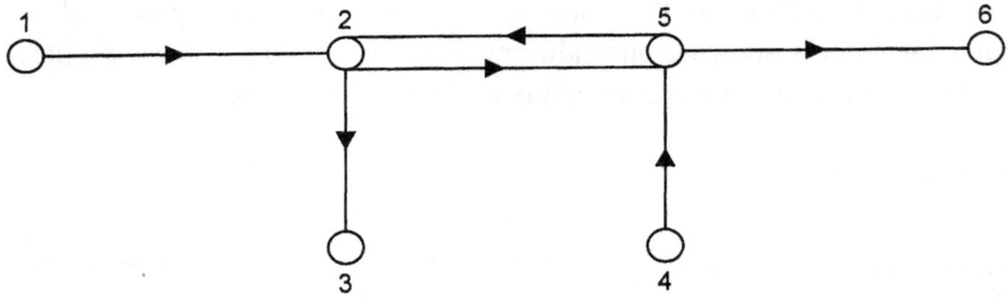

Figure 2. Network illustrating travel demand representations.

Table 1 summarizes the expected number of vehicles (rounded to the nearest integer) that originate at node 1 and follow various routes until they leave the network according to a simulation model that uses the above generation rates and turning percentages. On average, 25% of these vehicles cycle in the network causing overestimation of congestion in the network. Furthermore, the overestimation of congestion increases with the dimension of the network since the probability that a vehicle would leave the network decreases further. The result for vehicles originating at node 4 is similar.

Table 1. The expected number of vehicles following various routes in the network.

Route	Expected number of vehicles
1-2-3	50
1-2-5-6	25
1-2-5-2-3	13
1-2-5-2-5-6	6
1-2-5-2-5-2-3	3
1-2-5-2-5-2-5-6	2
1-2-5-2-5-2-5-2-3	1

In MITSIMLab time-dependent OD tables are used to input travel demand. OD tables may be defined for time intervals of varying duration. They may also be defined for up to 15 different user-defined vehicle types (e.g. cars, trucks, buses, SUVs) or generated vehicles may be randomly assigned a vehicle type based on an input fleet mix. Vehicle tables may also be used to specify the demand, replacing or complementing OD tables. Vehicle tables are lists of vehicles to be generated with pre-specified origins, destinations, departure times and optionally vehicle types and paths.

Routing behavior

The paths drivers follow in the network are determined by route choice models, which may be either path-based or link-based. Path-based models assume that drivers select between alternative routes to their destinations based on the attributes of the entire path. Link-based models assume that drivers only select the next link to travel on at each decision point. The main disadvantage of path-based models is that they require enumeration of all used paths, which may be prohibitively expensive in large networks. In contrast, link-based models do not require path enumeration but represent myopic behavior that may lead to unrealistic choices, such as a vehicle traveling on the freeway, taking an off-ramp and immediately after that an on-ramp back to the freeway. Moreover, link-based route choice may result in cyclic paths. The explicit path choice set defined for path-based models can easily exclude such paths.

MITSIMLab supports both route choice models. The path-based model utilizes a path-size logit model (Ramming 2002), which accounts for overlap between paths. The path choice set is specified by the user for each OD pair. Choice probabilities for the various paths are given by:

$$P(i \mid C_n) = \frac{\exp\left(V_{in} + \ln\left(PS_{in}\right)\right)}{\sum_{j \in C_n} \exp\left(V_{jn} + \ln\left(PS_{jn}\right)\right)} = \frac{PS_{in} \exp\left(V_{in}\right)}{\sum_{j \in C_n} PS_{jn} \exp\left(V_{jn}\right)} \quad (1)$$

V_{in} and PS_{in} are the systematic utility and the size of path *i* to driver *n*, respectively. C_n is the path choice set for driver *n*.

The systematic utilities V_{in} depend on perceived path travel times, which are calculated as the sum of perceived travel times on links that comprise the path. Drivers may perceive freeway travel times differently from travel times on urban streets. A freeway bias coefficient captures this behavior:

$$TT_i = \sum_{a \in \Gamma_i} \frac{TT_a}{\beta_a} \quad (2)$$

TT_i and TT_a are the travel times on path *i* and link *a*, respectively. Γ_i is the set of links in path i. β_a is a link-type variable defined by:

$$\beta_a = \begin{cases} \beta^{freeway} & \text{if link } a \text{ is freeway} \\ 1 & \text{otherwise} \end{cases} \quad (3)$$

$\beta^{freeway} (\geq 1)$ is a constant that captures drivers' preference to freeway travel.

The path-size term adjusts the path utility to account for overlapping links in alternative paths. It is bounded $0 < PS_{in} \leq 1$. Its value is 1 if the path does not overlap with other paths and decreases with the level of overlap. A number of path-size formulations have been proposed in the literature (see Ramming 2002 for a review). The formulation of Bekhor et al (2001) is adopted in MITSIMLab:

$$PS_{in} = \sum_{a \in \Gamma_i} \left(\frac{l_a}{L_i} \right) \frac{1}{\sum_{j \in C_n} \frac{L_i^{\gamma}}{L_j^{\gamma}} \delta_{aj}} \quad (4)$$

l_a and L_i are the lengths of link a and path i, respectively. δ_{aj} are elements of the link-path incidence matrix, i.e., $\delta_{aj} = 1$ if path j includes link a and 0 otherwise.

The link-based model uses a multinomial logit formulation:

$$P(l \mid L_{sn}) = \frac{\exp(V_{ln})}{\sum_{k \in L_{sn}} \exp(V_{kn})} \quad (5)$$

V_{ln} is the systematic utility, to driver n, of selecting link l as the next link. L_{sn} is the set of links (emanating from node s) that driver n may choose as the next link.

Systematic utilities V_{in} depend on the perceived travel time on link l, the perceived travel time on the shortest path from the downstream node of link l to the destination, and a freeway diversion dummy variable, which penalizes exiting a freeway (i.e. drivers traveling on a freeway and choosing a non-freeway next link).

The effect of traveler guidance and information on route choices may be captured in both the path-based and the link-based route choice models through link travel time tables that are updated according to the specifications of the guidance system. Drivers with no access to information use habitual link travel time tables, which represent prevailing traffic conditions. Drivers who have access to information (e.g. pre-trip, in-vehicle technologies and variable message signs) use updated link travel time tables, which incorporate real-time traffic conditions. In the path-based model route choices are reevaluated whenever new information is received. Drivers' preference to keep their previously assigned habitual paths is captured by diversion dummy variables, which penalize switching from the driver's habitual route.

Driving behavior

Traffic dynamics in micro-simulation tools is the emergent result of a set of driving behavior models, which determine the speed/acceleration and lane changes performed by vehicles under various conditions. Acceleration models include free-flow, car following and emergency behaviors. Lane changing behavior is most often classified as either mandatory (MLC) or discretionary (DLC). MLC applies when the vehicle must change lanes, for example, to follow the path or to avoid a blocked lane. DLC applies when the driver perceives that traffic conditions are better in other lanes. In most cases acceleration and lane changing behaviors are modeled separately.

The core model implemented in MITSIMLab is an integrated driving behavior model, which captures both lane changing and acceleration behaviors and accounts for inter-dependencies and correlations between these behaviors. The structure of the integrated model is shown in Figure 3. The model hypothesizes four levels of decision-making: target lane, gap acceptance, target gap and acceleration. This decision process is latent. The target lane and target gap choices are both unobservable. Only the driver's actions (lane changes and accelerations) are observed. Latent choices are shown as ovals in Figure 3. Observed choices are shown as rectangles. At the highest level the driver chooses a target lane. The target lane is the lane the driver perceives as best to be in. A utility maximization model is used to explain the target lane choice. Lane utilities integrate mandatory and discretionary lane changing considerations into a single model, and thus capture trade-offs between these considerations. Important variables that affect the target lane choice are the distance to the point where a the vehicle must be in a specific lane and the number of lane changes required to be in that lane, densities and speeds of the vehicles in the neighborhood of the subject, presence of heavy vehicles and tailgating behavior.

In the case that either the right lane or the left lane are chosen, the driver evaluates the adjacent gap in the target lane and decides whether this gap can be used to execute the lane change or not. If the gap is accepted the lane change is immediately executed. Gap acceptance decisions are based on comparing the available lead and lag space gaps, defined by the clear spacing between the subject and the intended leader and lag vehicles, respectively, with the corresponding critical gaps, which are the minimum acceptable gaps. Critical gaps are functions of explanatory variables, such as the relative speeds of the subject with respect to the lead and lag vehicles. In order to accept the gap and execute the lane change both the lead gap and the lag gap must be acceptable.

If the adjacent gap is rejected, the driver chooses a target gap in the target lane traffic. The target gap will be used to perform the desired lane change. In the current implementation the target gap choice set includes three alternatives, defined by their locations relative to the subject vehicle: the adjacent gap, the forward gap and the backward gap. The target gap choice is based on the lengths of the candidate gaps, relative speeds of the vehicles involved and the distance from the subject vehicle to an ideal position relative to the target gap, which would allow the driver to execute the lane change.

For each one of the situations described above, a corresponding acceleration behavior is defined. Vehicles that choose to stay in their current lanes would, depending on the time headway from the front vehicle (leader), either follow their leader or accelerate to attain their desired speeds. A similar behavior also applies for vehicles that are executing a lane change, but with respect to their leader in the new lane. The acceleration of vehicles that have chosen a target gap is determined such that to facilitate the execution of the desired lane change using the target gap (i.e. the driver tries to position the vehicle in a way that will increase the probability that the target gap will be acceptable). This acceleration may also be constrained by car following considerations, since the lane change is not immediate.

In addition to this core model, behaviors in several other situations are represented in the simulation. For example, the acceleration model also represents emergency braking, response to signals and signs and accounts for vehicle capabilities and driver aggressiveness. The lane changing model also

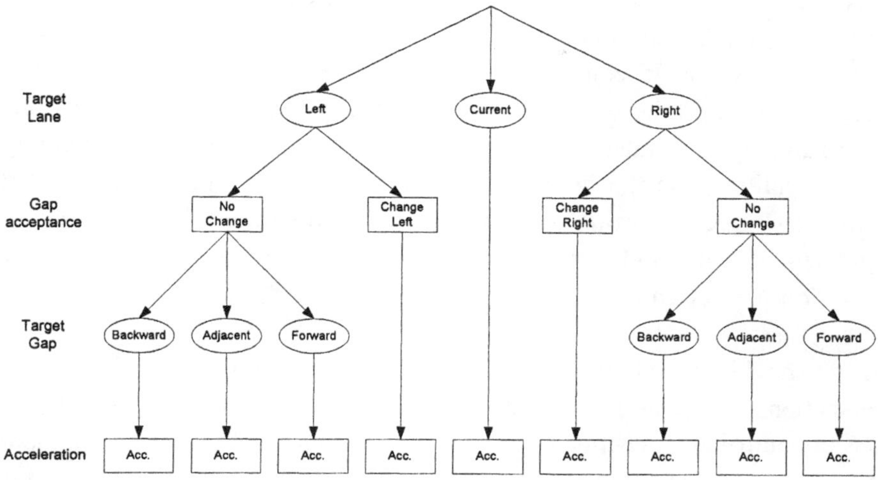

Figure 3. Structure of the integrated driving behavior model.

represents priority and non-priority merging behaviors and yielding and forced merging behaviors.

TRAFFIC MANAGEMENT SYSTEM REPRESENTATION

Accurate representation of the traffic management system is important to ensure the realism of simulation results, and is particularly essential when the simulation tool is used to evaluate the performance of various traffic control and guidance systems, which require support for a wide range of traffic control and route guidance systems, such as ramp control, freeway mainline control (e.g., lane control signs, variable speed limit signs, portal signals at tunnel entrances), intersection control, variable message signs and in-vehicle route guidance.

The traffic management simulator (TMS) module in MITSIMLab has a generic structure that can represent different designs of such systems with logic at varying levels of sophistication, from isolated pre-timed signals to real-time predictive systems. Figure 4 outlines the structure of TMS. Control strategies and routing information are generated using either reactive or proactive approaches. The reactive approach consists of pre-determined control laws that depend only on the current network state. In the proactive approach, the system predicts future traffic conditions and optimizes traffic control and routing strategies based on these predicted conditions. In this case, the generation of control and routing strategies is an iterative process. Given a proposed strategy, traffic conditions on the network are predicted and the performance of the candidate strategy is evaluated. If the strategy is found to be satisfactory, it is implemented; if additional strategies need to be tested, another generation-prediction iteration is performed.

The most widely used traffic management tool is traffic signal control at intersections. TMS supports modeling of intersection controls through three types of controllers: pre-timed, actuated and a generic controller. The logic for the pre-timed and actuated controllers, which are simpler to implement, require pre-specified phase plans and phase orders. In some cases, such as for pre-timed or four-phase actuated controllers, these types can simulate the control logic exactly. In other cases, they may be used to approximate more advanced control logic such as dual-ring controllers or European controllers in which phasing is not necessarily explicit.

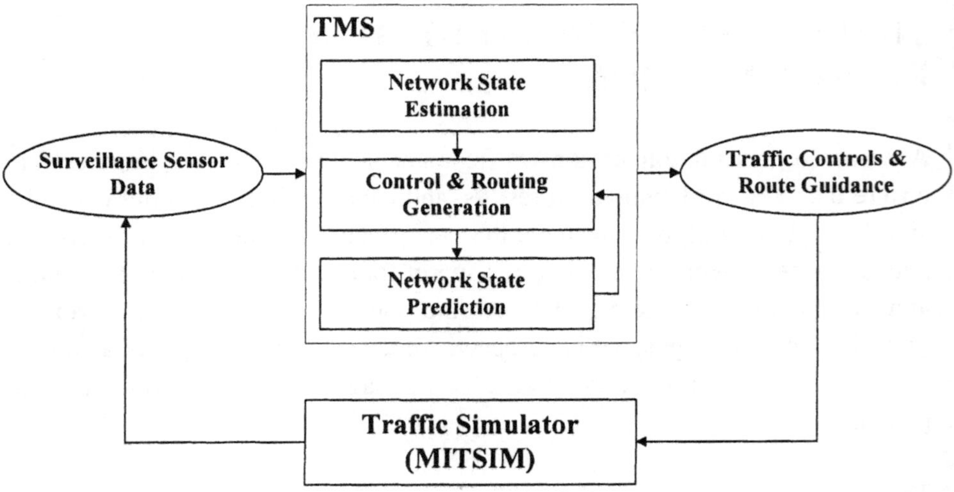

Figure 4. Structure of the traffic management simulator (TMS).

The generic controller (Davol 2001) addresses the limitations of the two controllers and allows the simulation of a wide range of advanced control strategies. Instead of requiring each phase to be specified explicitly, the generic logic allows each movement (signal group) to be controlled independently, thus providing full flexibility in the control logic. This logic is specified by means of detailed conditions that must be met before a signal indication changes, such as time, detector states, other signal states, and other controller states. By specifying these conditions, any logic can be modeled. For example, specifying only time constraints simulates a pre-timed controller. Constraints with respect to detector states add vehicle actuation, while constraints as to other signal states allow the specification of complementary or conflicting movements. Conditions on other controller states allow coordinated and area-wide adaptive control, which is necessary for the simulation of advanced control systems. Figure 5 summarizes the logic of the generic controller. The controller iterates through all the signal groups. For each signal group, it evaluates the logic conditions and determines whether the group's signal state should be updated. Because the group states may be inter-dependent, with the state of one group as an input to the logic of another group, this evaluation step must be iterative. Therefore, when the states of all the signal groups have been determined, the controller again cycles through the groups to check if the new states cause more signal groups to change. This process is repeated until the states of all signal groups are stable, at which point the controller displays the updated states of the traffic signals in TMS. This process is repeated each time step, typically 0.1 seconds of simulation time.

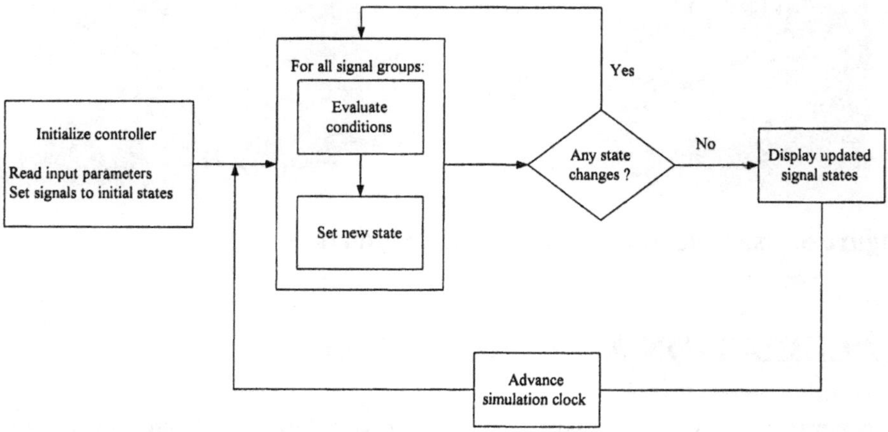

Figure 5. Logic of the generic controller

OUTPUT AND GRAPHICAL INTERFACE

Simulation tools are used to study complex systems, and therefore generate extensive sets of outputs. While very useful, this wealth of information can make it difficult to gain insights and analyze the system based only on the numerical outputs. To that end, most micro-simulation tools offer extensive graphical user interfaces (GUI). These interfaces are used for both debugging purposes and demonstration of traffic impacts through vehicle animation.
Figure 6 shows examples of the MITSIMLab GUI. The figure on the left illustrates a high level view of the network. Links are color coded to demonstrate congestion levels. The figure on the right focuses on the details of the operations of a complex interchange.

In addition to the graphical display of the network, MITSIMLab provides output data at system, link, segment (a part of a link with uniform geometry), lane, sensor, and vehicle levels. The level of detail of collected data at the vehicle level provides all the information necessary to develop measures of effectiveness (MOEs), such as travel times, delays and queue lengths, which may be used for evaluation.

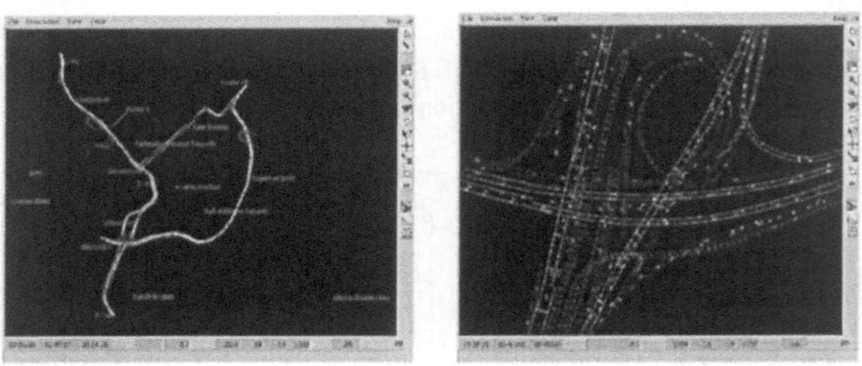

Figure 6. Example of the graphical user interface.

CALIBRATION AND VALIDATION

In general, calibration of microscopic traffic simulation tools should be based on the framework summarized in Figure 7. According to this framework the calibration process includes two steps: First, the parameters of the individual models the simulation consists of (e.g. driving behavior and route choice models) are estimated using disaggregate data, independent of the overall simulation framework. Disaggregate data includes detailed driver behavior information such as vehicle trajectories. In the second step, aggregate data (e.g. time headways, speeds, flows) is used to fine-tune parameters and calibrate general parameters in the simulator.

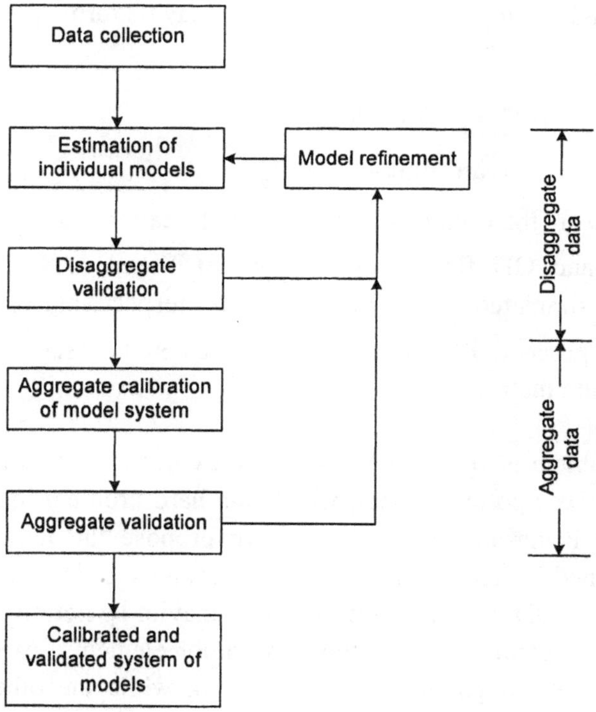

Figure 7. Overall calibration framework

While this two-step approach to calibration is desirable, data availability often dictates what steps are feasible. Most often, only aggregate data collected through loop detectors is available and therefore only aggregate calibration is possible. Aggregate calibration is formulated as an optimization problem, which seeks to minimize a measure of the deviation between observed and corresponding simulated measurements. The reason for this approach is that, in general, it is not feasible to isolate the contribution of individual models to the overall error. For example, OD estimation methods require an assignment matrix as input. The assignment matrix maps OD flows to counts at sensor locations. Usually the assignment matrix is not readily available and needs to be generated from the simulation model. Therefore, the assignment matrix is a function of the route choice and driving behavior models used. Similarly, important explanatory variables in route choice models are travel times, which are flow-dependent. Simulated flows are a function of the OD flows, driving behavior and the route choice model itself. Hence, the following optimization problem, which simultaneously calibrates the parameters of interest (OD flows, route choice and driving behavior parameters) may be formulated:

$$\begin{aligned} &\min_{\beta,\theta,OD} f\left(M^{obs}, M^{sim}\right) \\ &s.t. \quad M^{sim} = g\left(\beta,\theta,OD\right) \\ &\qquad OD = \arg\min_{X} \left\| AX - Y^{obs} \right\| \end{aligned} \tag{6}$$

β, θ and OD are the vectors of parameters to be calibrated: driving behavior, route choice and OD flows, respectively. M^{obs} and M^{sim} are vectors of observed and simulated traffic measurements, respectively. $g(\cdot)$ represents the simulation process. Y^{obs} are observed traffic counts at sensor locations. A is the assignment matrix.

The above problem is very difficult to solve exactly. The OD constraint, for example, is a fixed-point problem, which is a hard problem in its own merit (Cascetta and Postorino 2001). Hence, we propose the iterative heuristic approach outlined in Figure 8. This approach accounts for interactions between driving behavior, OD flows and route choice behavior by iteratively calibrating driving behavior parameters and travel behavior elements. At each step the corresponding sets of parameters is calibrated, while the other parameters remain fixed to their previous values. Calibration of the route choice model requires a set of reasonable paths for each OD and expected link travel times used as explanatory variables in the model. OD estimation requires generation of an assignment matrix. Hence, the travel behavior calibration step is also iterative: based on the existing OD flows, parameters of the route choice model are calibrated. The calibrated route choice model is used to generate an assignment matrix, and perform OD estimation. The new OD flows are used to re-calibrate route choice parameters and so on.

In summary the proposed calibration process proceeds as follows:
Step 1. Initialize parameters, β_0, θ_0 and OD_0.

Step 2. Estimate OD and calibrate route choice parameters assuming fixed driving behavior parameters.

Step 3. Calibrate driving behavior parameters assuming the OD matrix and route choice parameters estimated in Step 2.

Step 4. Update habitual travel times using the OD matrix, route choice and driving behavior parameters estimated in Steps 2 and 3.

Step 5. Check for convergence: If convergence, terminate.
Else, continue to step 2.

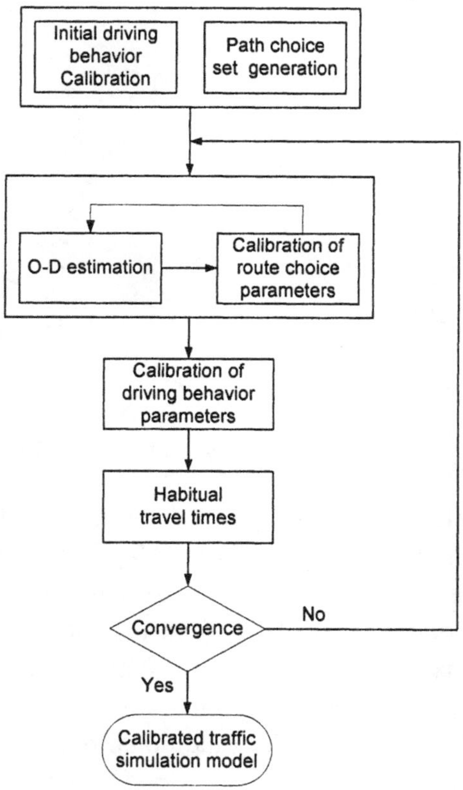

Figure 8. Methodology for aggregate calibration of micro-simulation models.

Case study

We now describe in more detail the application of the procedure described above to a case study in Stockholm, Sweden. The case study network is a mixed urban-freeway network in the Brunnsviken area, north of the Stockholm CBD, shown in Figure 9. The E4 corridor, on the west side of the network is the main freeway connecting the northern suburbs to the CBD. The east side of the network is a parallel arterial. These routes experience heavy southbound congestion during the AM peak period.

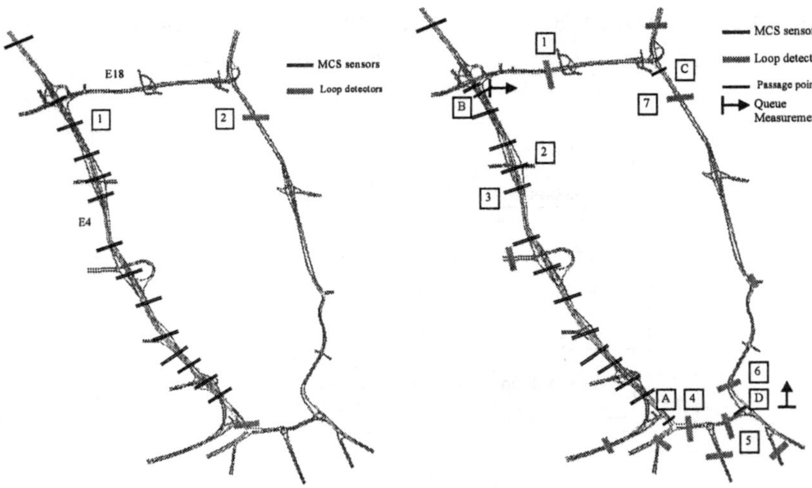

(a) May 1999 (b) May 2000

Figure 9. The case study network and measurement locations.

AM peak period traffic data from May 1999 was collected for calibration. Similar data was collected a year later, during May 2000, for validation. Sensor and other measurement locations for May 1999 and May 2000 are also shown in Figure 9. Sensor data was available from the Motorway Control System (MCS) and additional loop detectors. The validation data also included measurements of point-to-point travel times by probe vehicles. The probe vehicles also recorded queue lengths. Additional queue measurements were obtained from aerial photographs. A static AM peak OD flows matrix, previously developed for planning studies, was available for OD estimation. Additional information about vehicle mix by type (i.e. autos, buses, trucks etc.) and lane use privilege (i.e. permission to use bus lanes) was used to set corresponding input data for the simulation model. The type assigned to a simulated vehicle affects its physical properties (length and width) and performance capabilities (e.g. maximum speed, maximum acceleration and deceleration). The importance of lane use privileges in this application stems from the extensive bus lanes system in place. Only buses and taxis are allowed to use these lanes. The autos category was further split into two separate groups: high-performance and low-performance vehicles. Vehicles in these groups have different capabilities in terms of their maximum acceleration, deceleration and speeds.

Initial parameter calibration

Driving behavior parameters

The various parameters of the driving behavior models, with the exception of the desired speed parameters, were calibrated using the above procedure. The parameters of the distribution of desired speeds were estimated independently from speed data. The desired speed is the speed that the driver would choose in the absence of any restrictions imposed by other vehicles or by traffic control devices. This speed is affected by the geometry of the section and by driver characteristics. In MITSIMLab, the desired speed distribution is defined relative to the speed limit. This distribution was inferred from the speeds of unconstrained vehicles, which were defined as those vehicles that crossed the sensor stations at times when the flow rate was less than 600 veh/hr/lane. This threshold corresponds to the Highway Capacity Manual (HCM 2000) level of service A. An analysis of the sensitivity of the desired speed distribution with respect to the flow threshold was performed. Desired speed distributions were also developed assuming flow thresholds of 300 and 200 veh/hr/lane. The results were not significantly different from the one obtained for 600 veh/hr/lane.

A small sub-network of the Brunnsviken network was used to obtain initial values for other driving behavior parameters. Important considerations that led to adopting this approach were:

1. The calibration process is more manageable when performed on a sub-network.
2. The sub-network was chosen such that available sensor data can be used to generate accurate OD flows at 1 min. intervals. Moreover, for each OD pair in the sub-network only one path exists. Therefore most of the errors generated by OD estimation and route choice modeling were eliminated.

The sub-network and sensor locations within it are shown in Figure 10. This sub-network was chosen for several reasons including:

1. Minimal downstream effects – the location is far from possible spillbacks from the bottlenecks in the network that may affect the behavior but are not represented in the MITSIMLab sub-network model.
2. Representation of different behaviors – The sub-network contains on- and off-ramps, allowing capturing mandatory and discretionary lane changing and merging behavior, which are likely to be important behaviors in this case study.

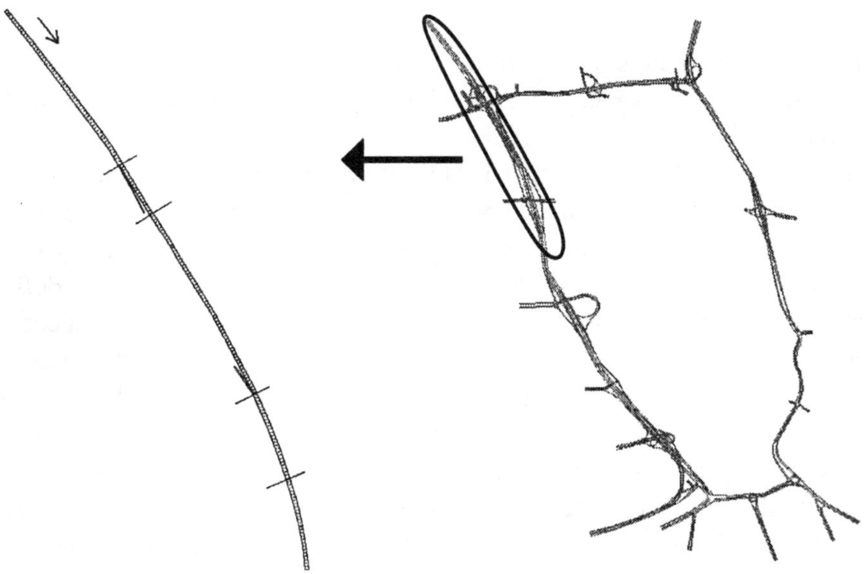

Figure 10. Location and details of the calibration sub-network.

Since sensor counts were used to extract OD information for the sub-network, the calibration objective function was to minimize the square deviations of simulated sensor speeds from the observed ones:

$$\min_{\beta} \sum_{t=1}^{T}\sum_{n=1}^{N}(V_{nt}^{sim} - V_{nt}^{obs})^2 \tag{7}$$

V_{nt}^{obs} and V_{nt}^{sim} are the observed and simulated speeds, respectively, measured at sensor n during time period t. N and T are the number of sensors and time periods, respectively.

Path choice set generation

The path-based route choice model requires a set of alternative paths for each

OD pair in the network. The following procedure was used to generate these sets:

1. Generation of a comprehensive path set – A comprehensive path choice set was generated using the link-based route choice model.
2. Unreasonable path elimination - The link-based route choice tends to generate a large number of paths. Unreasonable paths (e.g. paths using off-ramp and on-ramp immediately after that) were eliminated.

The path choice set may depend on traffic conditions, in which case the process should be repeated as OD flows and the parameters of the route choice model evolve to ensure that all reasonable paths are captured. The structure of the Brunnsviken network facilitates the generation of the path set, since only one or two reasonable paths exist for each OD pair. Therefore, the path generation exercise was preformed only once.

Overall Calibration

OD estimation

The OD estimation problem is often formulated as a generalized least squares (GLS) problem. The GLS formulation minimizes the deviations between estimated and observed sensor counts while also minimizing the deviation between the estimated OD flows and seed OD flows (Cascetta et al 1993). The corresponding optimization problem is:

$$\min_{X \geq 0} \left(AX - Y^H\right)^T W^{-1}\left(AX - Y^H\right) + \left(X - X^H\right)^T V^{-1}\left(X - X^H\right) \quad (8)$$

X and X^H are vectors of estimated and historical (seed) OD flows, respectively. Y^H are the historical (observed) sensor counts. A is the assignment matrix, which maps OD flows to sensor counts. W and V are the variance-covariance matrices of the sensor counts and OD flows, respectively.

However, in the problem discussed in this paper, the assignment matrix is not known. Hence, we propose the following iterative process: First, the simulation is run, using the calibrated parameters and a set of seed OD flows to generate an assignment matrix. This assignment matrix is in turn used for OD estimation.

Due to congestion effects, the assignment matrix generated from the seed OD may be inconsistent with the estimated OD. Therefore, the OD estimation process must be iterative.

The Brunnsviken network includes a large number of OD pairs and sensor locations, which made OD estimation computationally intensive. To overcome this limitation, a sequential estimation technique (Cascetta et al 1993), which exploits the sparse structure of the assignment matrix was employed. The sequential estimation process is as follows: The seed OD is taken as fixed for the first time period. An assignment matrix is generated and used to estimate the effect of the first period demand on sensor counts in subsequent periods. The demand in the second period is then estimated, based on the observed counts less the estimated contribution from OD flows in the first time period. The assignment matrix is used to estimate the effect of second period demand on subsequent periods. This process is continued until OD flows are estimated for all periods of interest.

Route Choice Parameter

Given the OD flows, path choice sets and habitual travel times, parameters of the route choice model were calibrated to match the split between the two sensors marked 1 and 2 in Figure 9a. These points were selected because the structure of the network ensures that all vehicles with a choice of paths pass exactly one of them. Splits rather than counts were used in order to reduce errors from inaccuracies in the scale of the OD matrix, especially at early stages of the estimation process.

Driving behavior parameters

While the initial calibration of driving models included a wide range of parameters, during this step only a limited set of parameters was calibrated. Sensitivity analysis indicated that the calibration could focus on the scale parameters of the various models. Scale parameters are the sensitivity constants in the acceleration models and alternative specific constants in the lane changing models. Hence, given OD flows and route choice parameters, scale parameters of the acceleration and lane changing models were calibrated using the formulation given in Equation (7), now applied to the entire network.

Habitual travel times

The route choice model uses habitual path travel times as explanatory variables. Calibration of the model parameters requires knowledge of these travel times. While planning studies can be used to provide initial values, further refinement is necessary to capture the time-dependent nature of travel times and improve the accuracy of the models. Therefore, an iterative day-to-day perception updating model was used to improve initial travel times estimates obtained from planning studies. At each iteration of this process, representing a day, habitual travel times were updated as follows:

$$TT_{it}^{k+1} = \lambda^k tt_{it}^k + \left(1 - \lambda^k\right) TT_{it}^k \quad (9)$$

TT_{it}^k and tt_{it}^k are the habitual and experienced travel times on link i, time period t on day k, respectively. λ^k is a weight parameter ($0 < \lambda^k < 1$).

Validation Results

Three types of measurements were used to validate the calibrated MITSIMLab model: traffic flows, travel times and queue lengths by comparing simulated measurements with the corresponding observed measurements. OD flows were estimated from the May 2000 traffic counts and the previously calibrated model parameters.

Traffic flows

Observed and simulated traffic flows at key sensor locations were compared using two hours of AM peak data at 15 min. intervals. The results are presented in Figure 11. Two measures of goodness-of-fit were used to quantify the relationship between observed and simulated measurements: The root mean square normalized error (RMSNE), which quantifies the total percentage error of the simulator and the mean normalized error (MNE), which indicates the existence of consistent under- or over-prediction in the simulated measurements. These measures are calculated by:

$$RMSNE = \sqrt{\frac{1}{N}\sum_{n=1}^{N}\left(\frac{m_n^{sim} - m_n^{obs}}{m_n^{obs}}\right)^2} \quad (10)$$

$$MNE = \frac{1}{N}\sum_{n=1}^{N}\frac{m_n^{sim} - m_n^{obs}}{m_n^{obs}} \qquad (11)$$

m_n^{obs} and m_n^{sim} are the observed and simulated measurements, respectively. N is the number of measurement points (over time in this case).

RMSNE for the different locations range from 5% to 17%. MNE values range from −12% to 14%. In general, simulated flows correspond well to the measurements and accurately capture temporal patterns. Note that sensor flows were used to estimate the OD flows. Hence, the results emphasize the importance of OD estimation.

Travel times

Point-to-point travel times measured by the probe vehicles were compared against average simulated travel times. Only few probe vehicle observations were available. Therefore, mean observed travel times could not be accurately estimated. Instead, Figure 12 compares average simulated travel times and individual probe vehicle observations. The figure also shows travel times values corresponding to the average ±2 standard deviations of the simulated travel time. Assuming that simulated travel times follow a normal distribution, these values define an interval containing 95% of simulated travel times. 80 out of the 120 (67%) probe vehicle observations are within these intervals.

Simulated travel times match the observed ones very in sections A-B, B-C and D-C, which are relatively uncongested during the AM peak period. Sections C-D, C-B and B-A are heavily congested, as indicated by the shapes of the travel time curves. These shapes are rather well replicated, although the simulation underestimates travel times. The largest incomparability between observed and simulated measurements is in sections A-D and D-A. These are short and congested sections dominated by traffic signals and roundabouts. Some of the error in these sections may be attributed to inconsistencies in the traffic counts in this area (e.g. entry flows to an intersection do not match exit flows) that led to poor OD estimation and to imperfections in the representation of traffic control (e.g. pedestrian and bicycle signals) in the simulation tool.

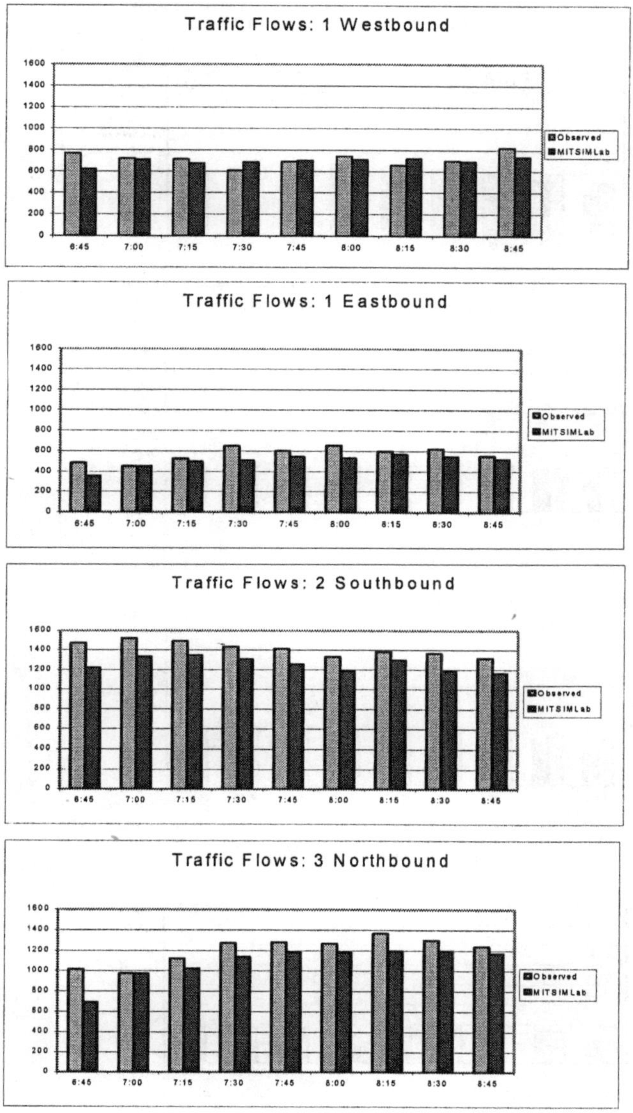

Figure 11. Comparison of observed and simulated flows at sensor locations.

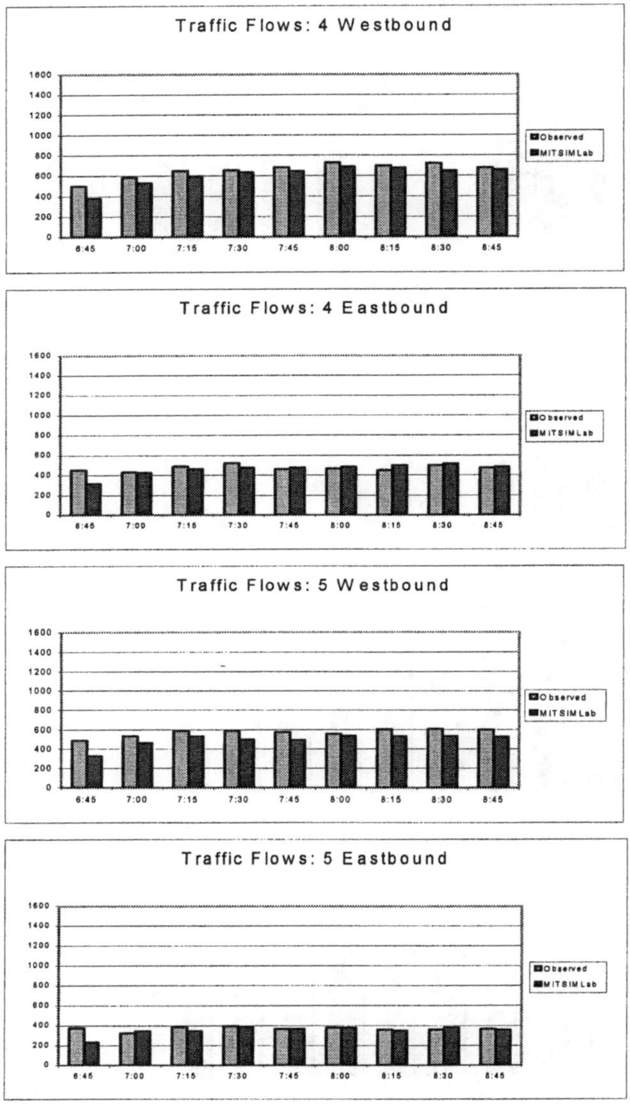

Figure 11 (cont'd.)

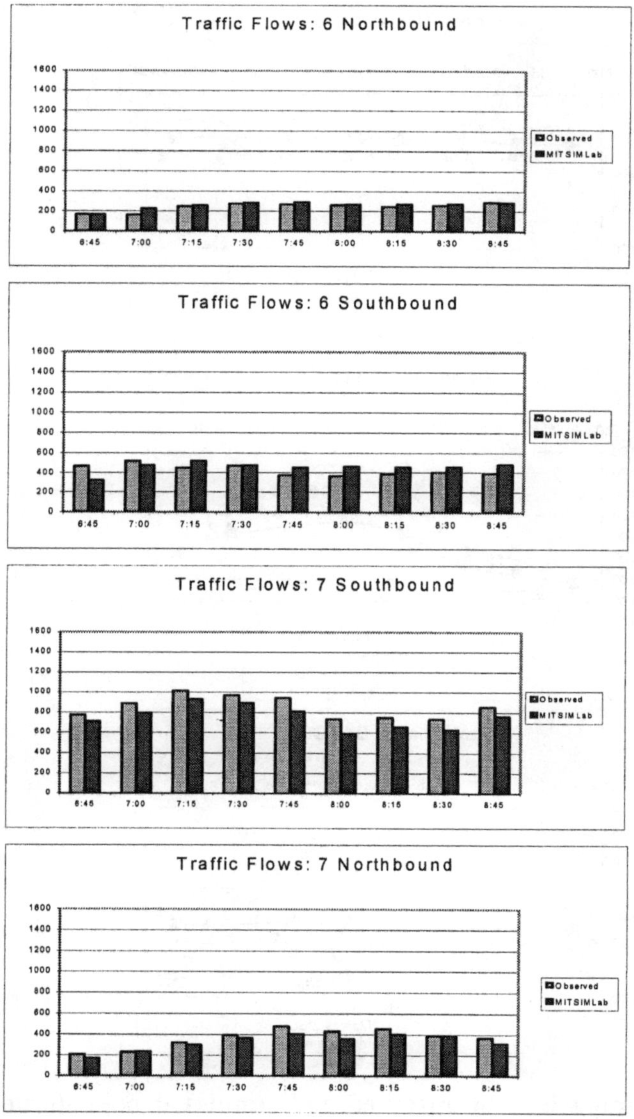

Figure 11 (cont'd).

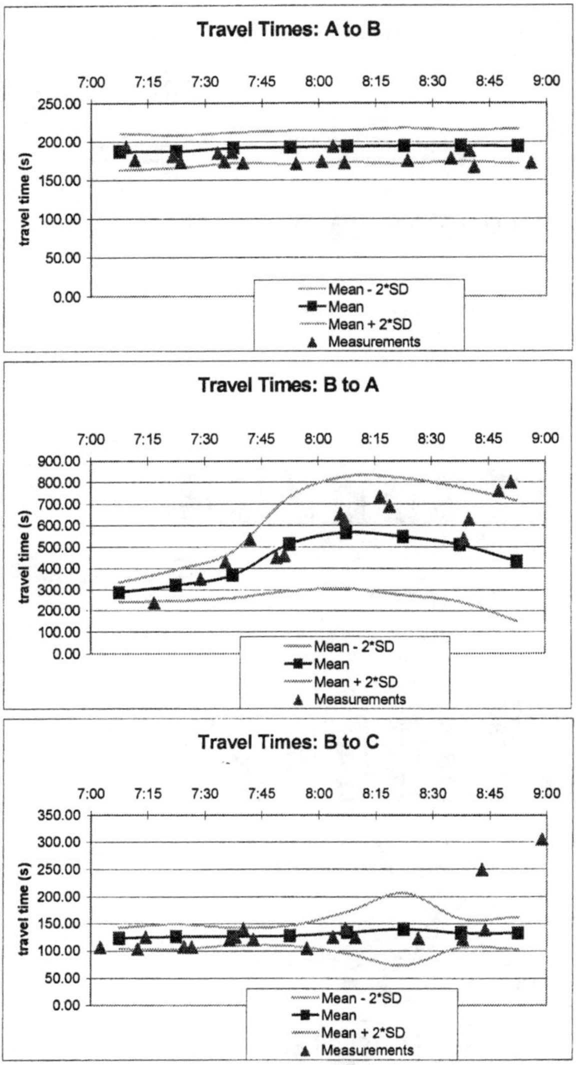

Figure 12. Comparison of observed and simulated point-to-point travel times.

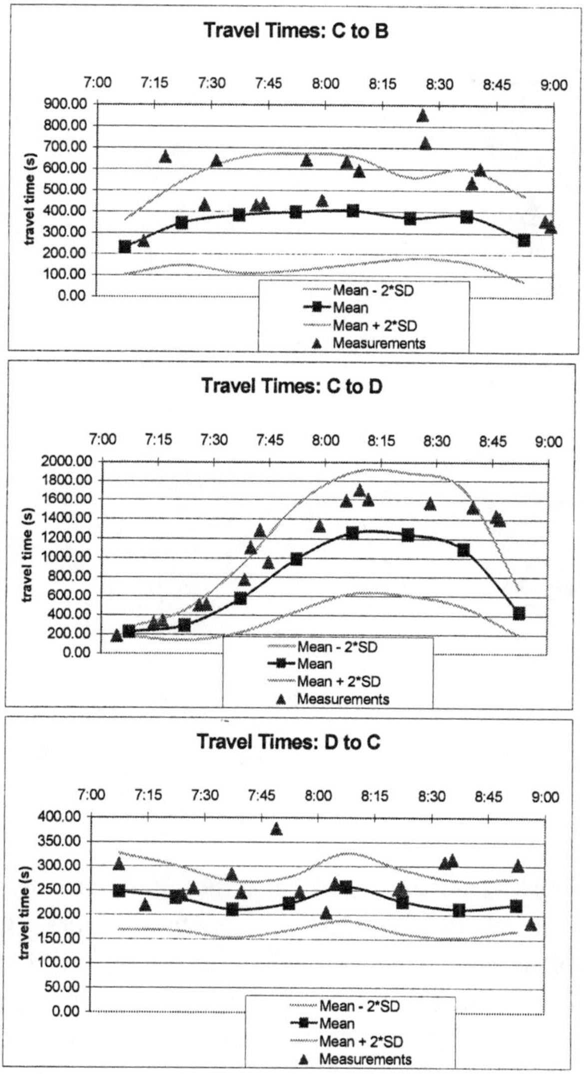

Figure 12 (cont'd).

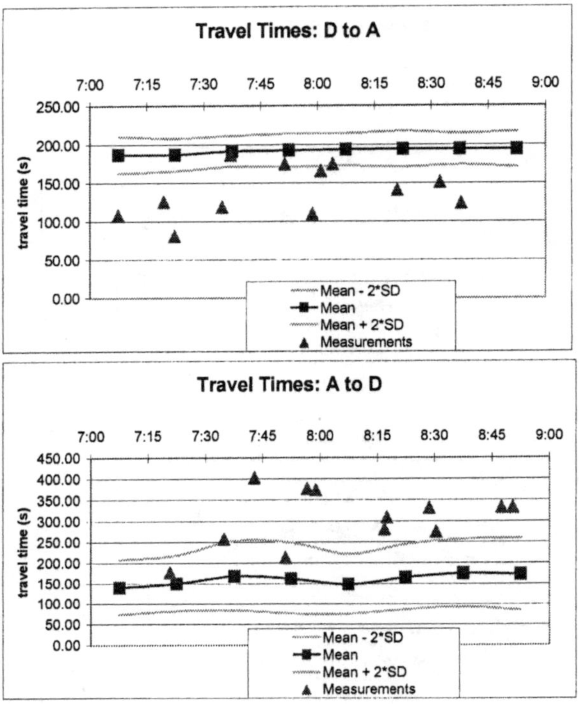

Figure 12 (cont'd).

<u>Queue lengths</u>

Simulated queue lengths were compared against those measured by the probe vehicles and from aerial photos. Both queues, shown in Figure 13, are very significant. At their peak they may interlock and grow beyond the Northern boundary of the network. They are well represented in the simulation both in terms of magnitude and time of occurrence. However, the number of observations is very limited, which forbids rigor statistical analysis of the results.

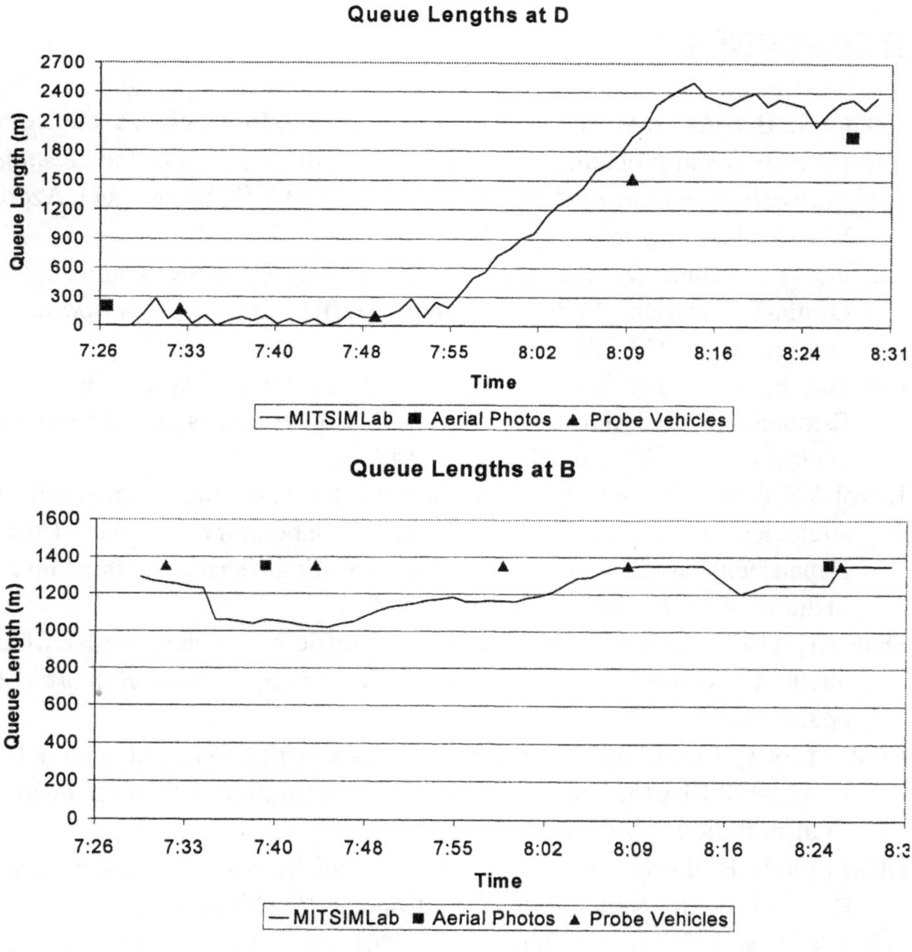

Figure 13. Comparison of observed and simulated queue lengths.

CONCLUSION

Microscopic traffic simulation is an important tool for traffic analysis, particularly in the presence of intelligent transportation systems (ITS). Microscopic traffic simulators analyze traffic phenomena through detailed representation of the behaviors of individual vehicles/drivers. Rigorous calibration and validation of these models are essential to the fidelity of micro-simulation tools.

REFERENCES

Bekhor S., Ben-Akiva M. and Ramming M.S. (2001). Route choice: choice set generation and probabilistic choice models, In *Preprints of the Triennial Symposium on Transportation Analysis, TRISTAN IV*, Sao Miguel, Azores Islands, Portugal, vol. **3**, pp459-464.

Cascetta E., Inaudi D. and Marquis G. (1993). Dynamic Estimators of Origin-Destination Matrices Using Traffic Counts, *Transportation Science*, **27**, pp363-373.

Cascetta E. and Postorino M.N. (2001). Fixed Point Approaches to the Estimation of O/D Matrices Using Traffic Counts on Congested Networks, *Transportation Science*, **35**, pp134 –147.

Davol A.P. (2001). Modeling of traffic signal control and transit signal priority strategies in a microscopic simulation laboratory, Master Thesis, Department of Civil and Environmental Engineering, Massachusetts Institute of Technology, Cambridge, MA.

Dial R.B. (1971). A probabilistic multipath traffic assignment model which obviates the need for path enumeration, *Transportation Research*, **5**, pp83-111.

FHWA (2001). Managing our congested streets and highways, publication No. FHWA-OP-01-018, Federal Highway Administration, US Department of Transportation, Washington DC.

HCM (2000). Highway Capacity Manual, *Special Report 209*, Transportation Research Board, National Research Council, Washington DC.

Schrank D. and T. Lomax (2001). The 2001 urban mobility report, Texas Transportation Institute Research Repor*t*, May 2001.

Ramming M.S. (2002). Network knowledge and route choice, PhD dissertation, Department of Civil and Environmental Engineering, Massachusetts Institute of Technology, Cambridge, MA.

Yang, Q. and Koutsopoulos H.N. (1996). A microscopic traffic simulator for evaluation of dynamic traffic management systems, *Transportation Research*, **4C**, pp113-129.

Yang, Q., Koutsopoulos H.N. and Ben-Akiva M.E. (2000). A Simulation Laboratory for Evaluating Dynamic Traffic Management Systems, *Transportation Research Record*, **1710**, pp122-130.

THE ART OF THE UTILIZATION OF TRAFFIC SIMULATION MODELS: HOW DO WE MAKE THEM RELIABLE TOOLS?

Ryota Horiguchi
i-Transport Lab. Co., Ltd., 2-12-404 Ageba-cho Shinjuku-ku Tokyo 162-0824 Japan
horiguchi@i-transportlab.jp

Masao Kuwahara
Institute of Industrial Science, Univ. of Tokyo, 4-6-1-C504 Komaba Meguro-ku Tokyo 153-8505 Japan
kuwahara@iis.u-tokyo.ac.jp

ABSTRACT

This paper, firstly, describes the current status of the utilization of traffic simulation models. Evaluation for the model's application in Japan was based on a questionnaire survey. Secondly, the Best Practice Manual for Simulation Application, which is currently being developed, is discussed. In addition some exerts from the manual in regards to addressing the issues of simulation application are presented: i.e. i) understanding the models' nature through verification and validation; ii) OD estimation from vehicle counts; iii) model parameter calibration; and iv) indices to measure the reproducibility. Finally this paper introduces the Clearing House of Traffic Simulation Models. These models promote simulation utilization.

INTRODUCTION

This paper describes the art of utilizing traffic simulation models in a practical scenario. In the last decade, quite a number of network traffic simulation models became available all over the world. Preceding these simulations, mathematical dynamic network modelling was studied. Probably, the first appreciable contribution was made by Merchant and Nemhouser (1978). Since then, several researchers have proposed dynamic assignment models, sometimes based upon the control theory and/or based on the queuing theory (Friesz *et. al.*, 1989; Wie *et. al.,* 1990; Boyce *et. al.*, 1993; Lam *et. al.*, 1995; Kuwahara and Akamatsu, 1993, 2001). These mathematical network models have greatly contributed to the dynamic traffic analysis by giving the theoretical background. However, the mathematical formulations have limitations in their flexibility to deal with detailed vehicle motions, user choice behaviour, and traffic operation schemes.

We now find many reports of simulation studies in the business scene. However the users, except for the model developers, have inadequate knowledge of simulation models due to the difficult nature in fully understanding the nature of such models by only reading literature or manuals. The simulation is sometimes criticized as a 'black box', and reconciles itself to unreliable techniques.

In order to cope with this criticism, the Japanese Society of Traffic Engineers' Simulation Committee (Sim@JSTE), has encouraged model developers to disclose the nature of their models through verification (Horiguchi, *et al.*, 2000) and validation (Horiguchi, *et al.*, 1998) with the purpose to promote the utilization of traffic simulation. The basic ideas of verification and validation are as follows:

> 'Verification' is a series of simple tests to confirm that fundamental model functions are properly programmed as in the specification. The simulated result is compared with the result obtained from the well-authorized theory. In order to individually examine each of the functions and also compare theoretical solutions, we should use virtual data on the simplest network.

'Validation' seems quite similar to 'Verification' in general. However, we clearly distinguish 'Validation' as the evaluation of model specification using real field data. Even if the model is verified as in the specification, the model specification itself may not be adequate to describe real traffic phenomena. The model cannot be practically applicable if actual traffic situations are not sufficiently reproduced due to the incomplete model specification. Furthermore, the model performance as a system should also be confirmed, such as whether the execution of the model can be finished within a practical computation time.

Rao and Owen (2000) proposed a similar concept on verification and validation. It is this paper's recommendation to concur with the following suggestion of these authors: that in order to understand a model characteristics 'Conceptual Validation', which requires the identification of the model's underlying theory, and 'Operational Validation', which compares the model outputs with actually measured behaviour, is needed. The importance of verification and validation has been shared amongst researchers as well as practitioners on different continents.

One of our outcomes is the verification manual (JSTE, 2001a), which describes the standard verification procedure, and can thus be applied to various different types of models. The manual contains a series of basic tasks to check the reproducibility of traffic conditions of a model by applying simple but ideal dataset. Each of the basic tasks evaluates: 1) vehicle generation pattern, 2) bottleneck capacity and saturation flow rate at an intersection, 3) shockwave propagation, 4) capacity of merging and diverging sections, 5) right (left) turn capacity decline at a signalised intersection, and 6) dynamic route choice behaviour. In each task simulation results are to be compared with "well-known" theories in traffic engineering. A theory, which sometimes over simplifies the traffic phenomena, can give us a good standing point to understand the models' behaviour. We do not therefore require a model that completely follows the theory, but when the simulation result is different from the theory, the verifier of the model could explain why the result shows such a discrepancy to vindicate the model itself.

Several simulation models that are practically used in Japan have been evaluated based upon the proposed verification process. We have verified seven pilot models, such as: AVENUE (AVENUE, WWW site), SOUND (Yoshii, *et al.*, 1995), tiss-NET (Sakamoto, *et al.*, 1998), Paramics (Paramics, WWW site), NETSIM (NETSIM, WWW site), REST (Yoshida, *et al.*, 1999), and SIPA (Yokochi, *et al.*, 1999) along with the verification manual.

Other output includes the benchmark datasets, which are real field data observed and processed appropriately (Hanabusa, *et al.* (2001)). Although verification can help us to comprehend the models' behaviour it does not tell us the applicability to the real world, where various traffic phenomena embed and affect each other. Therefore, the model developers have to show evidence that those models can reasonably reproduce such complex situations through validation with benchmark dataset. The benchmark datasets have been gathered so that developers can utilize them to validate their models, since generally, the acquisition of real field data is substantially more expensive and time consuming. Some of the pilot models were validated through the application of the benchmark dataset (Horiguchi, *et al.*, 1996) (Sawa and Yamamoto, 2002).

These outputs can be found in 'Clearing House of Traffic Simulation Models' at an Internet website (JSTE, 2001b). The developers and users of simulation models are also encouraged to publish their experiences of verification and validation through the Clearing House. Our activities for verification and validation make sense -as the criticism against the 'black-box' - only when the ability of the model is disclosed.

Another criticism that "*simulation is useful, but dangerous*" (Smartest, 1999) should be pointed out. Case studies of traffic simulation are sometimes only reported with their results. However, there must be a lot of room made for the users to calibrate the simulation models out of given conditions. Since a simulation model has the flexibility to reproduce traffic conditions by changing its model parameters, users may fit the simulation result to "any" conditions. Different ways of model calibration may lead to different results for the same case study. Therefore, unless the report tells us how the model is fitted to the desired condition, we barely believe the result is pertinent.

In order to overcome such a situation, Sim@JSTE is currently undertaking further exertions to establish the protocol for simulation studies as the 'Best Practice Manual of Traffic Simulations'. A simulation study consists of several phases to complete the application: e.g. data acquisition, model selection, models calibration, result interpretation, and so on. Firstly, we collected the case studies for business practice through a questionnaire survey, and then we extracted the generalized methodologies for each phase of application to be included in the manual.

In the following chapters, the current status of the utilization of traffic simulation models is first described by examining the results of a questionnaire survey for the model's application in Japan. Secondly, the best practice manual for simulation applications, which is currently under construction, is introduced. This manuscript summarizes major issues discussed in the manual, such as: the verification and validation processes in order to understand model characteristics, relevant practices of data acquisition, evaluation of reproducibility of models, and the interpretation of model outputs. Finally, the 'Clearing House of Traffic Simulation Models' that promotes simulation utilization is briefly explained.

CURRENT STATUS OF SIMULATION UTILIZATION

Questionnaire Survey for the Simulation Application in Japan

In order to understand the realities of the simulation model application, the questionnaire form shown in Figure 1 was delivered to the users who had much experience in simulation application (Horiguchi and Oneyama, 2002). A similar questionnaire survey was utilised in the 'Smartest Project' in Europe (Smartest, 1997). In the 'Smartest Project' survey, however, the major interest was to identify the requirements of the microscopic simulation models to be applied to the evaluation of advanced telematic services.

The questionnaire form included the following items to explain some details of each simulation study:

i) *Outline of the case study* -- purpose of the simulation, evaluated measures, etc.
ii) *Network information* -- size, area, shape, road type, etc.
iii) *Input data & model parameters* -- trip demand, link capacity, scan interval, etc.

iv) *Reproducibility* -- how to calibrate the model, what're the indices, etc.
v) *Output measurement* -- sort of measurement, etc.

There are 45 forms concerning the eight simulation models shown in Figure 2. AVENUE, tiss-NET, SOUND, VISITOK (VISITOK, WWW site), REST, and TRANDMEX (Suzuki, *et al.*, 2000) were the models developed in Japan. Paramics and NETSIM are overseas' models. The respondent refers to each model that he/she develops or owns.

The number of answered applications widely varied depending on the respondents' attitude to the questionnaire. This does not truly reflect the actual proportion of each model's usage in the business scene, but on the whole they might imply the activities of simulation application. Therefore, in the following sections, we do not distinguish specific models but consider all models.

Table 1 contains general information of each model (including SIPA, referred to in the next chapter). All models treat discrete image of a vehicle, i.e. no fluid approximation models, while there are two different types in the modelling of longitudinal vehicle movement. 'C-F' type represents the models based on car-following behaviour, while 'Q-K' type models use traffic flow characteristics such as 'flow-density' or 'speed-density' relationship. Some of them, the so-called 'microscopic' model, consider lane changing for lateral movement. Others are 'meso-scopic' models.

Although various functions could be considered to classify simulation models, one of the most influential model functions would be the capability of route choice. For models without route choice function, driving routes must be prefixed. Therefore, they cannot examine the policies affecting travellers' spatial behaviour; that is, these models generally cannot be applied to wide network areas. Most recent models however have this function.

There may be so many other functions that more or less determine application conditions. Those functions include signal control capability, several traffic operation' plans (i.e. the exclusive bus lane, reversible lanes, parking management, etc.), highway capacity reproducibility at merging and weaving sections, and so forth.

SIMULATION STUDY RECORD CARD

Code	Traffic Managemen	Application	TDM	Simulation Model	SOUND

Section	Item	Content
Outline	**Subject**	Evaluation of Road Pricing Scheme in Tokyo Central Area
	Purpose	The purpose of this case is to evaluate the mitigation of traffic congestion with the road pricing scheme in Tokyo Central Area, which is currently investigated by Tokyo Metropolytan Government.
	Note	Estimate the future trip demand by considering human behaviour for the road pricing scheme.

Section	Item	Content		
Network & Simulation Settings	**Area**	Tokyo central area (23 wards); Appx. 15km in radius	**Time Period**	24 hours
	When?	Present - 1996, Future - not specified		
	Type	Major arterial roads		

Size							
# of Nodes	942	# of Links	2952	# of Centroids	115	# of Trips	370 million
Arterial Road Networks		# of Intersections	392	# of Traffic Signals	0*	# of Roads	708
Expressway Networks		# of JCTs		# of Ramps		# of Sections	

Note * The capacity of a link is reduced by considering signal split.

Network Image

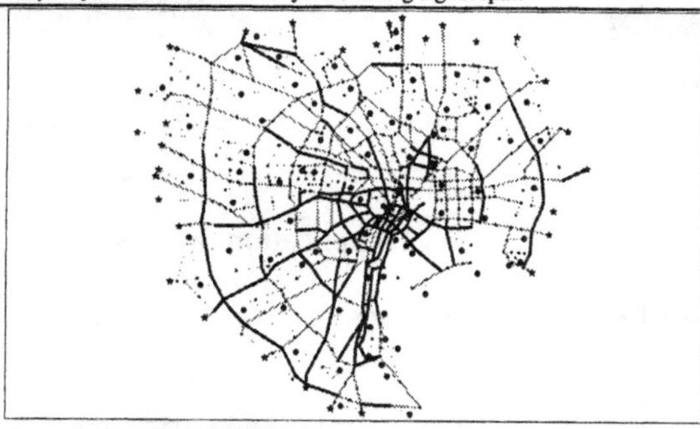

Section		Item	Content
Input & Parameters	Road Network	Link	Length, number of lanes, capacity of lanes, free flow speed.
		Intersection	Capacity for each stream direction
		Junctions	
	Signal	Cyc, Sp, Of	Signal control is not explicitly considered. Link capacities are reduced by the assumed signal split.
		ow do you knov	Based on assumption
	Demand	Type	Time varying O-D matrix
		ow do you knov	Present: based on the result of Vehicle O-D Survey in the Traffic Census in 1996. Future: modified the presend OD with some human behaviour models.
		pacial Resolutio	Middle Zone used in the Traffic Census.
		Time Resolution	1 hour.
		Vehicle Type	2 types (large / small)
	Note		

Section	Item	Content
Model Settings	**Scanning**	time periodic scanning (6 seconds for this case)
	Vehicle Size	10 vehicles for each packet.
	Route Choice behaviour	Stocastic route choice with Dial's algorithm. Travel time information is updated by 5 minutes. Pricing effect is incorporated into the generic cost.
	Note	

Section	Item		Content
Reproducibility	**Parameter Calibration**	What?	Capacity for each stream direction
		How?	To fit the average link travel speed in Tokyo Central Area to the reference data.
	Reference Data		Average travel speed in Tokyo Central Area
	Data Source		Traffic Census in 1996

Output

In this case, the following output is evaluated;

- average link travel speed, traffic count on link, areawise average travel speed.

Figure 1: An Example of Model Application Form

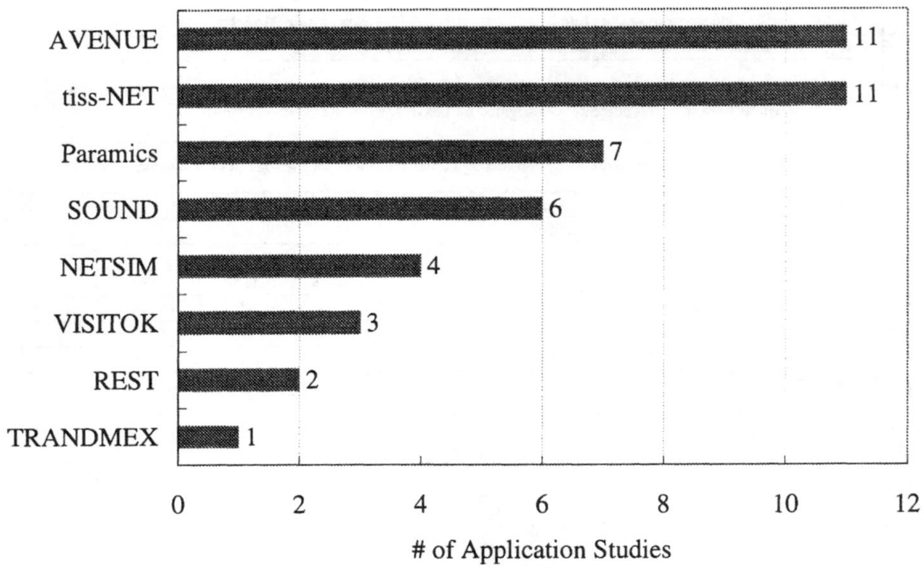

Figure 2: The Number of Collected Answers.

Table 1: General Feature of Listed Models

Model	Scanning Interval [sec]	Longitudinal movement	Lane changing behaviour	Route choice model	Target size of network
AVENUE	1	Q-K	Yes	Yes	Small
NETSIM	0.1-	C-F	Yes	No	Small
Paramics	0.1-	C-F	Yes	Yes	Small
REST	0.1-	C-F	Yes	Yes	Small
SIPA	0.1-	C-F	Yes	Yes	Small
SOUND	1-3	Q-K	No	Yes	Large
tiss-NET	Event	C-F	Yes	Yes	Small
TRANDMEX	1-3	Q-K	No	Yes	Large
VISITOK	0.1	C-F	Yes	No	Small

Outline of the Simulation Studies

Figure 3 illustrates the number of case studies in regards to the purpose of the simulation study. This tells us that the studies to be applied to smaller areas come to the forefront, such as 'Improve Bottlenecks', 'Inter-modal Junctions', or 'Building Shopping Malls'. This may result not only because there are more so-called microscopic simulation models in the answers, but also because of some technical difficulties in the studies with the larger areas, e.g. data acquisition, were prevented from the applications.

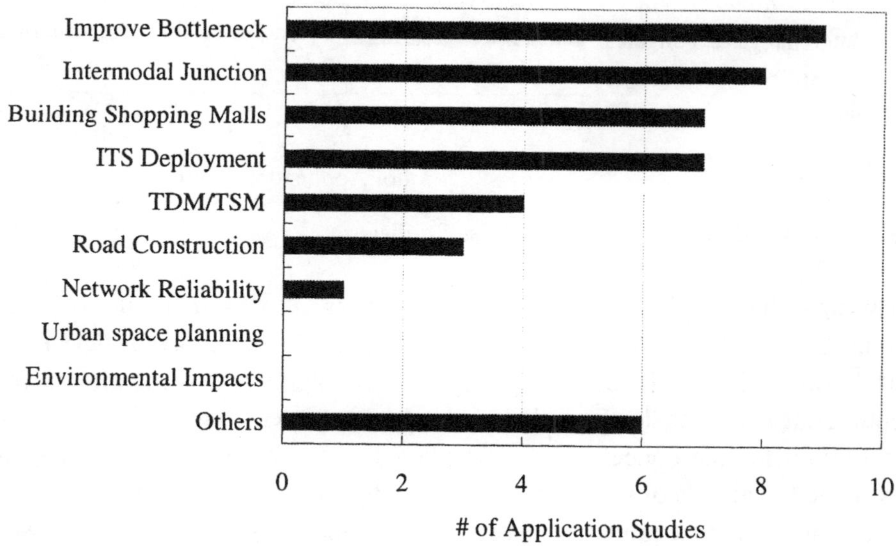

Figure 3: Purposes of the Case Studies.

Figure 4 sums up the countermeasures of case studies. The figure divides the countermeasures into 'Operation and Management' and 'Construction of Road and Parking Facilities'.

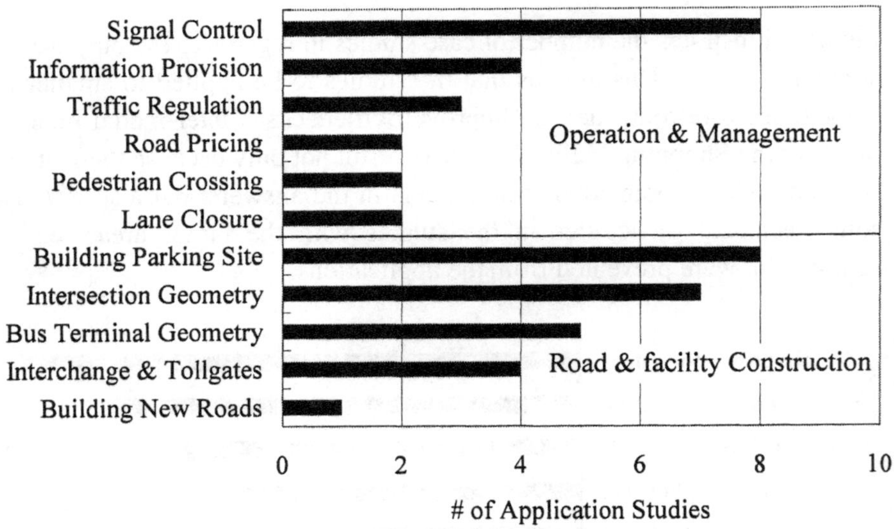

Figure 4: Countermeasures of the Case Studies.

Through the detailed pictures of both figures, it was difficult to find a remarkable difference between using the C-F type and the Q-K type. Even if we use a simulation model of the Q-K type, we can apply it to some microscopic countermeasures as long as the model has sufficient temporal and spatial resolution for the concerning traffic phenomena. This is because, when we evaluate the effects of the improvement on the signal control scheme, it needs to be enough for the model to reproduce the capacity as well as the shockwave propagation at every 1 or 2 seconds with the spatial resolution in several meters. Thus not only the C-F type model but also the Q-K type model which considers lane changing, can be regarded as a microscopic model.

Road Networks Applied in Simulation Studies

Figure 5 illustrates which road types are more commonly applied in the case studies. Most cases deal with the networks of surface streets so as to be applied with local countermeasures. This may come from a Japanese peculiarity where heavy congestions are frequently found on surface streets. However, we have to pay attention to the fact that there are no case studies with integrated networks of surface streets and expressways in practical business scenes. There must be requirements to the simulation to evaluate the macroscopic countermeasures like the road pricing scheme, where we have to apply the simulation to considerably large areas, including surface streets and expressways.

The biggest reason for this may come from the difficulty in knowing how traffic demands arise and how the drivers choose their route on large networks. These sorts of problems in data acquisition can be also implied to **Figure 6;** i.e. almost two thirds of the studies apply the simulation to the simple shape network that has no alternative route for each OD pair. The simulation users seem to be avoiding complicated network shapes in a practical scene. We will come back to the OD data acquisition problem in a subsequent section.

Figure 7 compares the number of nodes to the size of each network measured by the approximate length of the area boundary. It is natural that the numbers of nodes are getting larger as the size of networks becomes larger. In the mathematical expression of the network, not only the subjected intersections or junctions but also dummy junctions can be thought of as nodes.

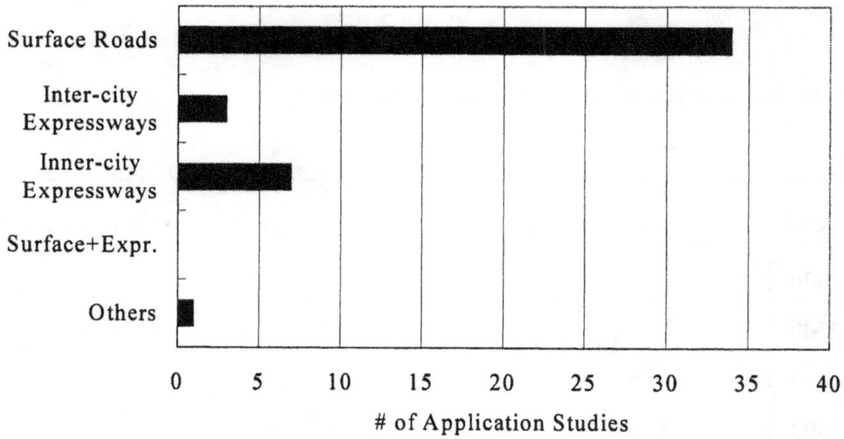

Figure 5: Road Types of the Simulation Studies.

Figure 8 now compares the number of actual intersections or junctions on the network. There are remarkable differences between the number of nodes and the number of actual intersections when the size of a network is fairly large. One reason for this is that many dummy nodes are inserted into the network to connect with the zone centroids at which the traffic is generated and/or absorbed. Another reason is that many minor intersections are still remaining, and neither of the crossing roads are included in the subjective network.

Increasing the number of nodes (= to the number of links) makes the simulation difficult in the model calibration, since most of the models have parameters that have an effect on the link capacity. It depends on the individual user's skill as to how to describe the network and how to calibrate the model. This may lead to the fact that the simulation result will be different for the same case with different users.

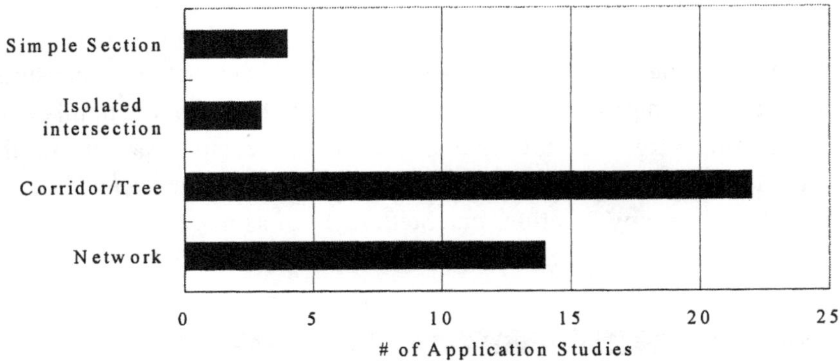

Figure 6: Network Shapes of the Simulation Studies.

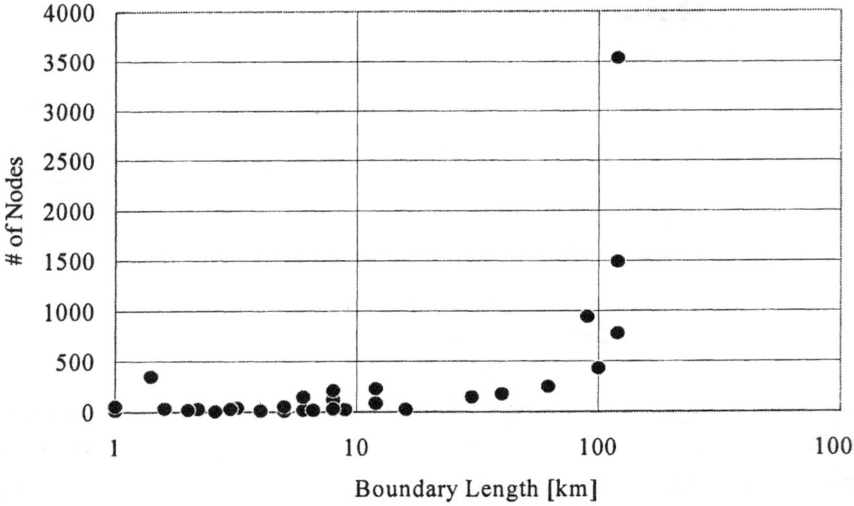

Figure 7: Network Size vs. Number of Nodes.

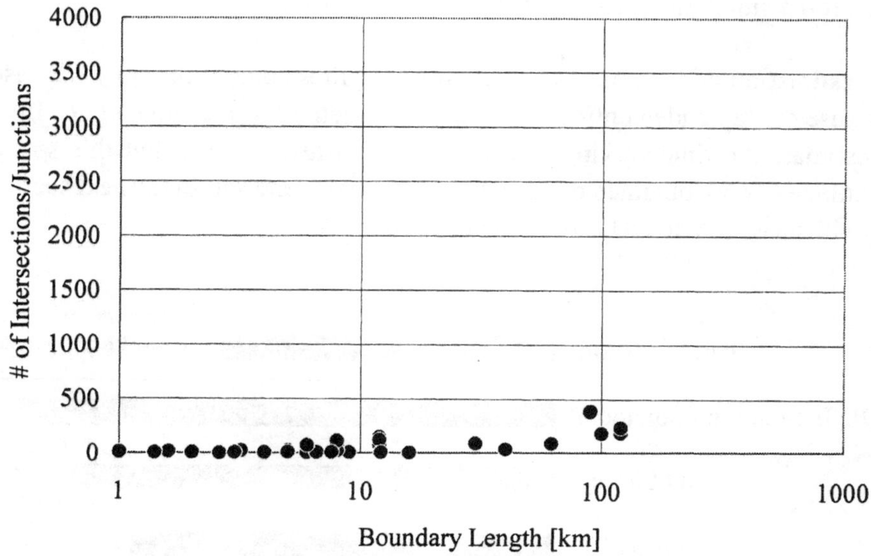

Figure 8: Network Size vs. Number of Intersections.

Traffic Demand Given to the Simulation Models

Figure 9 counts the data sources that the users used to obtain the traffic demand. Here two models (NETSIM and VISITOK) do not incorporate a drivers' route choice model, as a consequence they require the arrival flow rate at each network boundary and the turning ratio at an intersection.

The others accept time-varying OD matrices with an incorporated route choice model, but the data sources are different on a case-by-case basis. Some of the cases with small networks use OD matrices directly observed by licence plate matching, and some of them with large networks borrow results from conventional surveys, such as traffic census. The former can obtain somewhat precise OD matrices but it is quite a cost-consuming way. The latter is cost-effective but not as precise because the sampling rate of the census is normally very small.

There are two items of 'OD estimation' from vehicle counts, i.e. with the simple corridor networks and with complicated networks. For the corridor shape networks, we can calculate the OD matrix by using the turning ratios at intersections. This is completely equivalent to the cases of the 'turning ratio' even if the simulation incorporates the route choice model.

OD estimation with complicated networks requires some technique. All cases here use the 'extended entropy maximization method' (Oneyama, *et al.*, 1996) to estimate the time-varying OD matrix from traffic counts, but this sort of technique has not obtained popularity in practical scene yet. In the next section, we will survey some OD estimation techniques.

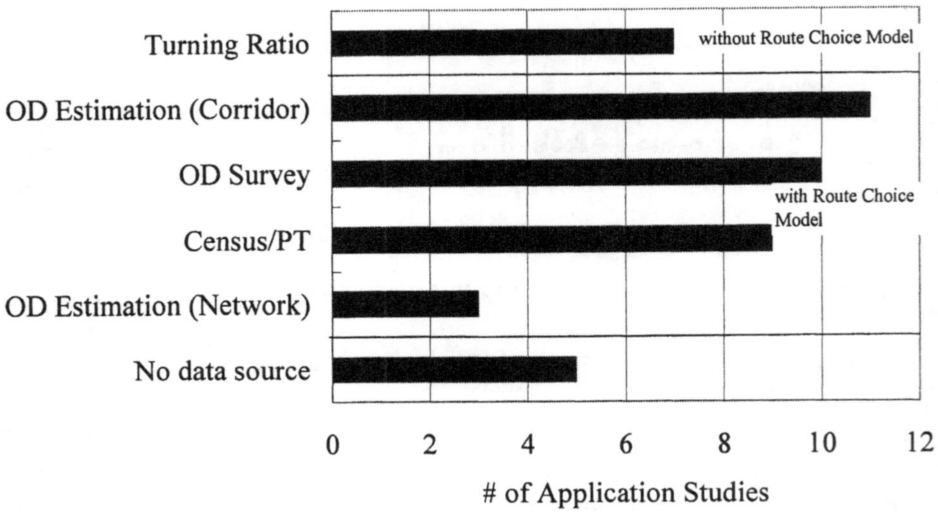

Figure 9: Data Sources of Traffic Demand.

Indices of Reproducibility of Traffic Conditions

Figure 10 shows what indices were employed to evaluate the reproducibility of traffic conditions. In general, traffic conditions on a link or a section must be explained with at least two measurements; i.e. traffic counts (TC) and travel time (TT), or throughput and queue length (QL). However, there are several cases using only one measurement because of limited available data. For these cases, it is suspicious whether the simulation really reproduces the present condition of traffic.

For the cases with more than two measurements, queue length is more popular than travel time. The reason for this may be inferred from the fact that it is hard to measure the travel time on the overall network with sufficiently short intervals, because the data collection mostly depends on a small number of floating cars. On the other hand, the queue length involves some uncertainty because the definition of queue length seems to be unclear.

Further problems could arise from how the users quantitatively estimate the reproducibility. A RMSE (root mean square error) or a correlation coefficient seems to be used out of habit, but how do we decide if the index value is sufficient or not? We will also consider this problem in the next chapter.

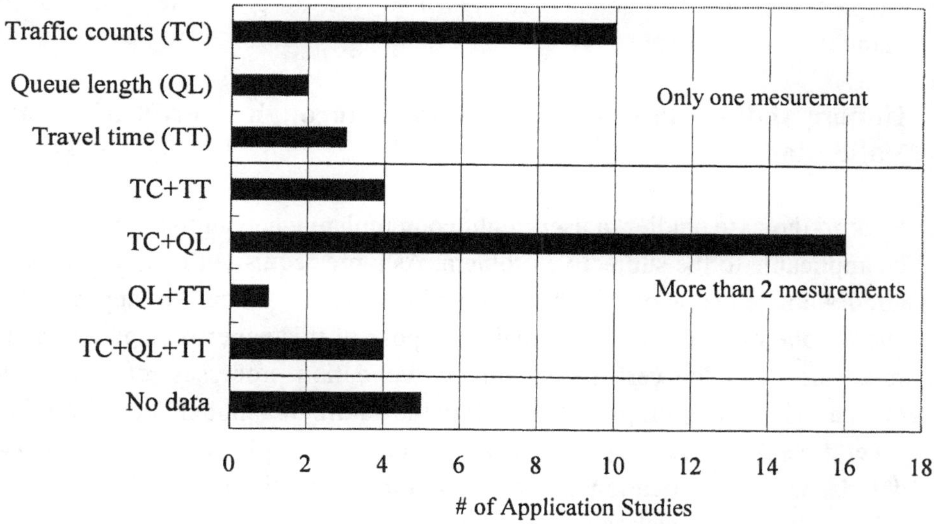

Figure 10: Indices of Reproducibility of Traffic Conditions.

BEST PRACTICE MANUAL FOR TRAFFIC SIMULATION APPLICATIONS

Sim@JSTE is now working on the best practice manual for traffic simulation applications, which contain useful sample procedures for each of the issues in practical case studies. There are several issues in typical simulation applications for example:

i) Taking advantage of the dynamic traffic simulation with the static analysis.
ii) Understanding the models' nature through verification and

validation.

iii) Traffic survey for dynamic traffic simulation.
iv) Trip demand estimation and network configuration.
v) Parameter calibration to reproduce present traffic conditions.
vi) Sensitivity to the analysis of each scenario.

For each issue, typical techniques that have been employed by the experts of simulation will be explained concretely. In addition, not only the conventional techniques but also the advanced ones will be included in the manual. In this chapter, some of the contents are presented so as to grasp the outline of the manual.

Understanding the Models' Nature through Verification and Validation

Prior to the case studies, a user might contemplate what simulation model can be applicable to the subjective problem. As a pre-requisite the user must have good knowledge about his/her available models' nature to appropriately choose one of them. Since the major purpose of this paper is to give a reader some idea how the verification and the validation processes referred in the manual will be helpful in understanding the nature of a model, let us introduce several results of verification and validation as examples in this section. The details of the verification process and further results of verification were referred from Horiguchi and Kuwahara (2002).

Generation of Vehicles

For the implementation of simulation, it is necessary to generate the traffic at the entry end according to the arrival distribution of vehicles from outside the study area. Most of the simulation models seem to assume random arrival at a network boundary section, but there might be some other arrival patterns to be adopted by considering the objective of the simulation study. For example, the uniform arrival may be assumed in some cases of the analysis for over-saturated traffic conditions, in order to avoid the undesirable tendency of pseudo-random series. The 'Standard Verification Process Manual' (JSTE. 2001a) examines whether the generation pattern assumed in the model was really achieved.

In addition to this, it should be also checked whether the number of vehicles generated in a certain time period are equal to or different to the given volume. Figure 11 and Figure 12 indicate the results with different random seeds for AVENUE (AVENUE, WWW site) and tiss-NET (Sakamoto, *et al.*, 1998), both of which assume random arrival in vehicle generation. AVENUE always generates the same number of vehicles as the given demand level (Q=500, 1000, 2000 [veh./hr]), on the other hand tiss-NET varies its results with each random seed.

The findings, resulting from a difference in the attitude of their 'specification' stages, can be realised only through qualifying tests in verification. It gives meaningful implications that literature would not reveal. For this case, a user of the simulation model that has the same nature as tiss-NET in vehicle generation should realize that he or she has to repeat the simulation for the same network and demand configuration with different random seeds. The user also has to be careful when choosing the set of random seeds not to be biased in regard to the number of generated vehicles against the given demand setting. Subsequently, the user must evaluate the variation of the number of generated vehicles for each calculation.

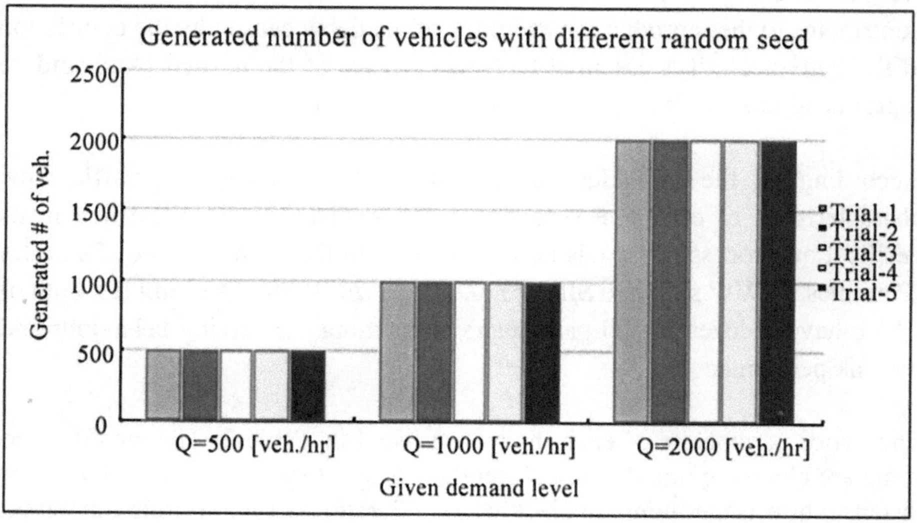

Figure 11: Total number of generated vehicles – AVENUE.

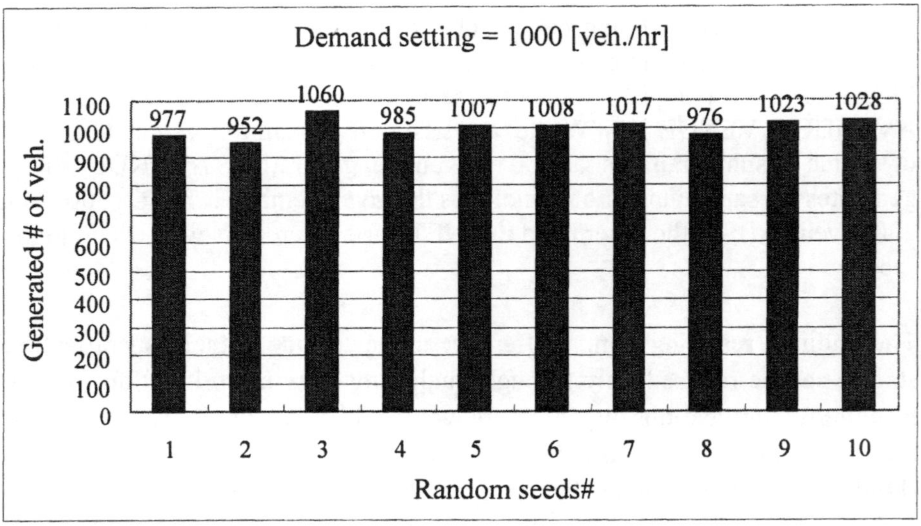

Figure 12: Total number of generated vehicles – tiss-NET.

Traffic Flow Characteristics and Bottleneck Capacity of C-F type Models

As the discharging flow rate from a bottleneck section sags or tunnels contributes to the reproduction accuracy of the delay caused by the congestion at the bottleneck, it is essential that the capacity of the bottleneck should be reproduced in a stable manner during the simulation.

According to the procedure described in the manual, the traffic flow characteristics of each C-F type simulation model must be identified in its verification process. Below is an introduction to the verification of Paramics (Paramics, WWW site) and SIPA (Yokochi, *et al.*, 1999), as examples, both of which have a dozen model parameters concerning the driving behaviour and the link performance.

The major parameters of each model are listed in Table 2. The meanings of some are clear, e.g. maximum acceleration or limit speed, but not all are. For instance: how is the 'minimum headway' of Paramics different from the 'target headway' of SIPA; what is the 'headway coefficient' of Paramics; or is the inverse of 'allowable minimum headway' of SIPA equal to the link capacity? Even if their software manuals or technical papers state the meanings of the parameters, they are mostly conceptual explanations. It is still mysterious how each of the parameters effects upon the bottleneck capacity.

Table 2: Major model parameters of Paramics and SIPA.

Model	Driving behaviour	Link performance
Paramics	Minimum headway, Maximum acceleration, Driving aggression, etc.	Headway coefficient, Limit speed, Gradient, etc.
SIPA	Target headway, Target speed, Maximum acceleration, Response delay, etc.	Allowable minimum headway, Limit speed, Gradient, etc.

Our interest here is to understand the quantitative relationships between the model parameters and the bottleneck capacity reproduced in the simulation. Furthermore, we would like to find the most sensitive parameters through the verification process, because it is the most efficient strategy to apply the simulation result to an actual traffic condition by changing the most sensitive parameters.

Figure 13 shows a portion of the results of Paramics. The dots in the figure indicate the volume-density plots observed with varying major parameters. The shape of the dots is associated with the sort of varied parameters. The remarkable point is that a decline of the flow rate is found only when the 'headway coefficient' of the bottleneck link is 1.5 (dots surrounded by the circle); otherwise there are no effects. This implies that only the changes on the 'headway coefficient' of the bottleneck link does affect to the bottleneck capacity while others have less influence.

Figure 14 shows the result of SIPA in a case where the 'minimum headway' of the bottleneck link changes from 2.0 seconds to 3.0 seconds. Theoretically, the minimum headway must be equal to the inverse of the capacity; as a consequence the bottleneck capacity must be 1200 pcu per hour if the minimum headway is 3.0 seconds. However, the bottleneck capacity reproduced in the verification is slightly greater than the theoretical value. We may realize that the 'minimum headway' of SIPA is similar but different in parameter from the link capacity.

There are common findings of the verification of bottleneck capacity for the C-F type models:

i) Most of them have parameters that affect the minimum headway of each link.
ii) Such parameters have strong influence on the bottleneck capacity but others have less influence.
iii) Such parameters are not exactly equivalent to the inverse of the bottleneck capacity.

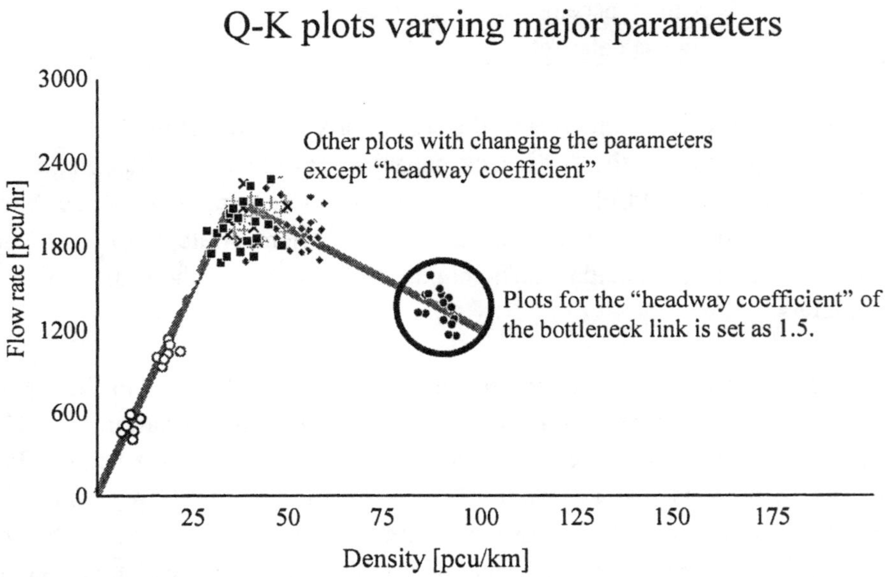

Figure 13: Bottleneck capacity and traffic flow characteristics of Paramics.

There are some implications obtained from i) and ii). Even if we use so-called microscopic simulation models, we have to be rather careful in calibrating the link parameters related to headways than those related to driving behaviours. In this sense, such microscopic simulation models are essentially equivalent to the macroscopic simulation models that require the capacities of links.

In addition, with iii), the tester of SIPA reports the study on the effects of the length of bottleneck sections. Figure 15 illustrates the reproduced bottleneck capacities that change as the length of the bottleneck link varies. The line with a symbol indicates the achieved flow rate along the distance from the bottleneck section. In this case, the bottleneck link for each line has the same

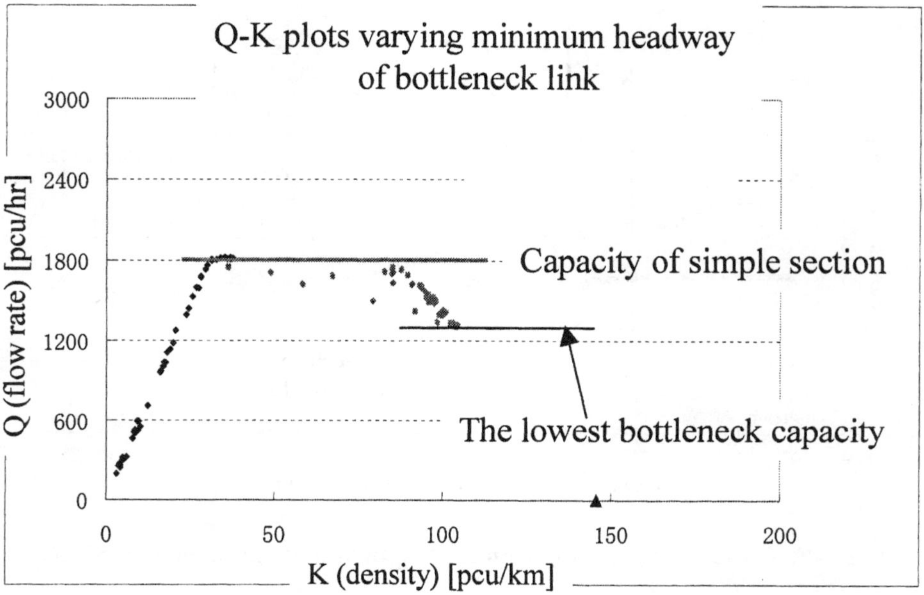

Figure 14: Bottleneck capacity and traffic flow characteristics of SIPA

'minimum headway on the link' value, but has a different length. The superimposed curve that lies on the flow rate at each downstream end of the bottleneck link shows that the capacity declines, as the length of the bottleneck link gets longer.

This phenomenon may be considered in the following: when a vehicle with the minimum headway allowed in the upstream section enters the bottleneck link, the headway achieved by the vehicle does not immediately shift to the new minimum headway. If the length of a bottleneck link is so short that a vehicle passes through the link before its headway will shift to the new headway, the capacity of the bottleneck link will not decline as expected. The users of C-F type models should realize this phenomenon because it is found among most of the models, not only in SIPA.

<u>Saturation Flow Rate at a Signal Intersection</u>

Even on surface streets in under-saturated conditions, a vehicle may have a delay caused at signal intersections. The outflow from an intersection continues at the saturation flow rate until a vehicle queue developed during the red vanishes. It is important to clarify how the saturation flow rate is

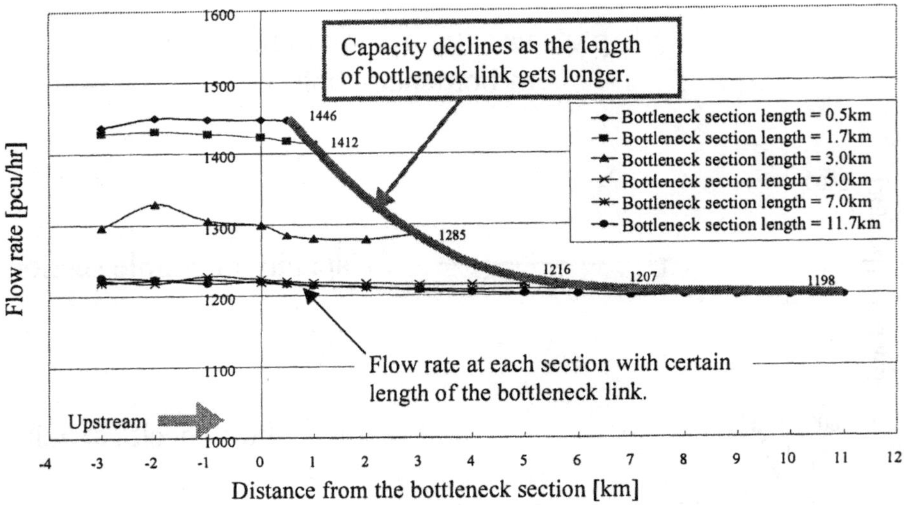

Figure 15: The capacity varying with the length of bottleneck section (SIPA).

reproduced in the simulation model in regards to the bottleneck capacity.

For the verification of the saturation flow rate at a signal intersection, the tester is required to show the profile of discharging traffic within a signal cycle. Let us introduce the result of SOUND (SOUND, WWW site), which has the combined flow model: car following for expressways (SOUND/express) and queuing vehicle lists for arterial roads (SOUND/A-21). The former calculates each vehicle speed according to the spacing-velocity (S-V) function given to each link. The S-V functions can be identified through the macroscopic surveys of traffic flows. On the other hand, the latter assumes the point-queue at the downstream end of each link. The point-queue of each link accepts vehicles up to the jam density and discharges them at the saturation flow rate of the link within the green signal.

Figure 16 illustrates the profile of vehicle discharging for SONUD/A-21. As SOUND/A-21 is a sort of 'Q-K type' model, the discharging flow rate at saturation is expected to strictly agree with the given saturation flow rate. Now we may confirm from the figure that the simulation result attains the given saturation flow rate as 1600 pcu/G1hr (an effective green hour) on average.

There is an additional point to be discussed in relation to Figure 16. The discharging flow rate of SOUND/A-21 immediately goes up to the saturation

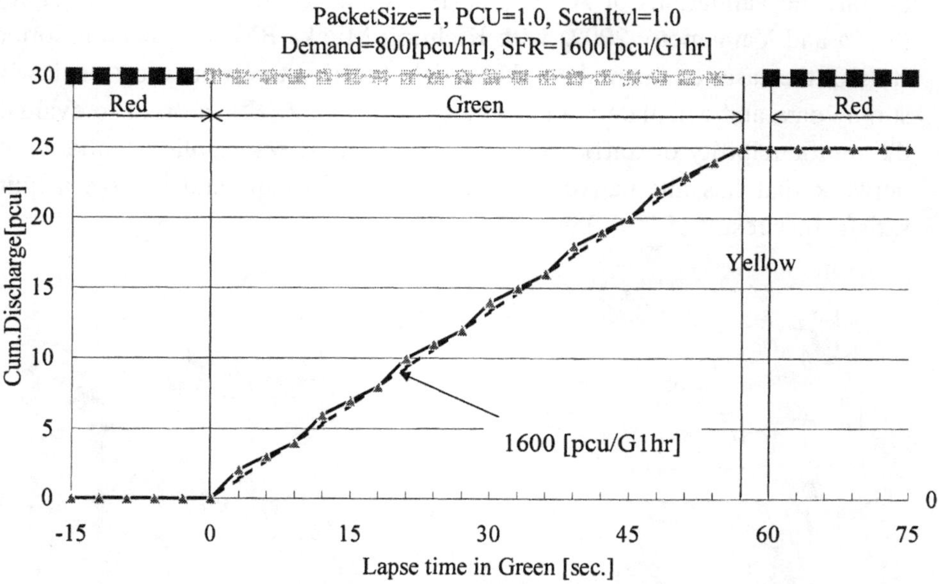

Figure 16: Link discharging profile at a signal intersection -- SOUND/A-21.

flow rate when the signal turns to green. In the actual situation, it takes some time to discharge the flow at the saturation flow rate because of the response delay of the drivers. The tester of SOUND gives the reason to this point as follows:

- Instead of neglecting the starting delay, a vehicle cannot flow out during yellow signals in order to adjust the effective green time.
- At normal intersections, the duration of the green signal is nearly equal to the effective green time so as to take the yellow interval as much as the starting delays.

Network Configuration to be Applicable for the Simulation Model

Not only the verification of simulation models but also the validation can give us useful information concerning the models' nature. Figure 17 illustrates the surface street network in the Kichijoji-Mitaka area in Tokyo, where precise OD trips were collected as well as travel time and signal settings. This data has been made public as the 'Benchmark Dataset (BM)' (Hanabusa, *et al.*, 2002) in order for it to be applied to model validations.

So far, the validations of AVENUE (Horiguchi, *et al.*, 1998) and NETSIM (Sawa and Yamamoto, 2002) with Kichijoji-Mitaka BM have been reported. Both of the cases compare the link throughputs from the simulation result with survey data and calculate the correlation coefficient (R^2) in order to evaluate the reproducibility of traffic conditions. AVENUE was applied to the whole network that has alternative routes for each OD pair, and it gave a quite satisfactory result of $R^2 = 0.98$.

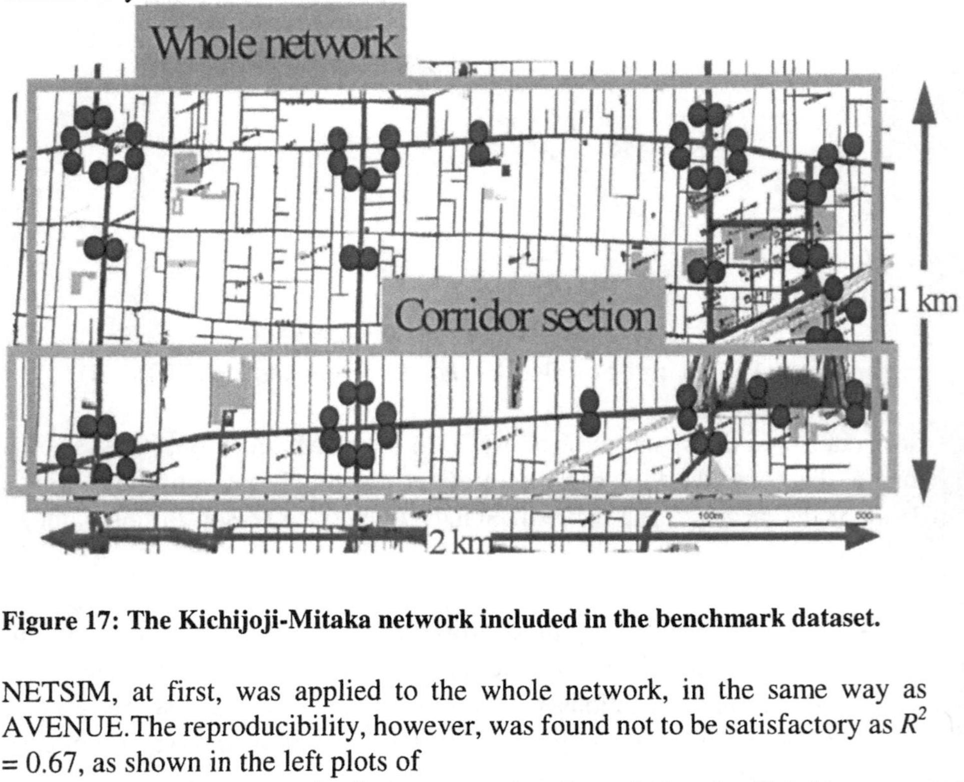

Figure 17: The Kichijoji-Mitaka network included in the benchmark dataset.

NETSIM, at first, was applied to the whole network, in the same way as AVENUE.The reproducibility, however, was found not to be satisfactory as $R^2 = 0.67$, as shown in the left plots of
Figure **18**. Furthermore, the linear regression line of plots is slightly steeper than the diagonal line. This means NETSIM tends to over estimate the traffic volume when it is applied to a network containing loops (Sawa and Yamamoto, 2002).

Subsequently, NETSIM was applied to a corridor section in the network that has no alternative route for each OD pair. In this case, the reproducibility was improved ($R^2 = 0.90$) and the regression line was found to lie along the diagonal line.

The reason for this problem can be explained as follows. Since NETSIM does not incorporate drivers' route choice model, the traffic demand is given as a turning volume ratio at each intersection. Thus, a vehicle may run along a looped route and use the same link more than twice within its trip. This leads to the over estimation of traffic volume. Therefore, the tester of NETSIM concludes that it should be applied only to corridor-shaped networks.

Trip Demand Estimation

How we obtain the trip demand of input traffic simulation is always problematic, especially for the model that accepts the trip demand in OD matrix. Some simulation models are combined with a software package to estimate OD trip demand, such as EMME/2 (EMME/2. WWW site). However, most of that software is based on the estimation procedure used in conventional transportation planning; i.e. the four-step estimation method. Traffic simulation has less compliance to accept such OD trip demand, because it might be estimated without considering the existing link flows. at present. As long as this sort of trip demand is used, it will be difficult for the traffic simulation to reproduce a present link throughput volume with satisfactory preciseness.

In an attempt to reduce the impedance of a mismatch between the traffic simulation and the trip demand estimation, some research exertions exist that estimate OD trip demand using vehicle counts on links. Cremar and Keller (Cremar and Keller, 1987) propose the 'entropy maximization' method that adjusts the OD matrix so as to maximize the entropy calculated with given vehicle counts on a link and assume link choice probabilities for each OD pair. However, their method did not consider the dynamic aspects of traffic conditions, there are still some mismatched to traffic simulation.

Oneyama, et al. (Oneyama, et al., 1996) extends the entropy maximization method to be applicable to dynamic traffic conditions. Their method re-constructs the road network representation along time-space axes as shown in Figure 19. According to the given traffic condition, the congested links are re-constructed as three-dimensional links with steep gradient in time-space. Observed vehicle count at each time slice on each link is assigned to each three dimensional link. By applying the entropy maximization method to this time-space network, we obtain the time-varying OD matrix taking into consideration the dynamic traffic conditions. Adding to this improvement, they incorporate the empirical OD pattern, which can be obtained from a conventional questionnaire, to the calculation of entropy. This empirical OD pattern may affect the estimated OD matrix with restrictive conditions. This 'Extended Entropy Maximization Method ($(EM)^2$)' is now implemented in a software package to work with SOUND/A-21 (SOUND, WWW site).

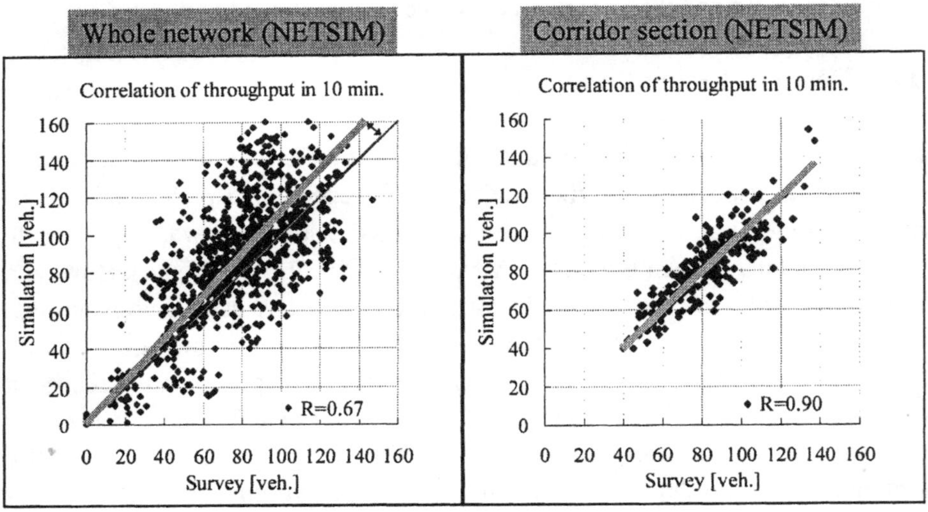

Figure 18: Link throughput reproducibility in the validation of NETSIM (Sawa and Yamamoto. 2002).

Yoshii and Kuwahara (1998) work for the further improvement on $(EM)^2$. Since $(EM)^2$ calculates the choice probability of each three dimensional link by assuming a certain stochastic route choice model, the reproduced link throughput in the simulation, which has a different drivers' route choice model, may be different from the traffic counts used for the OD estimation. In other words, $(EM)^2$ adopts a 'predictive' route choice model while most of the simulation models use 'reactive' route choice models.

The basic idea for their improvement is to replace 3-D 'time-space' network with traffic simulation itself. By applying the simulation with some assumed OD matrix, it can provide the link choice probability that can be used for the calculation of entropy. On the contrary, the link choice probabilities may change when the assumed OD matrix is different. Therefore, the simulation and the OD estimation are iteratively executed and converge on the OD matrix so as to fit the link throughputs in the simulation to the observed vehicle counts. Kitaoka, *et al.* (2002) employs a similar framework but decomposes the time series in order to reduce the calculation time and memory usage.

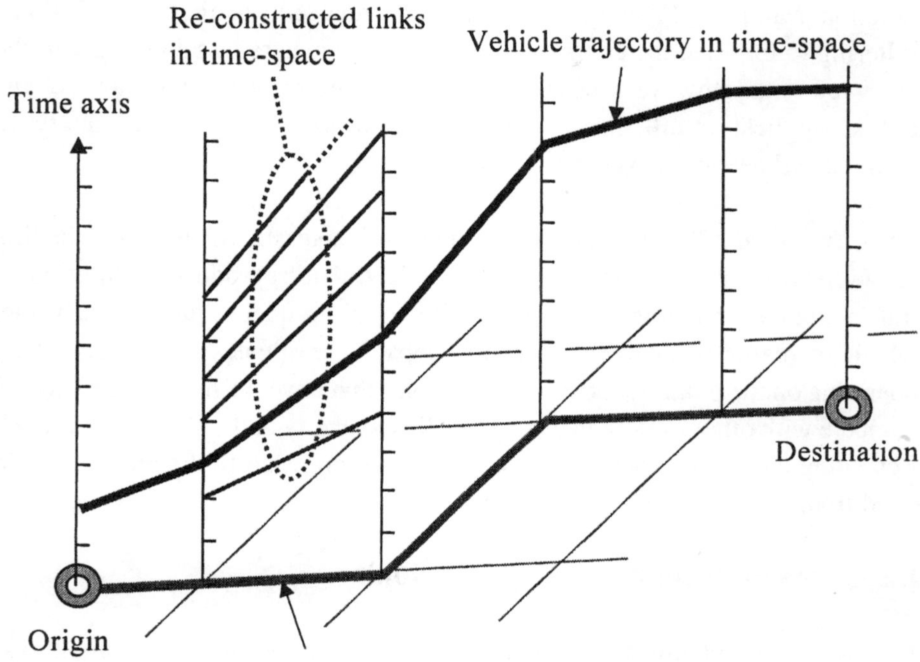

Figure 19: Re-constructed time-space network used for OD Estimation (Oneyama, *et al.* 1996).

Parameter Calibration to Reproduce Present Traffic Conditions

Most of the applications of traffic simulation include the phase to reproduce 'present' traffic conditions. The reason comes about not only because it is necessary to compare the basic case with future cases, but also because we need the evidence that the simulation model we use has the ability to reproduce desirable traffic conditions by adjusting model parameters.

For the microscopic, i.e. small case studies, it is enough to let the user calibrate the model parameters by hand, because there are not so many key parameters as to confuse the user. On the other hand, for the macroscopic, i.e. large case studies, there are too many model parameters to be adjusted by hand. We need some efficient methodology to tune these model parameters.

Cremar and Papageorgiou (1981) proposed the 'Box Complex Technique', which is based on a random search technique to solve a non-liner objective function. Nanthawichit and Nakatsuji (2001) proposed a method using Kalman Filtering to improve the estimation of model parameters by adjusting it to the observed data. However, these methodologies are strongly conscious of the online application of traffic simulation and seem difficult to apply to complicated networks except corridor shapes.

Furukawa, *et al.* (2000) proposed a method with heuristic rules to vary the link capacities of macroscopic simulation models. The reproduced condition in traffic simulation is transferred to the simple 'point-queue' model. Each rule, which increases or decreases the link capacity, evaluates its agreement by operating on the point-queue model. The rules that have the highest agreements compete with other rules then change the link capacity in the simulation model. Repeating these processes converge the simulation result to the present traffic condition.

Indicators to Evaluate the Reproducibility

An additional problem is also found in reproducing present traffic conditions; the question of how to evaluate the reproducibility arises. We frequently use RMSE or the correlation coefficient for the link throughputs and the travel times, but those indicators involve a couple of defects.

The first one is that they independently evaluate each time slice without considering time series. For the link throughput, the errors in earlier time slices may affect the reproducibility of the cumulative throughput to a greater degree than the errors in later time slices. The second is that their values include both the errors in the changes in the long term and the ones of short fluctuations.

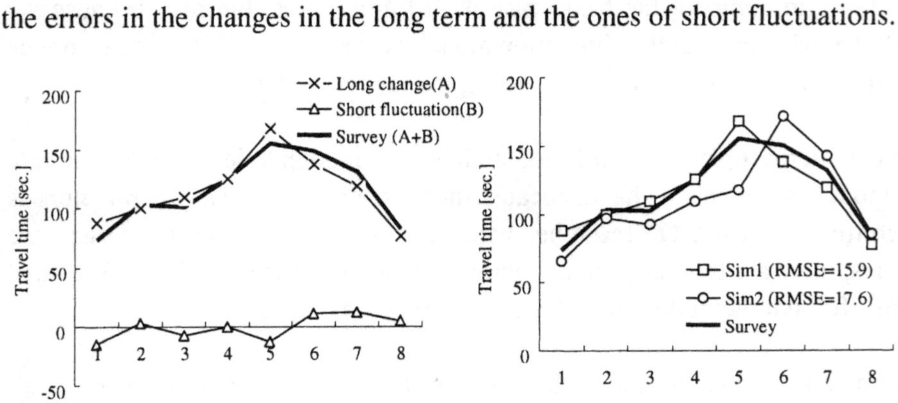

Figure 20: Problems for the evaluation of reproducibility of time series data with RMSE.

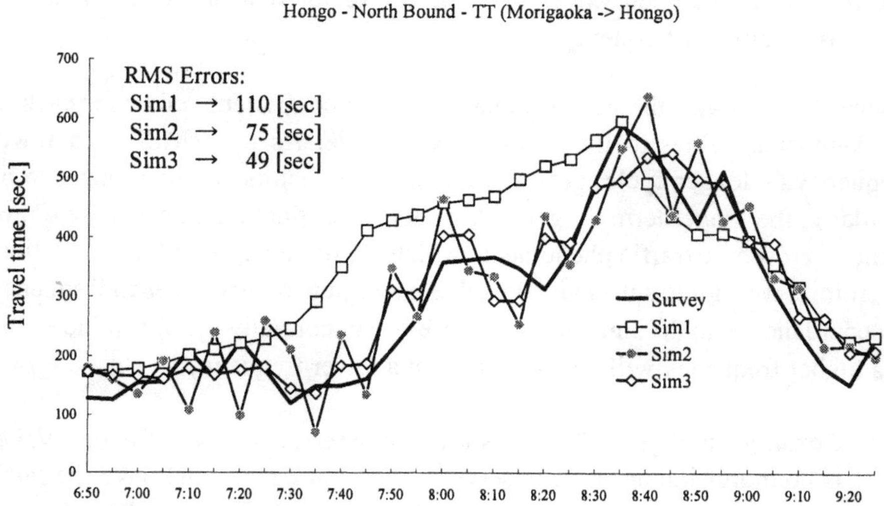

Figure 21: Examples of the Simulation Result.

For instance, let us assume the survey values to be reproduced by the simulation, as shown in Figure 20, contain long lengthy changes to be in subject and the short fluctuations not to be in subject. We have now two simulation results; Sim1 that completely agrees with the long changes in the survey data but does not agree with the short fluctuation, while Sim2 does not agree with the long changes but completely agrees with the short fluctuation. The RMSE value for the former is 15.9 and for the latter 17.6, but the question of how to judge if this tiny difference is significant or negligible arises. For this case, we finally have to approve the superior one by "seeing" its shape on the graph. As subjective judgement lies here, the result of approval may depend on individuals and this situation is not desirable from an engineering point of view.

Horiguchi (2002) proposes a newer measurement index based on wavelet analysis. This is a popular technique to evaluate the "similarity" of two vectors. In the case of evaluating the similarity of the three simulated travel times to the observed value, as shown in Figure 21, the proposed method calculates the error component (= wavelet coefficient vector) at each frequency through the wavelet analysis.

Figure 22 illustrates the basic idea of wavelet analysis. The original signal is decomposed into the approximation at lower frequency and the wavelet coefficient vector. Each wavelet coefficient vector represents the feature of changes in terms of frequency.

Figure 23 illustrates the RMSE of each wavelet coefficient vector for each of the simulated values to the one observed value. As the RMSE at a lower frequency (= longer cycle) gets smaller, the two vectors can be thought to be similar in the longer term. In general, a short-term fluctuation can arise out of some microscopic traffic phenomena, which are often neglected in simulation. Therefore, we should select an allowable magnitude of error at each frequency in accordance with the simulation purpose. Also, normally the allowable errors at a higher frequency will be looser than at a lower frequency.

For the example in Figure 22, Sim1 shows larger errors at lower frequency, i.e. Sim1 is no more similar to the observed value. Sim2 and Sim3 have the same errors except at the frequency of every 5 minutes. This means that the two values follow the observed value with the changes in longer terms so well that we may conclude Sim2 is satisfactory if we do not consider a great deal about every 5-minute change in the simulation.

Our final goal is to put the curves, as displayed in Figure 23, to indicate the allowable errors with a different grade. Here, the grade "A" means the strictest criterion while "C" means a looser one. What grade that is to be satisfied may depend upon the purpose of the simulation application and the preciseness of available data. Some applications that require reproducibility in short-term changes, such as signal parameter optimisation, should satisfy the strictest grade.

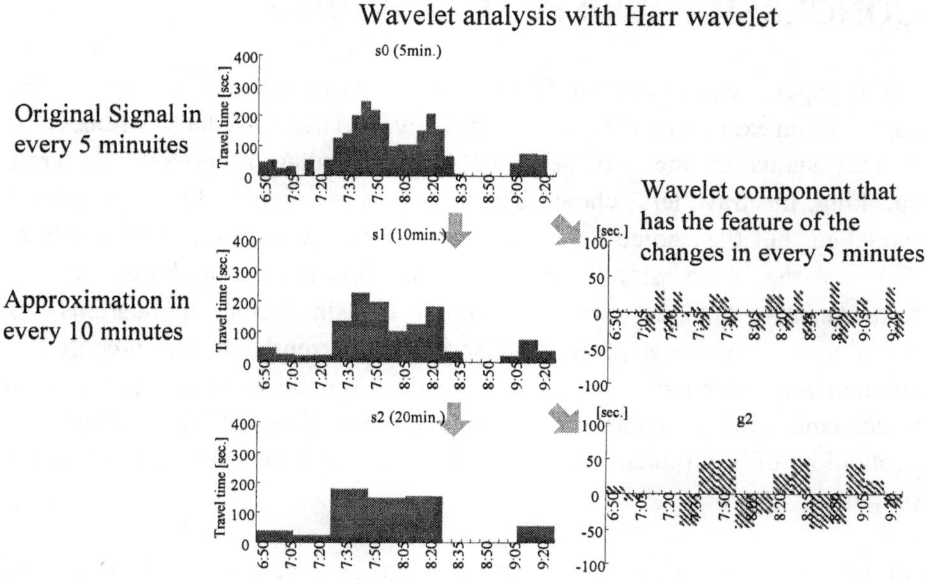

Figure 22: Basic Idea of Wavelet Analysis.

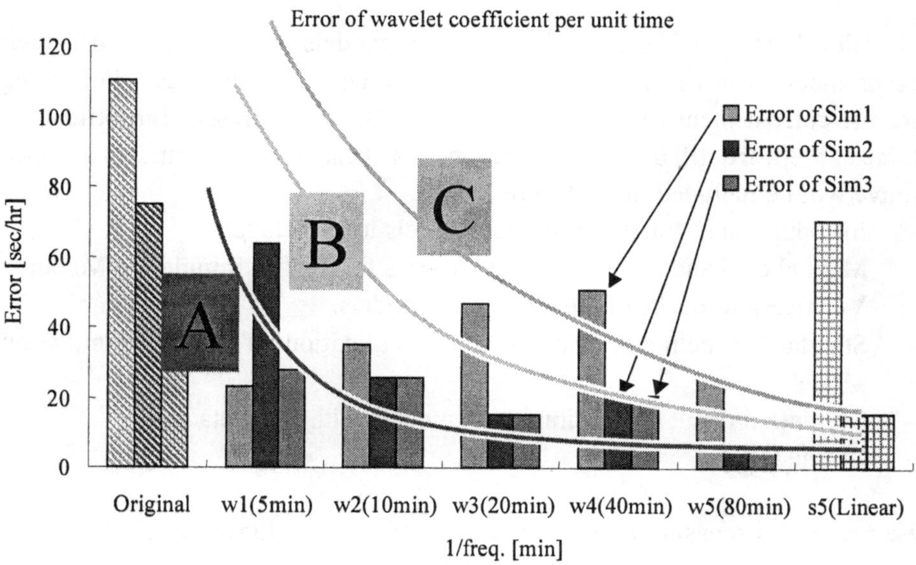

Figure 23: RMSE of Each Wavelet Coefficient Vectors at Each Frequency.

CONCLUSION AND FUTURE WORKS

In this paper, we have introduced recent activities in traffic simulation R&D in Japan. From examining the questionnaire given to users, we have discussed the current status of the utilization of traffic simulation models. Generally speaking, the travellers' choice behaviour is too complex to be sufficiently modelled, but the choice behaviour substantially affects the model outputs. This is the most significant limitation for most simulation models. Subsequently, the best practice manual for simulation applications was introduced. Although this manuscript is currently being prepared, it summarizes major issues such as: the verification and validation processes to understand model characteristics, relevant practices of data acquisition, evaluation of reproducibility of models, and the interpretation of model outputs.

So far, we have been working on standardizing several stages in the utilization of simulation as well as in the model development stage. These activities can be found in 'Clearing House' at the Internet website:

http://www.jste.or.jp/sim/ (in Japanese, Figure 24).

The developers and users of simulation models can also publish their experiences on verification and validation through Clearing House. Following are the current menus available at Clearing House at present (unfortunately some of them are still under construction). The best practice manual mentioned above will be included in the future.

- Introduction of traffic simulation models used in Japan.
- Manual of Standard Verification Process for Traffic Simulation Models.
- Verification results of the simulation models.
- Standard Benchmark Data Sets for Validation of Traffic Simulation Models.
- Validation result of the simulation models with BM data sets.
- Online Q&A.

We are now discussing how to encourage model developers to disclose their verification results to the public. Basically, we expect the verification process to be "de-facto standard" by educating about the necessity of verification to practitioners and also to people in public sectors who order consulting jobs using simulation models. Further discussion in our activity is expected in order to comprehend the results of the verification studies, and to estimate the characteristics of each model. Also, we will advocate the movement of this standard certification process for other simulation models worldwide.

Figure 24: The Clearing House of Traffic Simulation Models (in Japanese).

REFERENCES

AVENUE. *http://www.i-transportlab.jp/products/avenue/* (in Japanese).

Boyce D.E. , Ran B. , and LeBlanc L.J. (1993). Solving an Instantaneous Dynamic User-Optimal Route Choice Model, *Transportation Science*, **Vol. 29, No. 2**, pp128-142.

Cremar, M. and Papageorgiou, M. (1981). Parameter identification for a traffic flow model, *Automatica*, **Vol. 17, No. 6**, pp837-843.

Cremar, M. and Keller, H. (1987). A new class of dynamic methods for the identification of origin-destination flows, *Transportation Research*, **Vol. 21B, No. 2**, pp117-132.

EMME/2. *http://www.inro.ca/products/e2fea.html.*

Friesz T.L., Luque F.J., Tobin R.L., and Wie B.W. (1989). Dynamic Network Traffic Assignment Considered as a Continuous Time Optimal Control Problem, *Operations Research*, **Vol. 37**, pp 893-901.

Furukawa, M. *et al.* (2000). Automatic tuning of parameters in a network traffic simulation model, *Proceedings of 7th World Congress on Intelligent Transport Systems*, Turin, CD-ROM.

Hanabusa, H. *et al.* (2001). Construction of a data set for validation of traffic simulations, *Journal of Japan Society of Civil Engineers*, **No. 688/IV-53**, pp115-123.

Horiguchi, R. *et al.* (1996). A network simulation model for impact studies of traffic management 'AVENUE Ver. 2', *Proceedings of the Third Annual World Congress on Intelligent Transport Systems*, Orlando.

Horiguchi, R. *et al.* (1998). A benchmark data set for validity evaluation of road network simulation models, *Proceedings of 5th World Congress on Intelligent Transport Systems*, Seoul.

Horiguchi, R. *et al.* (2000). A manual of verification process for road network simulation models – an examination in Japan, *Proceedings of 7th World Congress on Intelligent Transport Systems*, Turin, CD-ROM.

Horiguchi, R. (2002). What should be proper indicators to evaluate the reproducibility of traffic conditions for dynamic traffic Simulation?, *Proceedings of Infrastructure Planning, June 2002*, Nagoya (in Japanese).

Horiguchi, R. and Kuwahara, M. (2002). Verification process and its application to network traffic simulation models, *Journal of Advanced Transportation*, **Vol. 35, No. 3**, Special Issue of ITS.

Horiguchi, R. and Oneyama, H. (2002). Future direction for R&D activities of traffic simulation through the analysis on practical applications in Japan, *Journal of Japan Society of Civil Engineers*, **Vol. 709, No. IV-56**, pp61-69 (in Japanese).

IATSS (2001). A manual of traffic assessment (in Japanese), *Report of International Association of Traffic and Safety Sciences*.

JSTE (2001a). A manual of standard verification process for traffic flow simulation model (Draft in English), *http://www.jste.or.jp/sim/verification/manual/VfyManE.html,* Japan Society of Traffic Engineers.

JSTE (2001b). Clearing house of traffic simulation models, *http://www.jste.or.jp/sim* (in Japanese).

Kitaoka, H. *et al.* (2002). Reproduction of traffic condition based on estimation of origin-destination flow, *Proceedings of Infrastructure Planning, June 2002*, Nagoya (in Japanese).

Kuwahara, M. and Akamatsu, T. (1993). Dynamic Equilibrium Assignment with Queues for a One-to-Many OD Pattern, *The 12th International Symposium on Transportation and Traffic Theory*, pp185-204, Elsevior, Berkeley.

Kuwahara, M. and Akamatsu, T. (2001). Dynamic user optimal assignment with physical queues for a many-to-many OD pattern, *Transportation Research*, **Vol. 35B, No. 5**, pp461-479.

Lam H.K.W. and Huang H-J. (1995). Dynamic User Optimal Traffic Assignment Model for Many to One Travel Demand, *Transportation Research*, **Vol. 29B, No. 4**, pp243-260.

Merchant D.K. and G.L.Nemhouser (1978). A Model and an Algorithm for the Dynamic Traffic Assignment Problems, *Transportation Science*, **Vol. 12**, pp183-199.

Nanthawichit, C. and Nakatsuji, T. (2001). Parameter estimation of macroscopic traffic simulation model, *Infrastructure Planning Review*, **Vol. 18, No. 4**, pp935-942.

NETSIM. *http://www.fhwa-tsis.com/.*

Oneyama, H., et al. (1996). Estimation of time dependent OD matrices from traffic counts, *Proceedings of the Third Annual World Congress on Intelligent Transport Systems*, Orlando.

Paramics. *http://www.paramics.com/.*

Rao L, Owen L (2000). Validation of high-fidelity traffic simulation models, *Transportation Research Record*, **Vol. 1710**, pp69-78.

Sakamoto, K. *et al.* (1998). Traffic assignment method considering car-by-car behaviour for traffic impact studies - development of the tiss-NET system, *Proceedings of 8th World Conference on Transport Research*, Antwerpen.

Sawa, M. and Yamamoto, F. (2002). Validation of NETSIM using Kichijoji Benchmark Dataset, *Proceedings of Infrastructure Planning*, June 2002, Nagoya (in Japanese).

Smartest (1997). Review of micro-simulation models, *http://www.its.leeds.ac.uk/smartest/.*

Smartest (1999). Best practice manual, *http://www.its.leeds.ac.uk/smartest/.*

SOUND. *http://www.i-transportlab.jp/products/sound/* (in Japanese).

Suzuki, H. *et al.* (2000). A model development for estimations of origin-destination travel time and flow on freeways, *Proceedings of 6th International Conference on Application of Advanced Technologies in Transportation Engineering*, ASCE, CD-ROM.

VISITOK. *http://www.visitok.gr.jp* (in Japanese).

Wie B.W., Friesz T.L., and Tobin R.L. (1990). Dynamic User Optimal Traffic Assignment on Congested Multidestination Networks, *Transportation Research,* **Vol.24B**, pp431-442.

Yokochi, K. *et al.* (1999). Development of microscopic traffic simulator for AHS Evaluation, *Proceedings of 6th World Congress on Intelligent Transport Systems*, Toronto.

Yoshida, T. *et al.* (1999). A basic study for the planning of ETC-dedicated highway interchange using a traffic simulator REST, *Proceedings of Infrastructure Planning*, **No.22 (2)**, pp231-234 (in Japanese).

Yoshii, T. *et al.* (1995). An evaluation of effects of dynamic route guidance on an urban expressway network, *Proceedings of the 2nd World Congress on Intelligent Transport Systems*, Yokohama.

Yoshii, T. and Kuwahara, M. (1998). Estimation of a time dependent OD matrix from traffic counts using dynamic traffic simulation, *8th World Conference on Transportation Research.*

Absorbing Markov Process OD Estimation and a Transportation Network Simulation Model

Jun-ichi Takayama and Shoichiro Nakayama
Dept. of Civil Engineering, Kanazawa University 2-40-20 Kodatsuno, Kanazawa, 920-8667, Japan
(takayma,snakayma)@t.kanazawa-u.ac.jp

ABSTRACT

Most studies of traffic network simulations aim to calculate travel times or traffic volumes as accurately as possible using origin-destination (OD) traffic volumes (i.e., an OD matrix). In general, estimating OD volumes is very difficult, and the performance of the simulation largely depends on OD estimation. This study proposes a transportation network simulation model that makes use of OD estimation. This simulation model estimates OD volumes using the absorbing Markov process, which can easily estimate OD volumes using only traffic counts at intersections, and which simulates the transportation network dynamically. This enables us to simulate the network states more closely to actual traffic counts.

INTRODUCTION

In the past, many researchers have studied network simulations or traffic flow simulations. Most of these simulate network states from given origin-destination (OD) traffic volumes. In general, it is very difficult to estimate OD volumes and, the accuracy of OD estimation is not necessarily sufficient to guarantee an accurate simulation. The performance of the simulation model seems to largely rely on OD estimation. Therefore, incorporating OD estimation in the traffic network simulation can be very useful even though the simulation itself is simple.

There have been several studies completed on the estimation of OD volumes or the OD matrix. Zuylen & Willumsen (1980) and Iida & Takayama (1986) proposed methods of estimating OD volumes more accurately using OD volumes that were already investigated or estimated. Low (1972), Hoberg (1976), and Holm et al. not wishing to use the existing OD volumes, estimated OD volumes using the gravity model. The gravity model is applicable to wide-area networks such as a city network. More specifically, the gravity model is not suited to an intra-city network or a small network.

The purpose of estimating OD volumes in most cases is to predict travel demand or to simulate traffic flow on a network, with traffic volumes calculated in some way. Also, in most OD estimation models, OD volumes are determined by adjusting calculated traffic volumes to observed traffic volumes (or traffic counts). Thus, traffic volumes on the links should be calculated when estimating OD volumes. As described earlier, OD estimation is not only important for the traffic network simulation, but also for calculating traffic volumes.

In this study, we constructed a traffic network simulation model that incorporates OD estimation. We adopted the OD estimation method that uses the absorbing Markov process. This method based on the absorbing Markov process enables us to estimate OD volumes using only the traffic volume of each lane at an intersection. In addition, is very convenient as traffic volumes are obtained relatively easily and accurately. The method of the absorbing

Markov process OD estimation enabled us to estimate OD volumes simply by using traffic counts. The simulation was then applied to the Kanazawa road network, and examined as to the validity of the simulation .

ABSORBING MARKOV PROCESS OD ESTIMATION

It seems, macroscopically, that traffic flow changes its direction with a certain probability at an intersection, and it then flows to the next intersection. Sasaki (1965) formulated traffic assignment as an absorbing Markov process. Sasaki's traffic assignment can also be interpreted as a kind of user stochastic equilibrium (Akamatsu, 1999).

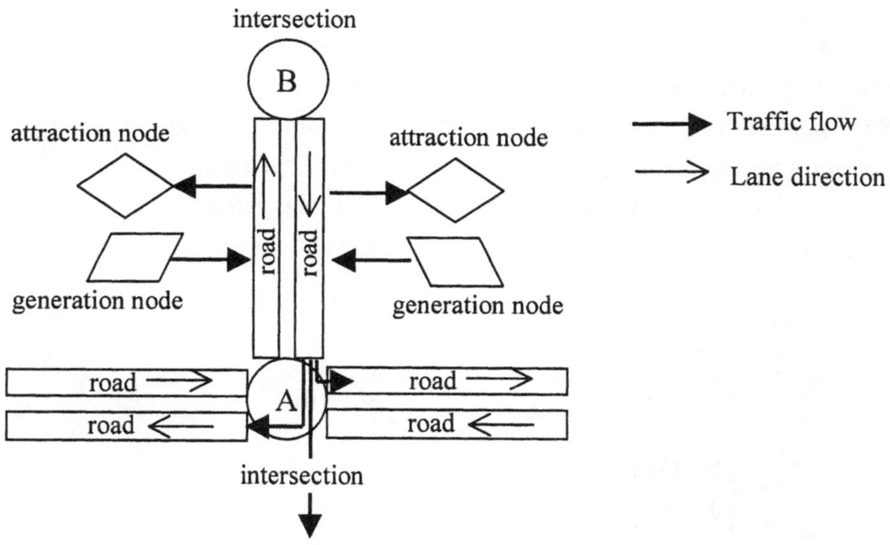

Figure 1. Generation and attraction nodes, roads, and traffic flows

In this study, an objective area is divided into several zones, each has a pair that consists of a generation node and an attraction node. It is assumed that there is no traffic flow that generates or attracts near intersections, and that the generation nodes and attraction nodes are set on the roads (links). Traffic flow is generated in the generation node, and the generation node is a source of traffic flow. On the other hand, traffic flow is attracted in the attraction node, and the attraction node is a sink of traffic flow. Figure 1 demonstrates the generation and attraction nodes and the roads.

Suppose that the network has r pairs of generation and attraction nodes and s roads (directed links). These nodes and roads are considered to represent the states in a network and have variables. The generation variable, the attraction variable, and the road variable represent generation volume, attraction volume and traffic volume on the road, respectively. The transition probability in the Markov process represents the probability of the movement of vehicles between two nodes. The transition probability matrix is expressed as:

$$\mathbf{P} = \begin{bmatrix} \mathbf{I} & \mathbf{O} \\ \mathbf{R} & \mathbf{Q} \end{bmatrix} \tag{1}$$

where **I** denotes a unit matrix, **O** represents a null matrix, **R** represents the attraction probability from a generation node or a road to an attraction node, ($(r+s) \times r$ matrix), and **Q** represents the transition probability between generation nodes or roads, ($(r+s) \times (r+s)$ matrix). There is no transition to the generation nodes and no direct transition from the generation node to the attraction node. Therefore, **R** and **Q** are as follows:

$$\mathbf{R} = \begin{bmatrix} \mathbf{0} \\ \mathbf{R}_2 \end{bmatrix} \tag{2}$$

$$\mathbf{Q} = \begin{bmatrix} \mathbf{0} & \mathbf{Q}_1 \\ \mathbf{0} & \mathbf{Q}_2 \end{bmatrix} \tag{3}$$

where $\mathbf{R}_2$ and $\mathbf{Q}_1$ are $s \times r$ matrices, and $\mathbf{Q}_2$ is an $s \times s$ matrix. The probability that the vehicle, which departed from node i will stay at node j after n transitions is given by the (i, j) element of the matrix $\mathbf{Q}^n$. Therefore, the expectation of the total number of times that the vehicles from the generation node pass through the nodes are $\mathbf{I} + \mathbf{Q}^1 + \mathbf{Q}^2 + \cdots = [\mathbf{I} - \mathbf{Q}]^{-1}$.

$$[\mathbf{I} - \mathbf{Q}]^{-1} = \begin{bmatrix} \mathbf{I} & \mathbf{Q}_1[\mathbf{I} - \mathbf{Q}_2]^{-1} \\ \mathbf{0} & [\mathbf{I} - \mathbf{Q}_2]^{-1} \end{bmatrix} \tag{4}$$

$[\mathbf{I}-\mathbf{Q}]^{-1}\mathbf{R}$ means the probabilities that the vehicle, which departs from each of $(r+s)$ generation nodes or roads, attracts to r attraction nodes. Therefore, given the $(r+s)$-dimension vector of the generation traffic volumes, $\boldsymbol{v} = (v_1, v_2,\ldots, v_r, 0,\ldots, 0)$, the attraction volumes, $\boldsymbol{u}$, and the traffic volumes on the roads (dim. s), $\boldsymbol{x}$, are calculated by:

$$\boldsymbol{u} = \boldsymbol{v}\,[\mathbf{I}-\mathbf{Q}]^{-1}\mathbf{R} \tag{5}$$

$$\boldsymbol{x} = \boldsymbol{v}\,\mathbf{Q}_1\,[\mathbf{I}-\mathbf{Q}_2]^{-1} \tag{6}$$

If the generation volumes are given, we can calculate the traffic volumes and attraction volumes using the above equations. When the traffic volumes are observed on several roads, we can estimate the generation traffic volumes by solving the following optimization problem:

$$\text{Min. } Z = (\boldsymbol{x}' - \boldsymbol{X}')\cdot(\boldsymbol{x}' - \boldsymbol{X}')^{\mathrm{T}} \tag{7}$$

where $\boldsymbol{X}'$ is the observed traffic volumes, $\boldsymbol{x}'$ is the traffic volumes (of the observed roads) that are calculated by Equation (6), and the exponent, T, denotes the transposition. As above, we can estimate OD volumes using equations (5) through (7) when the traffic volumes are observed. We shall call this the absorbing Markov process OD estimation.

When the matrices, **R** and **Q**, are known, we can solve the above minimization problem using conventional methods such as the modified Newton's method because Z in Equation (7) is quadratic of the unknown variables $\boldsymbol{v}$. In this case, the minimized function Z is convex. However, if the minimized function Z is not convex, for example, there are unknown variables in matrices **R** (or **Q**), the conventional methods cannot necessarily be applied. When Z is not convex, the conventional method may provide the local minimum, not the global minimum. In this study, we adopted genetic algorithms (Goldberg, 1989) to solve the minimization problem. The steps used to solve the problem are as follows:

1) Initial setting and coding
   Input the network and the known variables of the matrices **R** and **Q**. Create the population of chromosomes of the unknown variables (the

generation traffic volumes, $\boldsymbol{v}$ and other unknown variables of the matrices **R** and **Q** if they exist)

2) Calculate the traffic volumes
Calculate the traffic volume of each chromosome (the generation traffic volumes, $\boldsymbol{v}$) according to Equation (6).
3) Calculate the value of the minimization function
Substitute the traffic volumes calculated in Step 2) to Equation (7), to obtain the value of the minimization function of each chromosome.
4) GA operations
   i) Reproduction
   Individual chromosomes are copied according to their fitness values. In this study, the fitness value is the minimization function value in Step 3).
   ii) Crossover
   Portions of the chromosomes are combined at random to form new ones.
   iii) Mutation
   The genes of the chromosomes are altered occasionally (with a small probability)
5) Judgment of the termination conditions
If the termination conditions are met, this algorithm finishes; otherwise, repeat Step 2) through 4). In this study, termination occurs when all chromosomes become the same or the minimum fitness does not improve after 20 generations.

DYNAMIC NETWORK SIMULATION CONSIDERING CONGESTION AT TRAFFIC SIGNAL INTERSECTIONS

In the absorbing Markov process OD estimation, we can calculate the traffic volumes on the links using Equation (6). We can see the traffic volumes only on the links that are observed in the traffic count survey and those which are considered in the OD estimation. Traffic volumes that are not considered in the OD estimation, however, cannot be viewed. Using only Equation (6) would be insufficient. When the transportation network has plans, such as introducing new links, or when an analyzation of a network more detailed than the one in the OD estimation is required, we have to carry out the transportation network simulation based on the estimated OD volumes .

Our simulation model is specifically oriented towards OD estimation, and the network simulation is simple. It might be more accurate to call this a semi-dynamic traffic assignment rather than a dynamic simulation. The method we adopted is based on the all-or-nothing assignment of several pieces of OD traffic volumes.

Time is divided into periods of *T* minutes. According to the period, every OD traffic volume in each period is divided into several pieces, and each piece is assigned to the route, which had the minimal travel time in the previous period. Travel times on the links are calculated based on the traffic volume that remains from the previous period and the volumes that run into and out of the adjacent links. Figure 2 illustrates the basic concept of the traffic volume. Figure 2 shows a summation where, the gray parts of the traffic volumes are found on the traffic volume on Link 2.

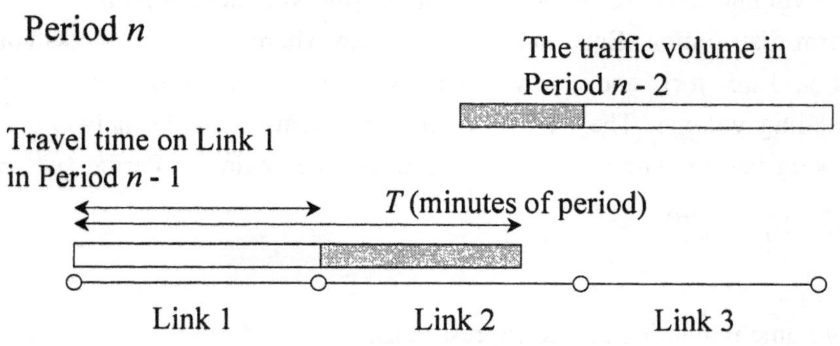

Figure 2. Calculation of link traffic volumes

It is assumed that the traffic volume (the vehicles) depart based on a uniform distribution in the period. The traffic volume of OD in the period, n, that flows through jth link of Route k, Link $j(k)$, is:

$$y_{ij(k)}^{n}=\begin{cases}\dfrac{q_i^n}{T}\cdot t_{j(k)}^{n-1} & \text{if } \sum_{j'=1}^{j(k)} t_{j'}^{n-1}\le T & (a)\\[2ex] \dfrac{q_i^n}{T}\cdot\left(T-\sum_{j'=1}^{j(k)-1} t_{j'}^{n-1}\right) & \text{if } \sum_{j'=1}^{j(k)-1} t_{j'}^{n-1}<T<\sum_{j'=1}^{j(k)} t_{j'}^{n-1} & (b)\\[2ex] 0 & \text{if } \sum_{j'=1}^{j(k)-1} t_{j'}^{n-1}\ge T & (c)\end{cases}\qquad (8)$$

where $y_{ij(k)}^{n}$ is the traffic volume of OD in the period, n, on Link $j(k)$; q_i^n is the traffic volume of OD and $t_{j(k)}^{n-1}$ is the travel time on Link $j(k)$ in the previous period. As described above, the volume from each period is assigned to the same route, and Route k is single. Equation (8*c*) is a case where the head of the volume in Period n does not reach Link $j(k)$ and the volume in Period n does not exist on Link $j(k)$ at all. Equation (8*a*) is a case where the head of the volume in Period n flows ahead of Link $j(k)$ and the volume on Link $j(k)$ is the product of the travel time on Link $j(k)$ and the constant density of the volume because the traffic volume (the vehicles) departs based on a uniform distribution. Equation (8*b*) is a case where the head of the volume runs on Link $j(k)$, and the volume on Link $j(k)$ is in proportion to the proceeding volume. The traffic volume may remain on the network in the following period. The volume of Period n that remains in Period $(n + m)$ is denoted by $y_{ij(k)}^{n,n+m}$.

The volume remaining can be expressed as:

$$
y_{ij(k)}^{n,n+m} = \begin{cases}
0 & if \sum_{j'=1}^{j(k)} t_{j'}^{n+m-1} \ge m \cdot T \quad (a) \\
\dfrac{q_i^n}{T} \cdot \left(\sum_{j'=1}^{j(k)} t_{j'}^{n+m-1} - m \cdot T \right) & if \sum_{j'=1}^{j(k)-1} t_{j'}^{n+m-1} < m \cdot T < \sum_{j'=1}^{j(k)} t_{j'}^{n+m-1} \quad (b) \\
\dfrac{q_i^n}{T} \cdot t_{j^k}^{n+m-1} & if \sum_{j'=1}^{j(k)} t_{j'}^{n+m-1} \le (m+1) \cdot T, \ \sum_{j'=1}^{j(k)-1} t_{j'}^{n+m-1} \ge m \cdot T \quad (c) \\
\dfrac{q_i^n}{T} \cdot \left\{ (m+1) \cdot T - \sum_{j'=1}^{j^k-1} t_{j'}^{n+m-1} \right\} & if \le \sum_{j'=1}^{j(k)-1} t_{j'}^{n+m-1} < (m+1) \cdot T < \sum_{j'=1}^{j(k)} t_{j'}^{n+m-1} \quad (d) \\
0 & if \sum_{j'=1}^{j(k)-1} t_{j'}^{n+m-1} \ge (m+1) \cdot T \quad (e)
\end{cases} \quad (9)
$$

Equation (9*a*) is a case where the last of the vehicles (the volume) passes over Link $j(k)$ and the volume in Period n does not remain in the link at all. Equation (9*b*) is a case where the head of the volume flows ahead of the link, however the last of the volume runs on the link. Equation (9*c*) is a case where the head of the volume flows ahead of the link and where the last of the volume does not reach the link. Equation (9*d*) is a case where the head of the volume flows on the link and where the last of the volume does not reach the link. Equation (9*e*) is a case where the head of the volume does not yet reach the link. If m is 0, Equation (9) is identical to Equation (8). According to Equation (9), the traffic volume on Link j, x_j^n, is:

$$
x_j^n = \sum_{n' \le n} \sum_i \sum_{j(k)} \sum_k \delta_{j,j(k)} \cdot y_{ij(k)}^{n',n} \quad (10)
$$

where $\delta_{j,j(k)}$ is 1 if Link j is Link $j(k)$, otherwise, $\delta_{j,j(k)}$ is 0.

In most cities in Japan, including Kanazawa, traffic congestion occurs due to waiting time at an intersection. Therefore, the travel time on the link is assumed to consist of running time (the time the vehicle runs), t^r, and waiting time at the intersection, t^w. Thus, the travel time, t, is $(t^r + t^w)$. The running time, t^r, is calculated using a formulation by the Bureau of Public Roads (BPR) as follows (some subscripts are abbreviated):

$$t^r = t_f \cdot \left\{ 1 + \alpha \cdot \left(\frac{x}{C} \right)^{\beta} \right\} \tag{11}$$

where t is the travel time on Link j in Period n, t_f is the free-flow link travel time, C is the link capacity, and α and β are constant parameters. As described above, we simplified the simulation for traffic flows. The waiting time, t^w, is also formulated as simple as possible, and is expressed as:

$$t^w = \frac{x^{\mathrm{int}}}{c^{\mathrm{int}}} \cdot s \tag{12}$$

where s is the length of the traffic signal cycle, x^{int} is the traffic volume that passes in the cycle, which includes the traffic volume that has remained form the previous period, and c^{int} is the capacity of the cycle.

As an index of the degree of congestion at an intersection, we utilize the length of congestion at the intersection. Traffic congestions rarely occurs from the neighborhood of the intersection in urban areas in Japan, and the lengths of congestions can express the network state roughly. The length of congestion at the intersection, L, is calculated as:

$$L = a \cdot (x^{\mathrm{int}} - c^{\mathrm{int}}): \tag{13}$$

where a is a positive parameter representing the length of an auto and the distance between autos. In the next section, a is set to about 6.0 meters.

RESULTS OF OD ESTIMATION AND SIMULATION

OD Estimation Results

The method of estimating OD volumes with the absorbing Markov process, described above, was applied to the road network in Kanazawa. Figure 3 shows a network that has been simplified for OD estimation. The data used is from the traffic count surveys obtained from the main intersections in Kanazawa. These were conducted on the 27th and 29th of April, 2000. The former day is a weekday and the latter is a holiday. The probabilities of turning right or left, or going straight, were calculated from this data, and the matrix **Q** was constructed. We estimated OD volumes once an hour from 9:00 to 18:00. Figure 4 and Figure 5 illustrate the scatter plots of the link traffic volumes that are estimated by Equation (6) and the observed ones on both days, respectively. Table 1 shows the correlation coefficients for every hour. The correlation coefficients on both days were more than 0.99 and the estimation is said to be sufficient.

Figures 6 and Figure 7 illustrate the scatter plots of the link traffic volumes that were estimated by the simulation in the previous section, and the observed traffic volumes on both days, respectively. Table 1 shows the correlation coefficients for every hour. The correlation coefficients are greater than 0.92 and the estimation is considered to be sufficient, although the correlation coefficients are slightly lower than the ones in Figures 4 and 5.

Simulation Results

Although the OD volumes are estimated every hour, the travel times and other indexes on the links are calculated every 20 minutes, that is, the period in the traffic simulation is 20 minutes. In Kanazawa, traffic congestion occurs mainly due to waiting at the intersections. So, our simulation displays congestion lengths using GIS as a result. Figure 8 through Figure 11 shows the congestion lengths in Kanazawa's road network at 9:00, 12:00, 15:00, and 17:00 on the 27th (weekday), respectively. In the figures, when congestion at the intersection is heavy, the link near the intersection is wide. It can be seen from these figures that the number of congested intersections and the degree of congestions increases in the evening. Detailed information at each intersection can also be displayed. Figure 12 and Figure 13 represent the exact congestion lengths and flow directions on the lanes at the Kata-machi intersection and the Kasamai intersection at 15:00 on the 27th (weekday), respectively. The density of color represents the congestion length in Figures 12 and 13 (in coloration on the real display). The lane is shown as middle gray (yellow) even if it is only slightly congested. As the congestion worsens, a peeper (redder) color is shown. Thus, the state of the intersection can be understood at a glance.

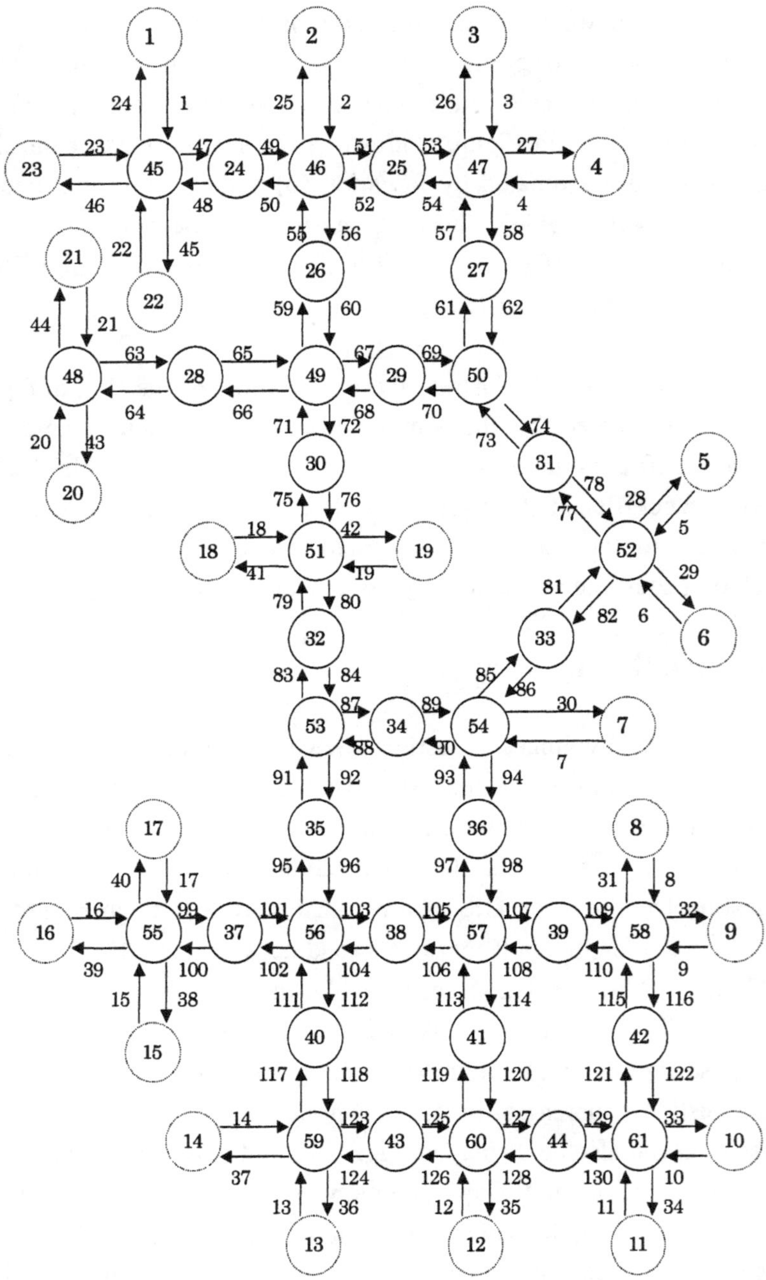

Figure 3. Kanazawa's road network for OD estimation

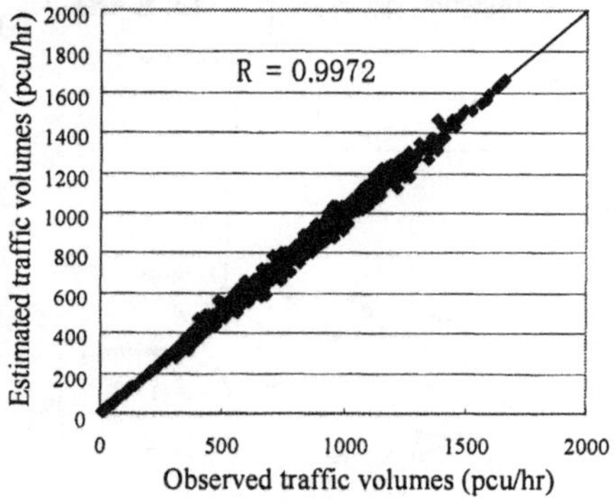

Figure 4. Scatter plots of the OD estimation on the 27th

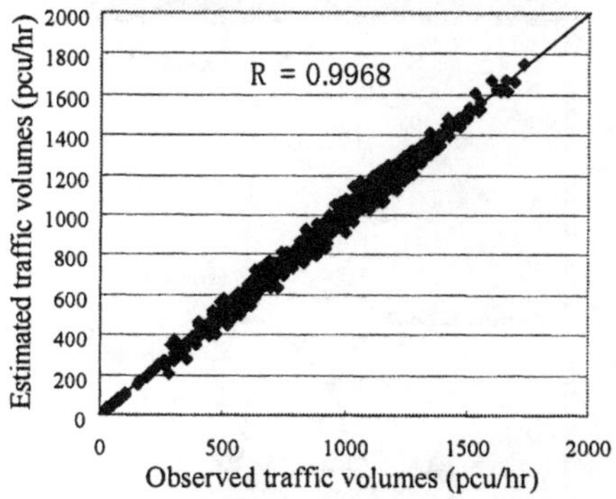

Figure 5. Scatter plots of OD estimation on the 29th

Table 1. Correlation coefficients in the OD estimation

	Weekday	Holiday
9:00-10:00	0.9967	0.9976
10:00-11:00	0.9970	0.9984
11:00-12:00	0.9969	0.9979
12:00-13:00	0.9972	0.9980
13:00-14:00	0.9977	0.9977
14:00-15:00	0.9973	0.9977
15:00-16:00	0.9979	0.9975
16:00-17:00	0.9978	0.9975
17:00-18:00	0.9980	0.9979

Table 2. Correlation coefficients in the network simulation

	Weekday	Holiday
9:00-10:00	0.9467	0.9506
10:00-11:00	0.9545	0.9577
11:00-12:00	0.9478	0.9564
12:00-13:00	0.9340	0.9555
13:00-14:00	0.9478	0.9573
14:00-15:00	0.9487	0.9595
15:00-16:00	0.9557	0.9589
16:00-17:00	0.9580	0.9532
17:00-18:00	0.9584	0.9594

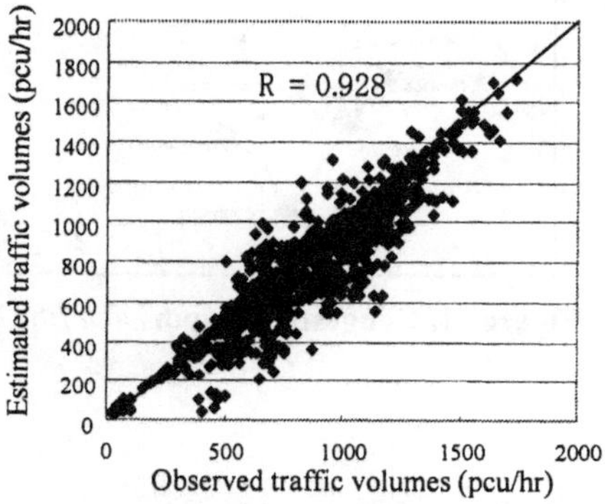

Figure 6. Scatter plots of the transportation simulation on the 27th

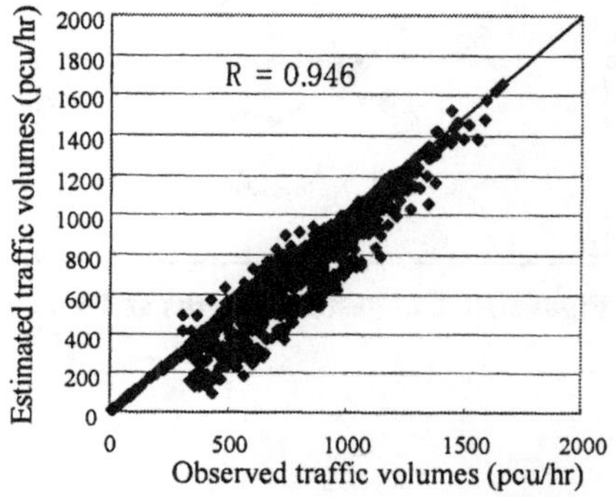

Figure 7. Scatter plots of the transportation simulation on the 29th

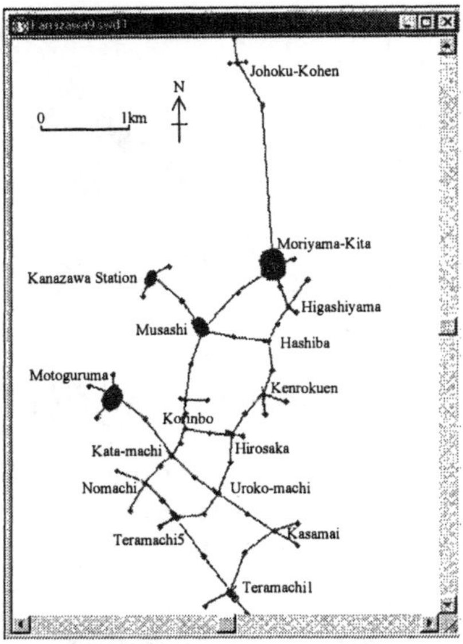

Figure 8. Congestion lengths at 9:00

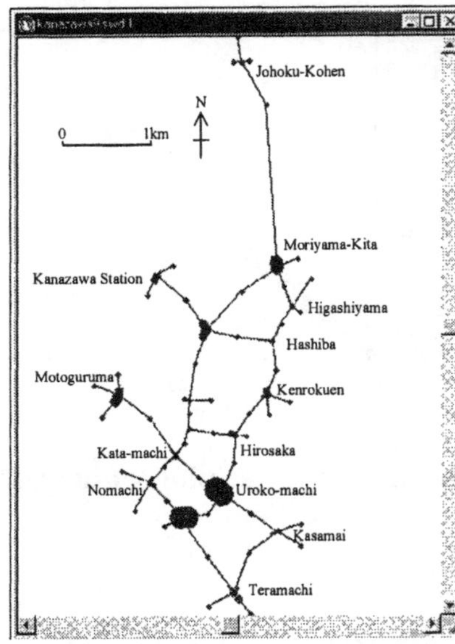

Figure 9. Congestion lengths at 12:00

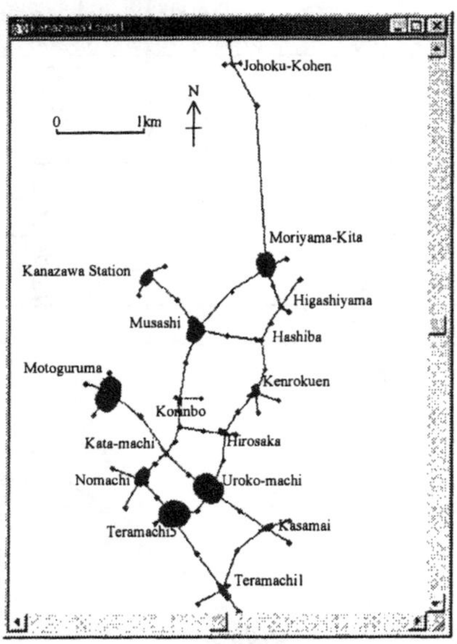

Figure 10. Congestion lengths at 15:00

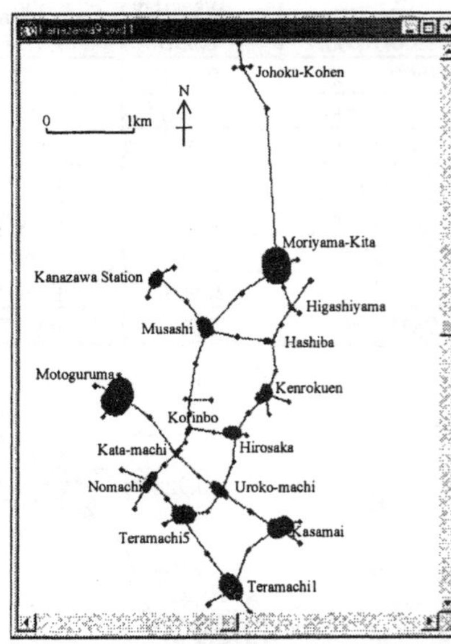

Figure 11. Congestion lengths at 17:00

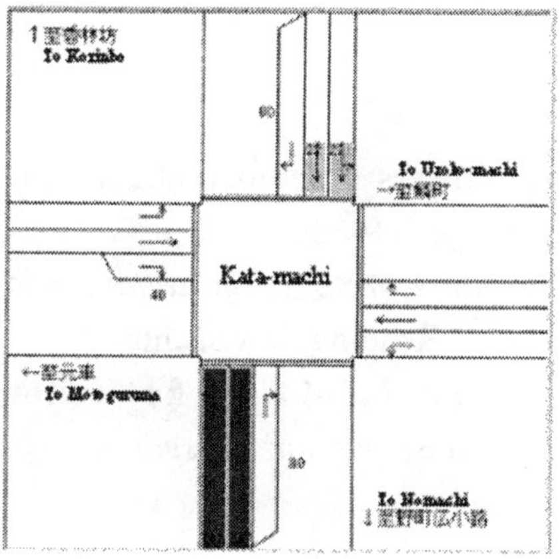

Figure 12. Detailed information at Kata-Kasamaimachi intersection (at 15:00)

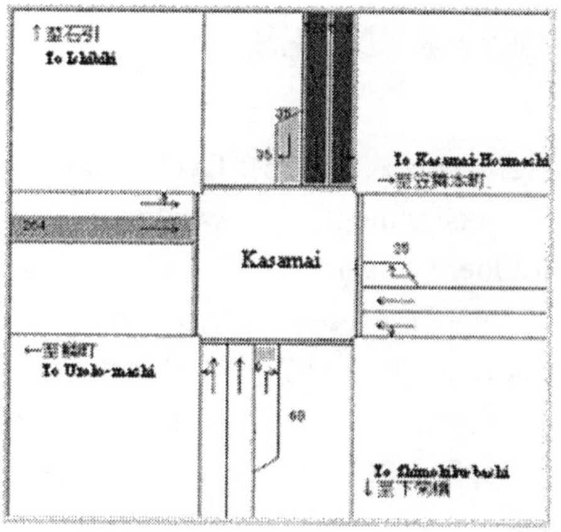

Figure 13. Detailed information at intersection (at 15:00)

CONCLUSIONS

This study, proposed a transportation network simulation model that makes uses of OD estimation. The simulation model estimates OD volumes using the absorbing Markov process, which can easily estimate OD volumes using only traffic counts, and which simulates the transportation network dynamically. This was then applied to Kanazawa's road network, and the validity of the simulation was examined. The simulation model focused on OD estimation, and the description of the network state was simplified. Despite its simplicity when compared to conventional traffic network models, the simulation is useful; it can be applied to the examination of traffic signal system controls and can be used to determine whether a change needs to be made in traffic regulations or lane direction. If a more elaborate network simulation is incorporated, the simulation becomes more useful and can be applied more openly.

REFERENCES

Akamatsu, T. (1996). Cyclic Flows, Markov Process and Stochastic Traffic Assignment, *Transportation Research*, **30B**, pp369-386.

Goldberg, D.G. (1989). *Genetic algorithms in search, optimization, and machine learning*, Addison-Wesley Pub. Co., Reading, Massachusetts.

Hogberg, P. (1976). Estimation of Parameters in Models for Traffic Prediction: A Non-Linear Approach, *Transportation Research*, **10**, pp263-265.

Holm, J., T. Jensen, S.K. Nielsen, A. Christensen, B. Johnsen, and G. Ronby (1976). Calibrating Traffic Models on Traffic Census Results Only, *Traffic Engineering and Control*, **17**, pp137-140.

Iida, Y. and J. Takayama (1986). Comparative Study of Model Formulations on OD Matrix Estimation from Observed Link Flows, *Proceedings of 4th World Conference on Transportation Research*, **2**, pp1570-1581.

Low, D.E. (1972). A New Approach to Transportation Systems Modeling, *Traffic Quarterly*, **26**, pp 391-404.

Sasaki, T. (1965). Theory of Traffic Assignment Through Absorbing Markov Process, *Proceedings of JSCE*, No. 125, pp28-32 (in Japanese).

Van Zuylen, H.J. and L.G. Willumsen (1980). The Most Likely Trip Matrix Estimated from Traffic Counts, *Transportation Research*, **14B**, pp281-293.

SIMULATING TRAVEL BEHAVIOUR USING LOCATION POSITIONING DATA COLLECTED WITH A MOBILE PHONE SYSTEM

Yasuo Asakura
Graduate School of Science & Technology, Kobe University, Kobe, Japan
asakura@kobe-u.ac.jp

Eiji Hato
Department of Civil & Environmental Engineering, Ehime University, Matsuyama, Japan
hato@eng.ehime-u.ac.jp
Katsutoshi Sugino
Institute of Urban Transport Planning Co. Ltd., Osaka, Japan
ks521@po.iijnet.or.jp

ABSTRACT

Mobile communication technology has recently become applicable in tracking surveys of individual travel behaviour in urban space. This is due to an increase in capacity to collect more precise time and location data of moving objects. This paper presents a study of a tracking survey system, which uses the location positioning function of a PHS (Personal Handy phone System) to collect travel data from 96 spectators at a Sumo tournament held in Osaka. It is shown that the space-time distribution map of location positioning data is useful in understanding the concentration and dispersion of travel demand in a day. A simulation-based methodology is proposed which involves multiplying the 96 samples by thousands of spectators. A number of profiles of hypothetical spectators were generated using the probability distribution of the sampled spectators. Another simulation model was developed to estimate the movement of spectators at a match held at a different venue. The probability distributions of the original samples were also used to generate spectator' profiles. These profiles were then put into a railway network simulator that described the movement of spectators and the quality of railway services in a congested situation. Case studies in the Osaka Nagai Stadium were examined to evaluate the congestion reduction policies at railway stations near the stadium. Along with guiding spectators to the less congested stations, sub-attractions in the stadium after the game were also found to be effective in reducing peak time congestion at popular stations.

Keywords: Travel Behaviour, Tracking Survey, Mobile Phone, Simulation

INTRODUCTION

A lot of interest has been directed at travel data collection methods using advanced information and communication technology. It has become possible to utilize various mobile instruments as data collection tools. For instance, GPS (Global Positioning Systems) have been applied to household travel surveys (Battele, 1997; Murakami and Wegner, 1999) in addition to vehicular traffic observation (Zito et al., 1995; D'Este et al., 1999). Mobile phones, including PHS (Personal Handy phone Systems) have also been utilized for collecting travel behaviour data (Asakura et al., 1999). An off-line travel data storage device was developed by Asakura et al. (2001). Furthermore, RF-ID (Radio Frequency Identification) tags used in logistics have been applied to travel data collection (Asakura and Hato, 2001.b).

The location positioning data of a mobile object, an individual traveler or a vehicle, are represented by a sequence of points in space-time dimensions. This is common in any survey system, mobile instrument and constitutes the general attributes of mobile objects. It is necessary to transfer the observed sequence of points when they are to be used for travel behavioural analysis. As such, algorithms were developed in order to identify each point and to indicate whether the tracked object was stationary or moving. The characteristics of each trip were then identified from the location positioning data of individual travelers (Asakura and Hato, 2000).

In order to describe the spatial movement of individual travelers, the location positioning data can be directly aggregated as the block-to-block flows in urban space are separated into several sub-blocks. Asakura and Hato (2001.a) applied the absorbing Markov-Chain model with the block-to-block transition probability matrices, which were estimated using location-positioning data. Although the spatial movement could be described well, the time attributes were not explicitly handled. Furthermore, the original time and location information of an individual object might not be fully utilized once the information has been aggregated.

The microscopic simulation approach has been studied recently in the area of travel behavioural analysis. It is recognized as a promising application for many areas from travel behavioural models to transport policy analysis and evaluation. For example, Fujii et al. (1997) used the sequential choice model of travel behaviour to develop a model named PCATS. This model can simulate the daily travel and activity of all inhabitants in a study area. The microscopic behavioural simulation approach seems effective when the sequential choice assumptions are validated, and reliable travel behavioural data is available for the specification of microscopic models.

The objective of this study is to show how location-positioning data, collected by a tracking survey using mobile instruments, can be utilized for simulating individual travel behaviour. One of the original ideas of the simulation model is to generate hypothetical individuals. The individuals' travel behavioural patterns are then created using the various probability distributions of the location positioning data samples that are obtained by a tracking survey. A simulator is interpreted as a sort of "incubator" generating a number of copied individuals with the attributes of travel behaviour similar to the samples. Instead of applying travel behavioural theories, we intend to utilize the observed location positioning data directly. This is not only because the observed location positioning data is not always sufficient for behavioural model specifications, but rather, it is because the accurate time and location data can be effectively used as they are observed.

In the second part of this paper, before developing the simulation model, the tracking survey method using mobile instruments is explained. The tracking survey method was used to collect location- positioning data from the spectators enjoying the Sumo tournament in Osaka. The data collection process is shown in chapter 3. The structure of the simulation model using the collected data is presented in chapter 4. Two types of simulation models are then discussed. Case studies using those simulation models are then shown in the same chapter with some transport policy experiments.

TRACKING SURVEY METHOD USING A MOBILE PHONE

Why Tracking?

Questionnaire surveys have been widely used for collecting travelers' behavioural data. A sampled individual is obliged to remember his travel activities and to answer a large number of detailed questions. However, it is not easy to provide the precise place and time of the activities. Axhausen (1998) discussed problems involving conventional questionnaire-type travel surveys. He examined the data collection methods of travel behaviour from the perspective of the validity and quality of travel and activity data.

Tracking surveys, on the other hand, are advantageous because they can measure the precise space-time attributes of an object directly when the appropriate survey instruments are available. However, tracking surveys have not been commonly utilized for travel data collection. They have been limited to observing the movement of a small number of pedestrians or cyclists in a narrow area. This is possibly because the cost of tracking human behaviour in urban space has been much higher than actually distributing a questionnaire survey.

Recently, mobile communication technologies have rapidly advanced to the point where they can be utilized as survey instruments for observing individual travel behaviour. As mentioned above, mobile communication systems such as GPS, mobile phone and RF-ID systems are available to accurately determine the place and time of a mobile object. These technologies are considered to be the core instruments of collecting precise travel behavioural data. Thus, a tracking survey using mobile instruments can compensate for the disadvantages of a questionnaire-type travel survey.

Tracking Survey Systems using a Mobile Phone

The number of cellular phone users has grown rapidly. The US Federal Communication Committee (FCC) determined that cellular phone carriers should improve the system to enable the location positioning of the average emergency call to be within 410 feet (125 meters) of the RMS error distance, by October 2001. Consequently, this has affected the rapid innovation of location positioning technology with regards to the use of cellular phones. In 1998, the earliest location positioning services using mobile phones were started by LOCUS, a private company in Osaka, Japan. LOCUS uses PHS, which is one of the cellular phone services. The PHS system emits a weaker signal than normal cellular phone systems, and therefore requires densely located base stations (antennas). As a result, the service carrier antennas are situated about every 100 meters in urban areas. The signal strength of an antenna decreases in direct proportion to its distance from the next antenna. The most useful characteristic of a PHS handset is that the handset usually measures the signal strength of multiple (up to seven) base stations even when the user is not making a call. The exact locations of the antennas are known, and the signal strength from each antenna can be measured. Thus, the position of the PHS handset can be calculated using the triangle survey method.

In spite of the effects of reflection and shielding by buildings and obstacles, the system gives a precise location positioning within a range of 20~150 meters. The error distance depends on the allocation of the base stations. The distinctive difference between GPS and PHS is that the latter is available even if a traveller is traveling or is inside of a building. Thus, the PHS based location-positioning system is suitable for the seamless tracking of travel behaviour in urban areas. The handy terminal for PHS is specified as 54 cc, 58 grams, which is very compact and its batteries last for 5-400 hours. The commercial PHS service was launched in Japan in April 1998. It has been used for various socio-economic activities, for example, personal welfare, security, amusement and so on.

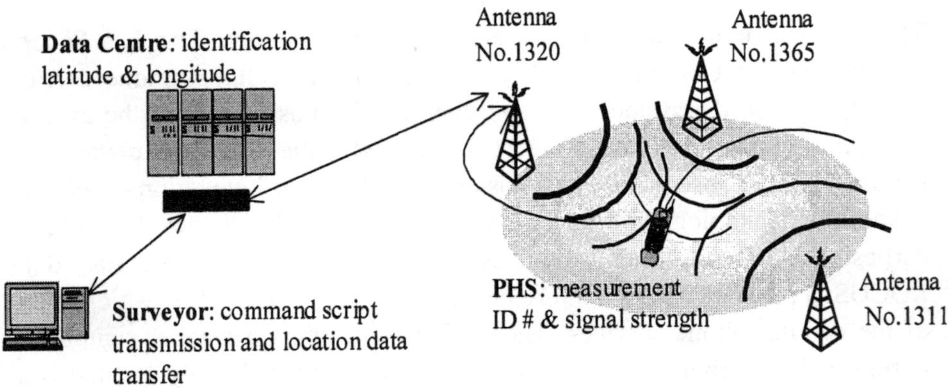

Figure 1. Data Collection System using LOCUS Personal Location Service.

Figure 1 shows an outline of the tracking survey system of travel behaviour using the personal location service of the PHS. In order to protect the privacy of a traveller, it is necessary to make an agreement with each traveller participating in the survey. A registered traveller only has to switch on his PHS handset and take it with him. The observer forwards the start script to the PHS handset of a traveller. The interval of data transmission, 60 seconds at a minimum, is included in the script. With the receipt of the script, the PHS transmits an ID number and the signal strength of neighbouring base stations to the LOCUS centre. The geometric coordinates of every base station are known, and the location position of the PHS can be calculated at the centre. The observer obtains the longitude and latitude of each PHS at every time interval via the Internet. Consequently, the movement of an individual traveler can be tracked.

DATA COLLECTION OF SUMO SPECTATORS

Outline of Tracking Survey

The performance of the tracking system using PHS was tested while observing the travel behaviour of a number of spectators at the grand Sumo tournament held at Osaka Castle Hall, on Saturday 3rd April 1999. The tournament started from 11:00 in the morning and the final match was held just before 17:00. The total number of spectators was estimated at about 10,000. Before the tournament, we collected 100 volunteers who agreed to join the tracking survey. A PHS handset was mailed to each of them with an admission ticket and a program of the tournament. In order to supplement the tracking survey, they were also asked to report the details of their travel and other activities on that day.

The tracking survey was scheduled between 7:30 and 24:00, and the time interval for collecting positioning data was set at two minutes. However, the average number of positioning data was 279 per person. This was mainly because of a delay in switching on the PHS handset in the morning. The loss of PHS signals also occurred when a traveller was in a suburban area without sufficient base stations. Four of the 100 registered volunteers could not participate in the survey due to various reasons, so the number of effective samples was 96.

Data Accuracy

Because of the fluctuation and the reflection of signals, the observed positioning data inevitably recorded some location errors. The accuracy of location positioning was examined using the positioning data at different stationary conditions where the exact locations were known. The following index was used for evaluating the data accuracy. It is the distance of an observed point from the position of a stationary condition. The distance of the i-th observed location at a stationary point n is defined as $d_{ni} = \sqrt{(x_{ni} - \bar{x}_n)^2 + (y_{ni} - \bar{y}_n)^2}$, where x_{ni} and y_{ni} denotes the UTM (Universal Traverse of Mercator) coordinates of the i-th observation at the stationary point n. $\bar{x}_n$ and $\bar{y}_n$ are the corresponding coordinates of the stationary point n.

Figure 2 shows the distribution of the error distance for two stationary sites near Osaka Castle Hall. There were 30 points observed for each site. Due to the fluctuation of signal strength and the multi-path effects, observed points distributed contained some errors at both sites. However, all of the observed points remained in the circle of 100 meters from the exact stationary point. The values of the average distance at each site were 36 meters and 55 meters, respectively. This amount of positioning error may not be sufficient if we intend to identify an individual object precisely. For example, it is not possible to identify a person at either side of a street of 20 meters in width. However, it is possible to track the movement of a person from one place to another when the distance of the two places is not so small. Although it is true that the positioning errors have location-specific tendencies, it seems difficult to completely remove these tendencies. Further analysis might be needed to improve the accuracy of positioning.

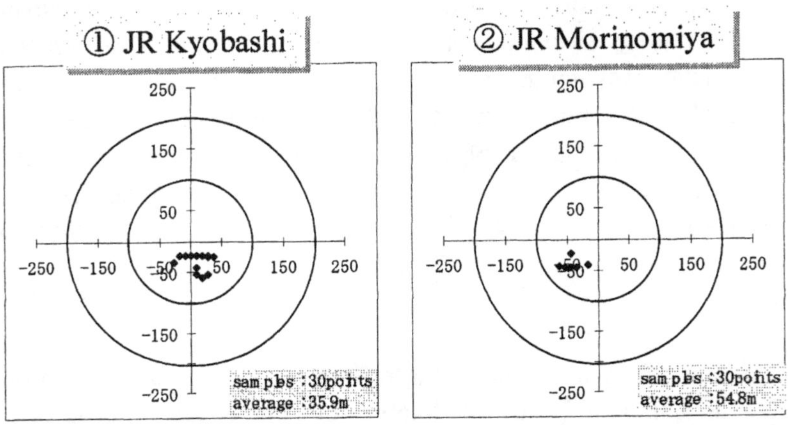

Figure 2. Error Distributions at Two Different Sites near Osaka Castle Hall.

Figure 3 shows the comparison of the exact trail of an individual and the observed location data. A research student was required to report his position as precisely as possible and the exact trail represented by his movement. Volunteers were required to report their position and the exact trail represented by their movements as precisely as possible. Though a few positioning errors were found, the tracking route, with PHS, was mostly consistent with the actual route. Thus, the movement of an individual as a whole could be successfully traced using the location positioning data obtained via PHS.

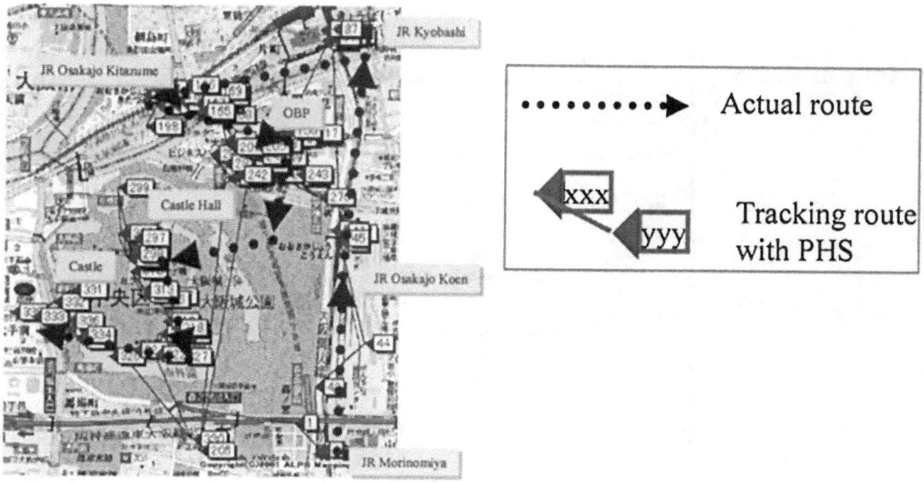

Figure 3. Tracking an Individual around Osaka Castle Hall.

Space-time Distribution of Spectators

When the observed location positioning data of 96 spectators in a day are plotted in space-time dimensions, a hyperbola shaped figure can be depicted. The X–axis and the Y-axis are the direction of longitude and latitude, respectively. The Osaka Castle Hall is placed at the centre of the coordinates. The location position in the longitudinal and latitudinal angler coordinates is converted to the UTM (Universal Traverse Mercator) grid coordinates. The Z-axis represents the time of the day.

Figure 4 shows the spatial concentration and dispersion of spectators for four selected time intervals in the day. Spectators were distributed within an area of 60km from the hall in the morning. Then, they moved towards the hall, and the distribution area at 13:00 was reduced to within 20km from the hall. The main tournament started at 15:00, at this time the spectators were concentrated near the hall. This condition continued until 17:00 when the last match was held. Dispersion of spectators was not very fast and they were still distributed within a 20km area from the hall 2 hours after the tournament. The distribution at 22:00 was almost the same as the first set of data collected in the morning. This diagram is useful to understand the change of the space-time distribution of spectators in large-scale areas such as 100 square km.

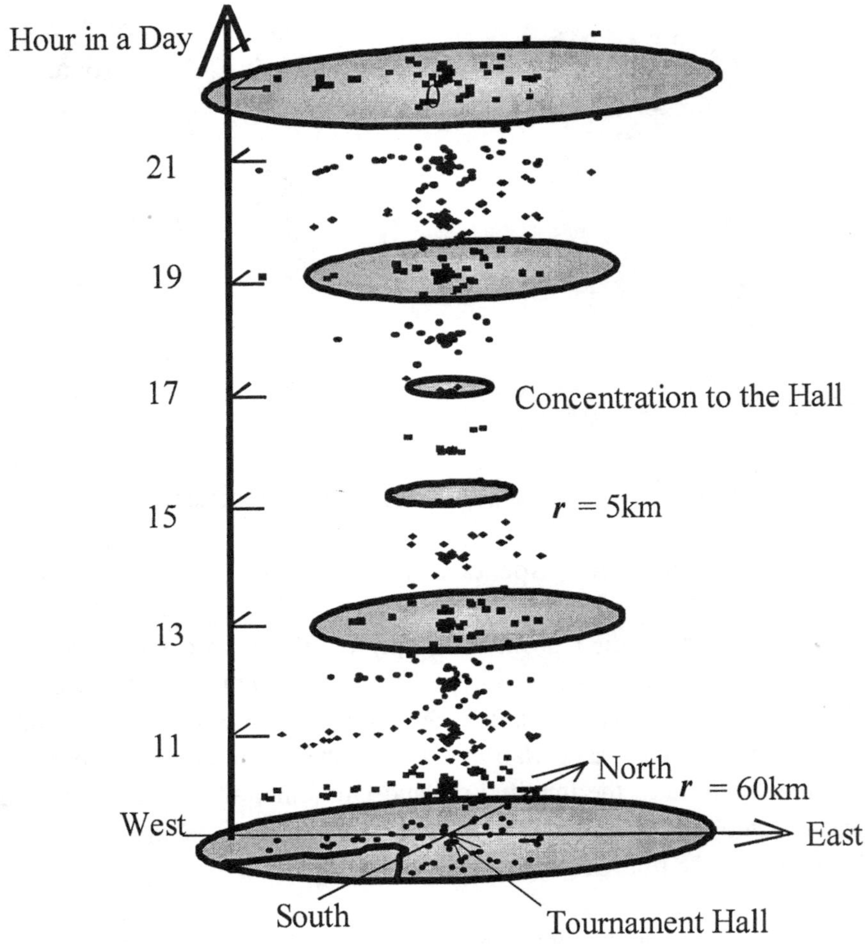

Figure 4. Space-Time Distribution of Location Positioning Data of Spectators.

In order to analyze the movement of spectators around the hall, a rectangle area of 2.08km by 1.36km near the hall was cut off and divided into seven blocks. As the averaged error distance was about 50 meters, the side length of each block was about 500 meters. The block-to-block movement of spectators after the game can be represented using the transition matrices shown in Table 1. The hall is located in block 7, and block 8 corresponds to the area outside of the rectangle. The location of a spectator observed with PHS was aggregated into

Table 1. Block-to-Block Movement of Spectators around the Hall.

16:00 → 16:30

O \ D	1	2	3	4	5	6	7	8	Total
1									0
2						1			1
3							1		1
4									0
5									0
6						1	1		2
7		1	1		1		86	1	90
8								2	2
Total	0	1	1	0	1	2	88	3	96

16:30 → 17:00

O \ D	1	2	3	4	5	6	7	8	Total
1									0
2								1	1
3								1	1
4									0
5								1	1
6						2			2
7		1	2		4	3	78		88
8								3	3
Total	0	1	2	0	4	5	78	6	96

17:00 → 17:30

O \ D	1	2	3	4	5	6	7	8	Total
1									0
2		1							1
3		1						1	2
4									0
5					2			2	4
6				1		3	1		5
7	6	13	9	3	8	7	17	15	78
8								6	6
Total	6	15	9	4	10	10	18	24	96

17:30 → 18:00

O \ D	1	2	3	4	5	6	7	8	Total
1	5							1	6
2		4						11	15
3					1			8	9
4				2		1		1	4
5	3				2			5	10
6			1		3	5	1		10
7	1		2	1		5	4	5	18
8								24	24
Total	9	4	3	3	6	11	5	55	96

one of seven blocks for every 30 minute interval. Then, the movement of a spectator was assigned to a cell of the matrix. The figure in a cell indicates the number of spectators who moved from a block to another. The small number of samples in this study might not be sufficient to discuss the aggregated zonal movement. However, the matrices are useful to understand the tendency of the dynamic change in the spatial distribution of individuals.

From 17:00 to 17:30, spectators dispersed in different directions around the hall. Some of them visited Osaka castle (block 2,3 and 6) to enjoy cherry blossoms in full bloom. One of the streams of spectators went to block 2 where the nearest train station is located. The number of spectators in block 8 (outside of the area) was 24 at 17:30, the rest of the spectators still remained in the rectangle. The percentage of those spectators was 75% (72/96). At 18:00 the number of spectators in block 8 increased to 55 indicating that 57% (55/96) of them had left the rectangle. These transition matrices can be used to understand the main flows of spectators around the hall. An absorbed Markov Chain model has been applied to analyze the movement. The results were reported by Asakura and Hato (2001.b).

SIMULATION MODELS

Two Simulators

The collected location positioning data with mobile instruments gave us valuable information for analyzing individual travel behaviour. Although the location positioning data is more accurate in space-time dimensions, than data obtained with a conventional paper and pencil based questionnaire survey, the number of individual samples may be limited due to constraints in the number of available instruments. One of the possible ways to overcome this shortcoming will be from the simulation of travel behaviour using location-positioning data.

The coverage of the word of "simulation" may be a little different from the ones which have been used for simulating traffic flow and travel behaviour. In this paper, we intend to create a number of individual travelers whose travel attributes are the same as the samples. A more precise definition <is generated> from a set of travel profiles of an imaginary traveller using various probability distributions obtained through the time-space distributions of the location-positioning data from the samples. In this sense, the simulation model used in this study could be interpreted as an "incubator" or "profile generator". For example, the profiles of 100 samples are multiplied to the profiles of 10,000 clone travellers. Generated travel profiles can work as if a number of travelers were observed by a tracking survey. Details of the simulation with the function of "proliferation" will be explained in the next section.

The other simulation model discussed in this paper aims to transplant the observed location positioning data to a different place in urban space. For example, the location positioning data collected at the Sumo tournament is processed to simulate the movement of spectators at a football match held at a different stadium. The objective of developing the "transplant" simulation is to estimate the effects of transport policies adopted at a different event site and its surroundings. It is the same as the calculation used where a number of imaginary spectators are created using information from the original location positioning data. Details of the "transplant" simulation will be explained in the later section.

Profile Generator

When we generate a number of profiles of hypothetical travellers, no interaction is assumed between individual travel behaviour and system performance. In other words, there is no congestion during the profile generation. The level of transport service is kept unchanged, and a traveller's behaviour has no affect on it. This assumption makes it possible to produce a number of travellers' profiles, one by one. Beckman et al. (1996) studied a methodology for the creation of a synthetic baseline population of individuals and households, which could be used for activity-based models. They used census data and focused mainly on the socio-demographic characteristics.

It is not always necessary to describe all the details of travel behaviour when the profile of an individual traveller is generated. There are a number of technical methods for generating profiles. The only common concept among the different methods is to utilize the time-space probability distribution of the sample data. The random number and the inverse of the probability distribution can determine the time and space attributes of travel and activity of an individual. The following procedure is a typical example of generating a set of travel profiles of imaginary individuals in a day.

Step 0: An imaginary individual is generated. Socio-economic characteristics are given if it is necessary.

Step 1: The residential location of this individual is determined using the spatial distribution of the residential place of the original data.

Step 2: The departure time of the first trip of the day is determined using the departure time distribution of the first trip. The scheduled departure time of the last trip is also determined using the departure time distributions of the last trip. The counter of trips in a day, n, is set to equal 1.

Step 3: The location of the destination of the n-th trip is determined using the spatial distribution of the destination of the n-th trip. Travel time to the destination is calculated in a network, and arrival time is then determined.

Step 4: The departure time of the next trip is determined using the duration time distribution at the destination.

Step 5: If the departure time in Step 4 is before the scheduled departure time in Step 2, up-date n=n+1 and go back to Step 3. Otherwise, the travel of this individual is over. Return back to Step 0 and generate the next individual.

When the time of a day and the location in urban space are discretely defined, a generated profile is described as a series of discrete points in space-time dimensions. Thus, the profile of an imaginary individual has the same data format as the profile of an observed sample. For example, an additional label denoting either "move" or "stay" can be attached to each point as well as the time and the location attributes. When a point is categorized as "move", either a link number or a location index is given to the point. The profile data can be managed in the same manner as the sample data. They are depicted on a map or aggregated at any given time-space dimension.

The simple profile generator shown here seems effective when a large number of hypothetical individuals are required for analyzing the aggregate effects of travel behavior, such as congestion. There might be several variations in generating profiles. The trip-to-trip relations should be considered when trip-chaining behavior is essential for the analysis. The transport policy variables could be reflected as constraints in the above procedure. The original characteristics of the source samples have a great impact on the generated profiles. Thus, it is very important to design the generating procedure carefully and to understand the limitations of the generated profiles when a large number of source samples may not be available.

Data Transplant Simulation

When a large-scale event is held at a central area in a city, access by private cars may be restricted and spectators may be advised to use public transport. The location positioning data obtained at the original event site could then be transferred to a different event site in order to evaluate several transport options held at the site. This means that the original data is transplanted as if the movement of spectators was observed at the different site. The data transplant simulation model is designed to describe the movement of spectators who will use railway networks. The model consists of two sub-models: the profile generator and the railway network simulator. This profile generator has the same function as the one explained above. The difference is that generated imaginary spectators are assumed to gather at a different event site using railway networks. The output of the profile generator is put into the railway network simulator. The railway network simulator aims to describe the movement of spectators on railway networks. The congestion due to demand concentration at railway stations is taken into consideration, and the effects of congestion control policies existing at those stations can be estimated.

Figure 5 represents the range of the profile generator. Using the population distribution in a city, the location of residency of an imaginary spectator is determined. The shortest path is calculated from his/her home to the event site, and the railway stations near the home and near the event site are determined along the path. The activities before the event such as shopping are not considered. The arrival-time distribution at the event site is assumed to be the same as the original data. This implies that the original characteristics, such as arrival times can work if they are transferred to a different site. When the opening time (t_4) is given by the event program, the arrival time at the event site (t_3) is estimated using arrival time distribution. The departure time from home (t_1) and the arrival time at the station near the event site (t_2) are then calculated using the travel times along the shortest path.

The profiles after the event can be generated in similar procedures. The closing time of the event (t_5) is given, and the departure time distribution from the event site is also assumed as the same as the original. Using this distribution, the departure time (t_6) is determined. When an attraction is scheduled after the main event, the effect of the post-attraction can be considered in the departure time distribution. The walking time from the event site to the station near-by (t_7-t_6) is calculated, and the expected arrival time (t_7) at the station is determined. This time may not be able to be calculated due to the concentration congestion near the station. When there is no congestion, the expected waiting time at the station (t_8-t_7) can be calculated using train headway. However, the expected departure time (t_8) may not be realized due to concentrated congestion in the station.

Figure 6 shows the flowchart of the railway network simulator. The input data of the simulator are the profiles of imaginary spectators, the timetable of railway systems, and the conditions of the station facility and train cargo, such as platform capacity and train capacity. When a spectator arrives at a station, he/she may not be able to take the train or may even have difficulty reaching the platform due to congestion. The difference between the expected departure time determined in the profile generator and the actual departure time

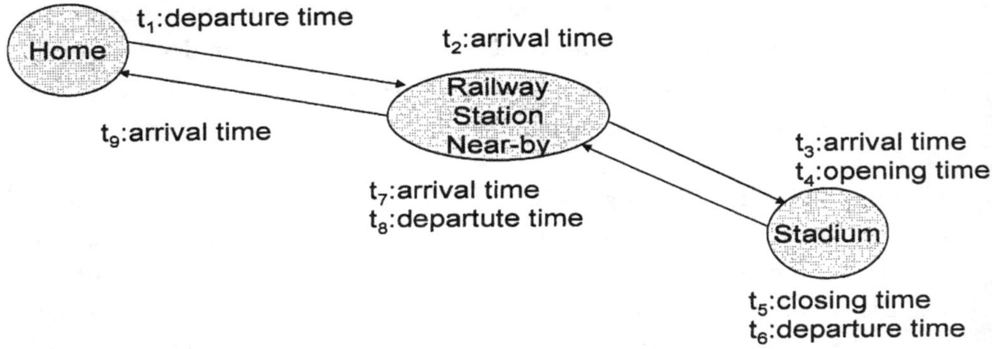

Figure 5. Coverage of the Profile Generator.

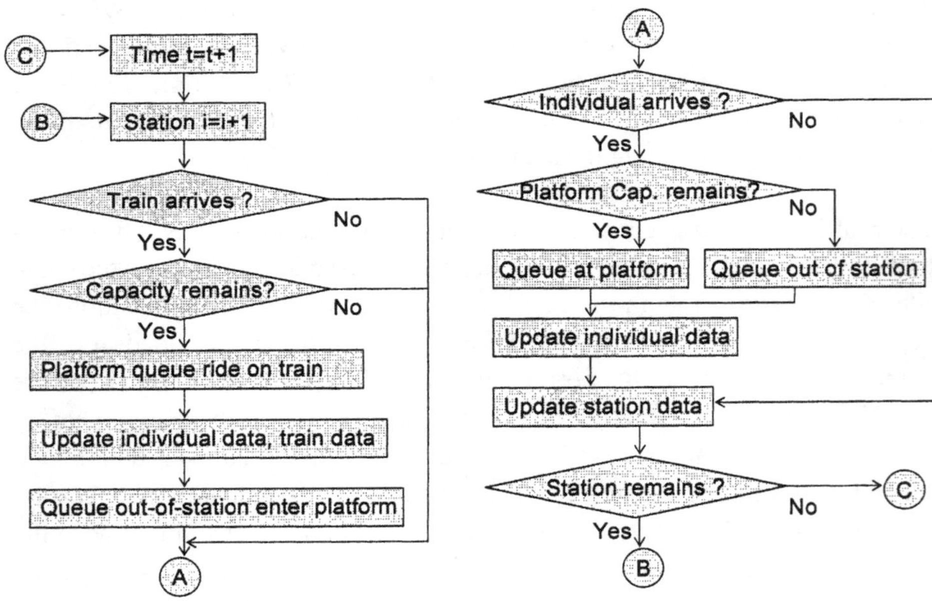

Figure 6. Railway Network Simulator Flowchart.

calculated in the railway network simulator corresponds to the congestion delay at the station. Train headway and capacity are the control variables of the simulation. The effects of extra trains for reducing congestion can be calculated. The guidance of spectators to different stations could be evaluated in the event that some of the spectators were guided to less congested railway stations.

In the transplant simulation model, we assumed that the data obtained at some specific event site could be transferred to a different site in a city. The transplant simulation can work well when the type of event, the transport environment in the application site and the characteristics of the spectators are similar to those of the original site. The profile generator in a different transport environment should be carefully treated when it is applied to areas with different levels of transport services. When it is applied to a case with different spectator characteristics, the expected percentages of the group of spectators with the same socio-economic characteristics could be used for reducing the estimation bias in the profile generator.

Case Studies

The transplant simulation model is calculated to examine the effects of a congestion management scheme for a football game held at Nagai stadium Osaka, Japan. The input conditions are as follows:

a) Date and game: FIFA world cup game held on the 12th June 2002 from 15:30-17:30.
b) Spectators: 50,000 spectators are generated, which is equal to the capacity of the stadium. The home distribution of spectators was limited to within Osaka city and surroundings.
c) Probability distribution: The arrival time distribution to the stadium and the departure time distribution from the stadium were assumed to be the same as those of the probability distribution of the sampled spectators at the Sumo tournament held at Osaka Castle Hall.
d) Railway network: The railway network consisted of Japan Railway network, private railways and Osaka municipal subways. Link impedance was defined as the travel time between adjacent stations, and travel fare was not considered.
e) Stations near-by: Seven railway stations were considered to be accessible on foot from the stadium. For each station, two or three access routes were assumed.

The results of the profile generator and the railway network simulator are shown in Figure 7. Each figure shows the conditions near the stadium after the game. Queues were generated at stations near by, and then disappeared. Each point represents an imaginary spectator and has its own location coordinates. Thus, the movement of spectators can be aggregated at any space-time unit.

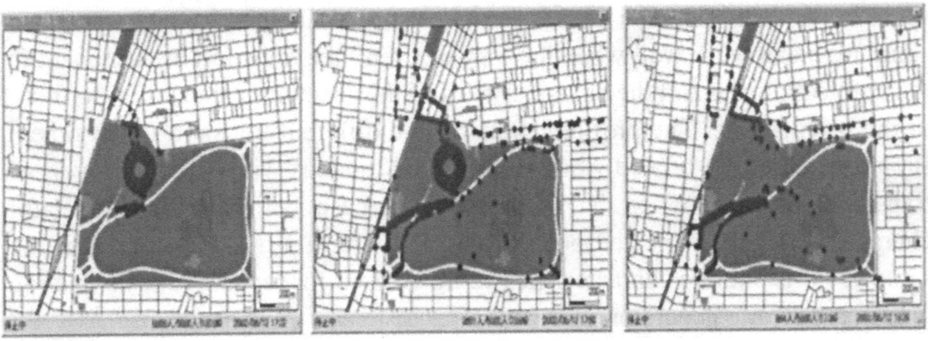

Figure 7. Dispersion of Spectators after the Match.

The following scenarios for reducing congestion after the football game can be examined using the railway network simulator.

a) Supply side improvement: The effects of improving capacities of the railway network systems can be evaluated, such as additional train frequency, additional train cargo and so on.

b) Spatial redistribution scenario of demand (Figure 8-1): This corresponds to the guidance of spectators to less congested stations. A spectator is guided to the down stream station in his/her home direction.

c) Time redistribution of demand (Figure 8-2): This corresponds to the detention of spectators in the stadium for a period of time. For example, small sub-attractions may be used for detaining some spectators. The length of the sub-attraction and the percentage of the detained spectators represent the detention scenario. It is assumed that the departure time distribution of spectators is then shifted.

Figure 9 shows the effects of the sub-attractions and the guidance of spectators. A 30-minute attraction is held after the match, and the spectators are guided to the downstream stations. The effects of guidance without sub-attraction were not so large and the results of the case were not shown. In comparison to the case without post attractions and guidance, the total number of spectators in a queue and the averaged waiting time were not so different. The maximum waiting time just after the game with the guidance scenario would be reduced for almost half of the cases without guidance. The guidance scenario seems effective in delaying the appearance of congestion. However, the waiting times

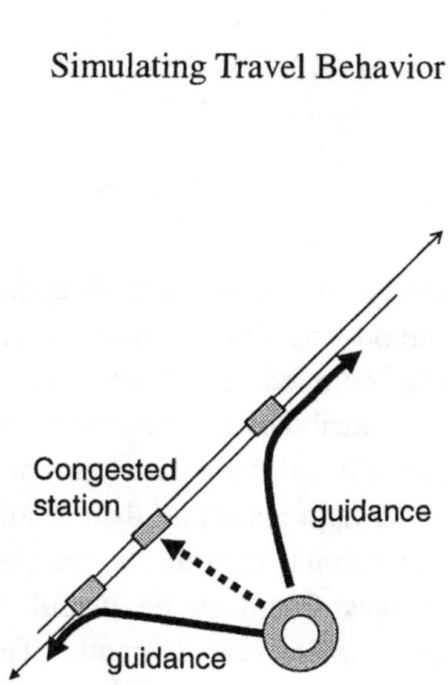

Figure 8-1 Spatial Redistribution Scenario.

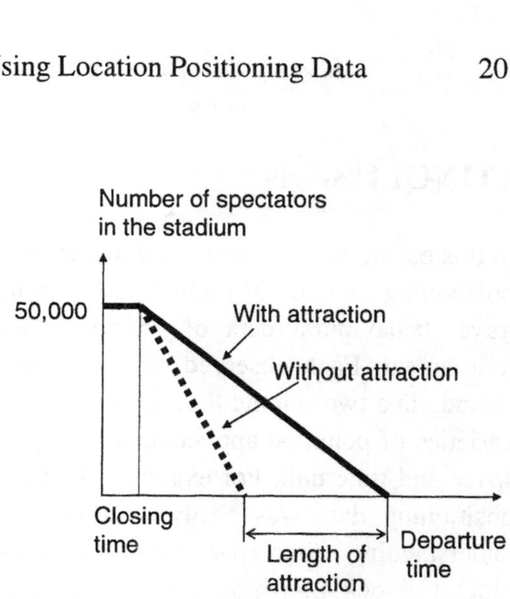

Figure 8-2 Time Redistribution Scenario.

of the guidance scenario greatly increased and the peak of the average waiting time was worse than that of the without-guidance case. Due to the capacity constraint of the railway systems, the guidance scenario might not show a remarkable effect. It was found through further numerical tests that the queue at the railway stations could disappear if more than 90 minutes of sub-attractions were held after the match.

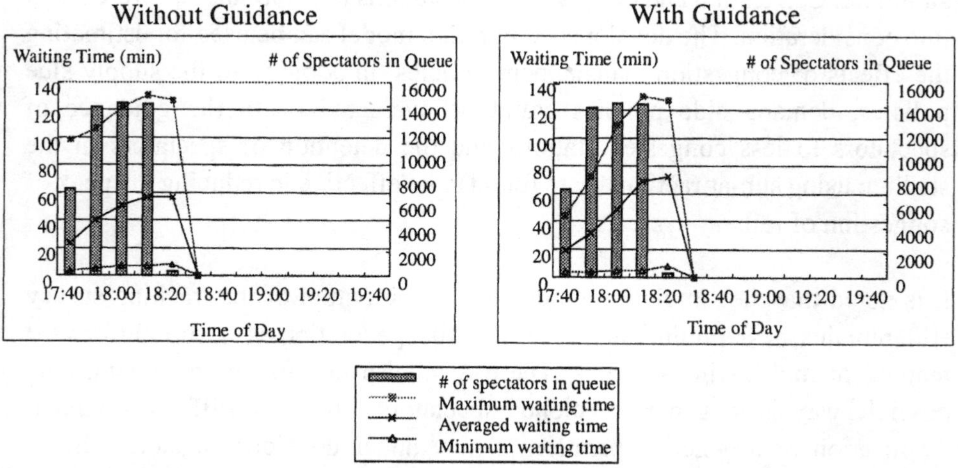

Figure 9. Effects of 30 Minutes of Attractions and Guiding Spectators to Less Congested Stations.

CONCLUSION

In this paper, we have presented the tracking survey method using the location positioning function of mobile phones. The method was applied to collecting travel behavioural data of 96 spectators who enjoyed the Osaka Sumo tournament. Each observed point has the space attributes of longitude and latitude in a two-minute time interval. The location positioning data has many varieties of potential applications. It is possible to aggregate that data at any space and time unit. For example, the three dimensional map of the location positioning data was easily obtained, and it was found to be useful in understanding the space-time characteristics of travel demand. The block-to-block transition matrices around the hall were also available for extracting the main stream of spectators after the game.

Using the probability distribution of sampled spectators, thousands of profiles of imaginary spectators can be generated. This is a type of multiplying simulation of sampled spectators. The generated profiles have the same location characteristics as the original ones, and they can be aggregated the same way as the original data. Another simulation model was developed to describe the movement of spectators at a different stadium in Osaka. As well as generating profiles of spectators using the probability distribution of the original samples, generated profiles were also put into the railway network simulator. Congestion and delay at popular stations near the stadium were taken into consideration. The developed simulation model can be used for estimating the effects of congestion management policies. In addition to the supply side policies, demand side policies could be evaluated. Both the guidance of spectators to less congested stations and the detention of spectators in the stadium using sub-attractions were found to be effective in reducing the peak of congestion of railway systems.

It is noted that the simulation models using the output of the tracking survey still remains at a preliminary level. Profile generation is a straightforward method of multiplying samples. There is much room for improvement. One possible way is to generate "clone" spectators through a different manner. Application of a genetic operation is now under development, and will be presented elsewhere in the near future.

Although, the output of the profile generation was not verified in this paper, verification of the simulation models is another important topic to discuss. It is possible to count the number of passing spectators at some observation points around the event site. For example, the number of passengers who used the railway stations can be used to verify the simulation results. Combinations of location positioning data and traffic count data are another interesting topic to study.

ACKNOWLEGEMENTS

The authors would like to express their sincere appreciation to the research staff of the Urban Transport Planning Co. Ltd collaborated with e-TSU project. This study was supported by the Grant-in-Aid for Scientific Research (#12450207) of Ministry of Education and Science. The research sub-committee on ITS in Japan Society of Civil Engineering is also acknowledged.

REFERENCES

Asakura, Y., Hato, E. Nishibe, Y., Daito, T., Tanabe, J. and Koshima, H. (1999). Monitoring Travel Behaviour using PHS Based Location Positioning Service System, *Proc. of the 6th ITS World Congress* in Toronto, CD-ROM.

Asakura, Y. and Hato, E. (2000). Analysis of Travel Behaviour using Positioning Function of Mobile Communication Devices, *Travel Behaviour Research; the Leading Edge*, Hensher, D. (ed.) Pergamon, pp885-899.

Asakura, Y., Okamoto, A., Suzuki, A., Lee, Y.H. and Tanabe, J. (2001). Monitoring Individual Travel Behaviour using PEAMON: a Cellular Phone Based Location Positioning Instrument Combined with Acceleration Sensor, Paper submitted to the *8th ITS World Congress* in Sydney.

Asakura, Y. and Hato, E. (2001.a). Possibility of Using Location Positioning Data with Mobile Phones in Management of Concentrating Travel Demand, Paper presented *the Symposium on Infrastructure Planning* held at Tsukuba, 9 pages (in Japanese).

Asakura, Y. and Hato, E. (2001.b). Rethinking Urban Travel Measurement, *Proc. of Asia GIS Symposium* in Tokyo.

Axhausen, K. (1998). Can We Ever Obtain the Data We Would Like to Have?, *Theoretical Foundations of Travel Choice Modeling(A)*, Pergamon Press, pp305-334.

Battelle Transportation Division (1997). Lexington Area Travel Data Collection Test: Global Positioning Systems for Personal Travel Surveys, Final Report to FHWA (Federal Highway Administration). http://www.fhwa.dot.gov/ohim/trb/reports.htm

Beckman, R.J., Baggerly, K.A. and McKay M.D. (1996). Creating Synthetic Baseline Populations, *Transportation Research A,* **Vol.30, No.6,** pp415-429.

D'Este, G.M., Zito, R. and Taylor, M.A.P. (1999). Using GPS to Measure Traffic System Performance, *Journal of Computer-Aided Civil and Infrastructure Engineering,* **14**, pp273-283.

Fijii, S., Otsuka, Y., Kitamura, R. and Monma, T. (1997). A Micro-Simulation Model System of Individual's Daily Activity Behaviour that Incorporates Spatial, Temporal and Coupling Constraints. *Infrastructure Planning Review*, **No.14**, pp643-652.

LOCUS(1999). Location Information Services by Portable Phone Networks, http://www.locus.ne.jp/regular/e/locus2.system.htm

Murakami, E. and Wagner, D.P. (1999). Can Using Global Positioning System (GPS) Improve Trip Reporting?, *Transportation Research C*, **Vol.7**, pp149-165.

Zito, R., D'Este, G. M. and Taylor, M.A.P. (1995). Global Positioning Systems in the Time Domain: How Useful a Tool for Intelligent Vehicle-Highway Systems?, *Transportation Research C*. **Vol.3**, No.4, pp193-209.

SIMULATION OF THE AUTOBAHN TRAFFIC IN NORTH RHINE-WESTPHALIA

Michael Schreckenberg, Andreas Pottmeier, Sigurður F. Hafstein, and Roland Chrobok
University of Duisburg-Essen, Physics of Transport and Traffic, 47048 Duisburg, Germany
(schreckenberg, pottmeier, hafstein, chrobok)@traffic.uni-duisburg.de

Joachim Wahle
TraffGo GmbH, Grabenstr. 132, 47057 Duisburg, Germany
wahle@traffgo.com

ABSTRACT

The amount of vehicular traffic has reached the capacity of road networks in many densely populated regions world-wide. Especially in North Rhine-Westphalia, situated in the western part of Germany, growing traffic demand gives rise to more and more congestion on the autobahn network. Therefore, the need for intelligent information systems has become increasingly important. Here we present a combination of real time traffic data analysis and microscopic traffic simulations as the basis of an online-tool that provides full information of the traffic state on the complete autobahn network. Using a java applet the results are made topical minute by minute in the internet (*www.autobahn.nrw.de*).

INTRODUCTION

Since the construction of the first autobahn in Germany in the early 30's between Bonn and Cologne (today known as A555), the vehicular traffic has risen continuously all over, but in some areas in a dramatic manner. Whereas in the early days the traffic demand could easily be satisfied by simply expanding the road system this is no longer possible for several reasons. Therefore nowadays, particularly in densely populated regions, the existing autobahn network has reached its natural capacity limit. The daily occurring traffic jams cause significant economic damage. And in fact, in these areas it is usually hardly possible and socially untenable to enlarge the existing network infrastructure.

This is in any case true for the German state of North Rhine-Westphalia. The network is not able to cope with the demand in the daily rush hours and the drivers have to deal with the traffic jams in the Rhine-Ruhr region (Dortmund, Duisburg, Düsseldorf, Essen, Mülheim, etc.) or around Cologne (Bonn, Leverkusen, etc.). In addition the prognosis for the near future paints an even worse picture as the demand will increase significantly further on. Therefore new information systems and traffic management concepts are necessary.

One obvious precondition for these goals is the detailed knowledge of the current traffic state in the *whole* autobahn network. But today traffic flows are measured only at separated locations in the network. These locations are quite densely in the highest loaded regions, but sparsely or even nonexistent in the other ones. To straight out this lack of information we propose the use of online-simulations for the traffic flow to derive the complete information for the whole autobahn network from the locally measured traffic data.

The outline of this contribution is as follows: In the next section we briefly describe the general concept of an online-simulation. The traffic data analysis will be discussed in the following section. Then a more detailed discussion of the simulation model proposed and its capability to reproduce measured data are presented. The network topology is discussed afterwards. The application of the online-simulation is shown in the next section and a summary and an outlook finally close the article.

GENERAL CONCEPT OF ONLINE-SIMULATIONS

The general concept of an online-simulation is to map the real traffic one-to-one, i.e., car by car, into the computer. The data, e.g., the flow and the speed, are measured by detection devices like loop or radar detectors and transferred as aggregated sets, e.g., minute by minute. The data are collected into a 'historical' database to evaluate the consistency of the sets and to generate a 'background'.

A very efficient microscopic simulation model, based on the cellular automata technique, which was improved step by step and has reached now a very realistic and high level of description, is used to incorporate the online-data to generate line information from the locally measured point information (Rickert *et al.*, 1996; Nagel *et al.*, 2000; Kaumann *et al.*, 2000; Knospe *et al.*, 2000; Schreckenberg *et al.*, 2001).

While the simulation is running and producing results, i.e., the traffic flow, the mean speed, etc., the outputs are analyzed. Upon these data the traffic state of the whole system, not only at the positions of the detection devices, is generated (Esser *et al.*, 1997). Therefore information about the traffic state can also be derived for those regions that are sparsely supplied or even not covered by detection devices.

Besides, relevant 'indirect' quantities are achieved, e.g., travel times. By implementing virtual floating cars into the simulation, travel times can be measured via the time needed by the driving (virtual floating) cars. The adjustment of the simulation with real world data is performed every minute. Therefore, virtual loop detectors at the locations of the real detection devices measure the local traffic quantities inside the simulation and adapt, if required, the number of cars and their characteristics, into the running simulation.

Accelerating the simulation, which runs in multiple real-time, and incorporating a combination of the last measured actual data set and historical data from the database instead of the current data leads to a smart tool to predict the traffic and acquire traffic information in a wide time horizon (Chrobok *et al.*, 2000). This is briefly discussed in the next section.

Traffic Forecast

Different approaches to predict traffic states have been proposed in the past, e.g., Box-Jenkins techniques (Ahmed, 1983; Van der Voort *et al.*, 1996; Williams, 2001), nonparametric regression (Nair *et al.*, 2001; Smith *et al.*, 2002; Sun *et al.*, 2003), neural networks (Barceló *et al.*, 1999; Dia, 2001), and heuristics (Wild, 1997; Van Iseghem *et al.*, 1999). Although there are many different ways to predict traffic states some general parameters influence the results of all approaches significantly in the same manor. Which procedure proves to be the most efficient forecast algorithm strongly depends on the prognosis horizon. The second important point concerns the input data, i.e., the number and the locations of the data sources.

For an appropriate long-term prediction it is necessary to use experience about the past, i.e., heuristics, in form of a statistical database consisting of traffic time series (Chrobok *et al.*, 2000). To forecast the evolution of traffic jams, spatial correlations can be used in a combination with a model of the dynamics of a moving jam (Kerner *et al.*, 2000).

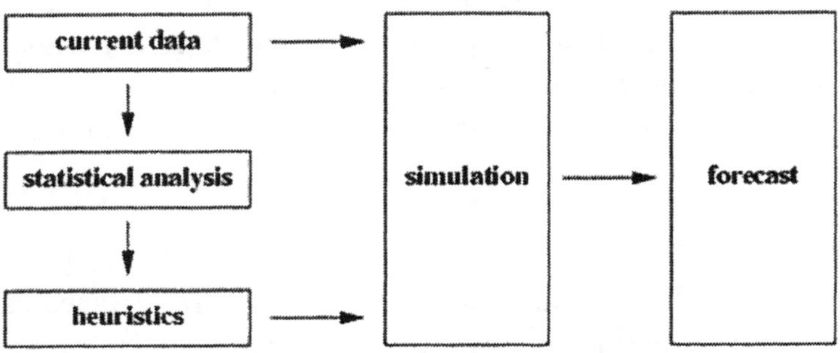

Figure 1. Schematic sketch of the forecast procedure. The input of the simulator are current and, provided in form of heuristics, i.e., historical data. A forecast is carried out by using both simultaneously.

Note that it is important for a reliable forecast that the driver's reactions are taken into account and the process of prediction is iterated until a stationary state is reached.

The coupling of the online simulation with current and statistical data allows for both: short- and long-term forecasts of the traffic state in the whole network. The basic idea is to create the current traffic state by combining traffic data directly with the simulation. To obtain an elementary prognosis, the simulation has to be performed faster than real-time.

One important problem is how to consider the traffic demand at the boundaries of the network in the simulation. This is resolved by collecting data at the boundaries offline and analyzing them to produce complete traffic states in form of heuristics. All together these data than serve as the basis for the predictions. With regard to traffic forecasts the simulation tool serves as a connector between current and historical data (see Fig. 1).

TRAFFIC DATA

As stated before, one important module of a realistic traffic simulation are current traffic data. On the one hand these data serve as the input for the online-simulation, on the other hand a data base for the further analysis is required.

Data Acquisition/Data Transfer

It is very important that the traffic data are provided to the simulation frequently. The traffic patterns in North Rhine-Westphalia are transmitted to the simulation minute by minute. In the first step, the data are collected in two local traffic data centers: in Leverkusen and in Recklinghausen. Using a permanent line these data are transferred to the computer in the University of Duisburg and fed into the simulation. Additionally, all data are aggregated in a large data base and analyzed to get detailed information about the characteristics of the daily traffic demand. Thereby, we can distinguish between different classes in the daily traffic as well as seasonal differences like a reduced demand on holidays as described in the next section.

Examples of Traffic Data

The data provided by the loop detectors are minutely aggregated data of the flow of all vehicles, the flow of the trucks, the speed of the private cars, the speed of the trucks, the time headway between vehicles, the occupancy, the deviation of the velocity and the exponentially smoothened velocity. It is quite obvious, that some of the data are redundant. In the following we focus on the first four values, because they are used for the feedback control of the simulation. In Figure 2 a data set of a particular Monday is shown. The time dependent evolution of the traffic demand can clearly be seen, as well as the synchronized traffic during the rush-hours in the morning and in the evening.

Since Figure 2 shows traffic data for the right lane, there is a relatively high fraction of trucks. This kind of contemplation makes sense to distinguish the scenarios on the different lanes. But to consider the whole traffic demand in one direction or to investigate the truck fraction on an autobahn, it is useful to aggregate the data of one cross-section. In Figure 3 the traffic demand of a cross-section on the BAB 40 is shown. The contingent of trucks, which is at about 10% at this point, is displayed. The diagram shows the characteristic commuter behavior in the traffic patterns. In Figure 3(a) (driving direction Dortmund) the traffic demand is significantly higher in the evening, whereas in Figure 3(b) (driving direction Duisburg) the traffic demand is higher in the morning.

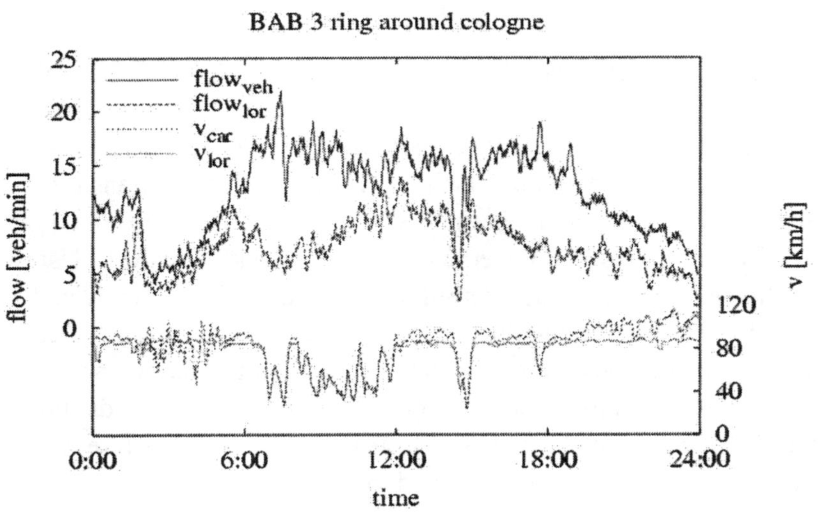

Figure 2. Traffic data provided by the loop detectors. The traffic data of a particular Monday of a loop detector on the BAB 3 at the ring around cologne is shown. In this region the highest traffic demand in Germany is measured. As an example, the flow of all vehicles and the trucks as well as the average velocity of cars and trucks is shown. The loop detector is placed on the right lane in north direction. To keep track of the different curves moving averages of the raw data are used.

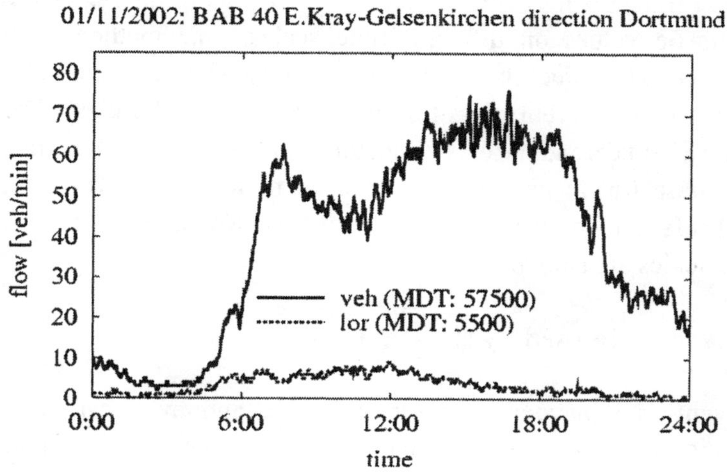

(a)

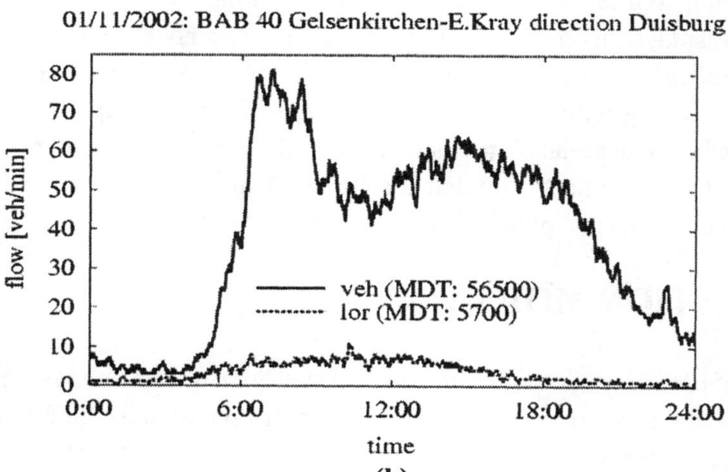

(b)

Figure 3. Daily traffic demand on a cross-section of the BAB 40 between junction Essen-Kray and junction Gelsenkirchen: (a) direction Dortmund, (b) direction Duisburg. The traffic demand of all vehicles measured per minute and the contingent of trucks is displayed. The numbers inside the graph are the daily traffic of vehicles and trucks. The commuter behavior can be seen as well: There is higher traffic flow in direction Duisburg in the morning and in direction Dortmund in the evening.

To use these traffic patterns for a prediction one has to distinguish different kinds of traffic values on different time scales. One method to find the heuristics needed by forecasts is to use the calendar as an external data source. In Figure 4(a) two different classified days are shown, the class 'Fri' which represents the Fridays and the class 'SunHol', which represents the Sundays and holidays. These traffic patterns can be used to forecast traffic states of a particular Friday and Sunday respectively. In Figure 4(b) seasonal differences on larger time scales are shown.

Special Events Mapped by the Data

Special events can have a dramatic influence and even can disturb the characteristics of the daily traffic demand described above. A very impressing example is the impact of the soccer games during the world championship 2002 in Japan/South Korea. A comparison of the relative traffic demand, i.e., the demand compared to the average of the corresponding heuristics, which is set to 100%, displays this influence (Fig. 5). Already two hours in advance of the kickoff the traffic patterns show much higher values of the flow. This is followed by a negligible traffic during the game only interrupted by a short peak in the halftime break. After the final whistle the traffic demand rises again and meets the heuristic traffic demand just an hour later. The knowledge of these special events is a prerequisite of a valid traffic forecast.

SIMULATION MODEL

The mathematical model is the most important part of the online-simulation tool. As the data are provided more or less in real-time to the simulation, it has to calculate the current traffic state also at least in real-time. Due to their design cellular-automaton models are very efficient in large-scale network simulations (Esser *et al.*, 1997; Nagel *et al.*, 2000; Kaumann *et al.*, 2000; Rickert *et al.*, 1996; Schreckenberg *et al.*, 2001).

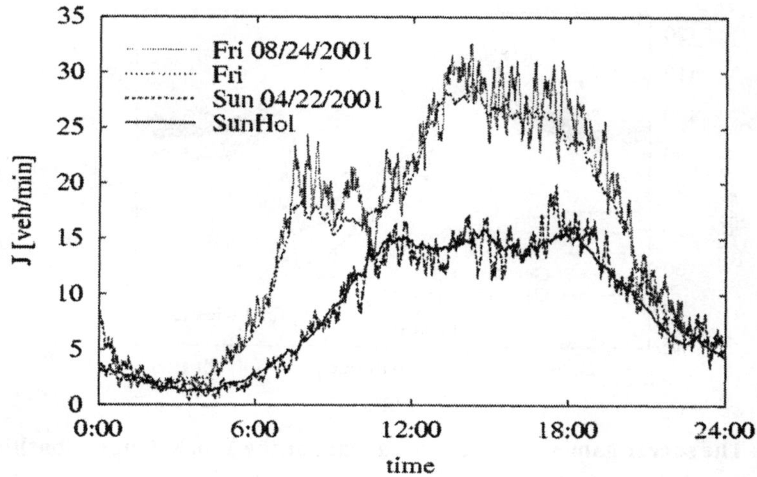

(a)

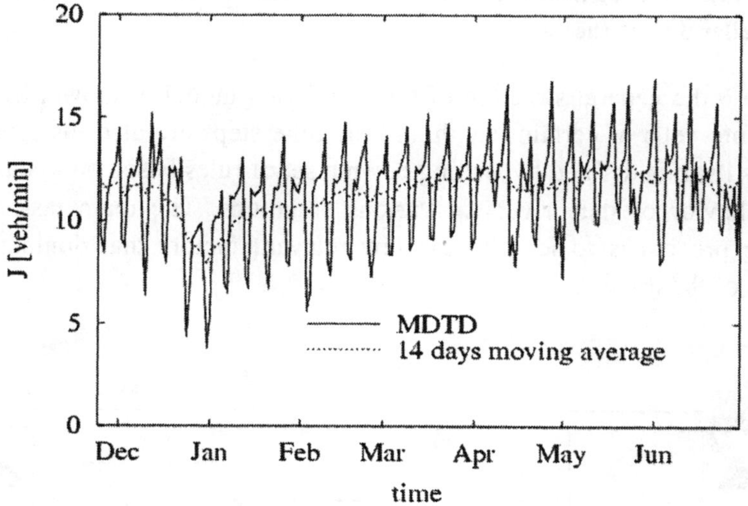

(b)

Figure 4. Different traffic patterns on different time scales. Each traffic pattern is a fingerprint of a special kind of the driver behavior. (a) Classified daily traffic patterns. The difference between the class Fri, which represents the Fridays, and SunHol, which represents the Sundays and holidays, is obvious. (b) Long term traffic patterns: One can see the Mean Daily Traffic Demand (MDTD) and its 14-day moving average. The highest peaks are on Fridays, as the most vehicles are measured generally on this particular day in the week. The low dips are the weekends. The averaged data have their minimum in the yuletide at the end of December.

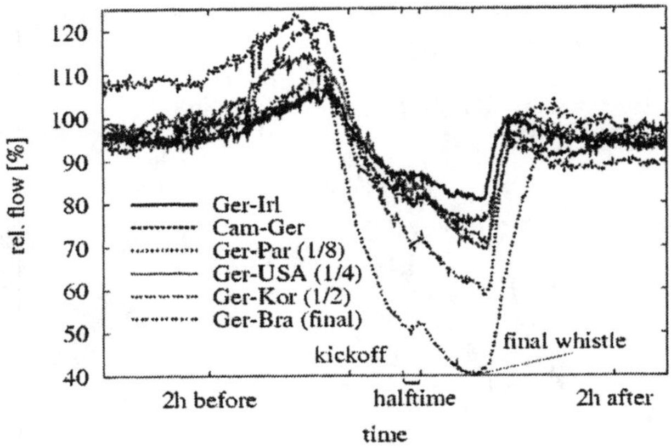

Figure 5. The soccer games of the German team at the World Championship 2002 in Japan/South Korea had serious impact on the traffic patterns. Right before and just after the game the traffic demand is much higher than normal, but naturally much smaller during the game.

In Figure 6 the general structure of the simulation model is shown. Space is divided into cells of certain length and the time steps are also discrete. The dynamics itself is defined by writing down a set of rules based on assumption about behavior of the drivers in various situations. The main task in the modeling process is to identify the few relevant factors that dominate the reactions of the drivers.

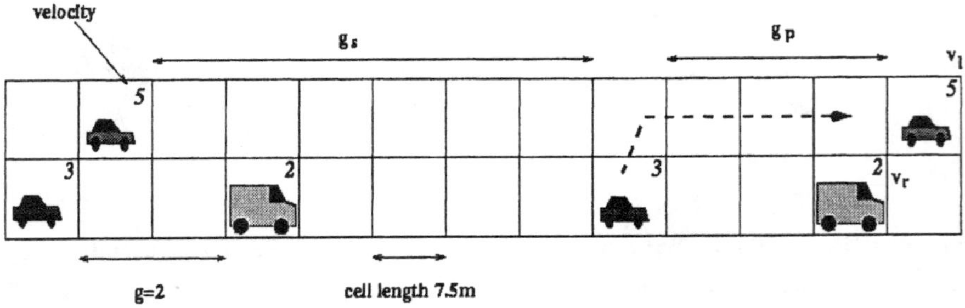

Figure 6. Schematic sketch of the cellular-automaton model. It is discrete in time and space. The dynamics base on a few update rules that are applied in parallel. The algorithms are fast and efficient.

Basic Model and its Extensions

Models which reproduce the dynamical phases of traffic are still under vital debate. The first cellular-automaton model for traffic flow that was able to reproduce some of the characteristics of real traffic like probabilistic jam formation, was introduced by Nagel and Schreckenberg (Nagel *et al.*, 1992).

In the scope of the network wide simulation a more realistic traffic flow model is required that rebuilds the microscopic properties in a greater detail. This is obtained by using *smaller cells* and by extending the rules with *velocity dependent randomization*, *anticipation*, and *brake lights* as shown in the so called "brake light model" (Knospe *et al.*, 2000).

Smaller cells allow for a more realistic acceleration and more speed bins. Currently an elementary cell size of 1.5m is used, in contrast to the 7.5m in the original Nagel-Schreckenberg model. This corresponds to speed bins of 5.4km/h and an acceleration of 1.5m/s² (0-100km/h in 19s) which is of the same order of the "comfortable" acceleration of somewhere about 1m/s². A vehicle occupies 2-5 consequent cells. By using velocity dependent randomization (Barlovic *et al.*, 1998), realized through the introduction of 'slow-to-start rules', meta stable traffic flows can be reproduced in the simulation, a phenomenon observed in empirical studies of real traffic data (Treiterer, 1975; Kerner *et al.*, 1997; Helbing, 1997). The inclusion of anticipation and brake lights (Barrett *et al.*, 2000; Knospe *et al.*, 2000) in the modeling leads to a more realistic driving, i.e., the cars no longer determine their velocity solely in dependency of the distance to the next car in front, but also take regard to its speed and whether it is reducing its speed or not.

In the following algorithmic description of the dynamics, $x_n(t)$ is the position, $v_n(t) \in \{0, 1, \dots, v_{max}\}$ the velocity, and $d_{n,m}(t)$ the number of free cells between cars n and m (in front of n) at time t. When the simulation algorithm decides about braking of car n, it does not only consider the distance to the next car m in front, but also makes an estimate of how far car m will move in the next time step ('anticipation'). Note, that the moves are performed in parallel, so that the model remains free of collisions. This leads to the effective gap

$$d_{n,m}^{eff}(t) := d_{n,m}(t) + \max(v_m^{\min}(t) - d_s, 0)$$

relevant for car n at time t. In this formula d_S is a safety distance and

$$v_m^{\min}(t) := \min(d_{m,l}(t), v_m(t)) - 1,$$

where $d_{m,l}(t)$ is the number of free cells between car m and car l in front of it. Brake lights are further components of the anticipating driving. They allow drivers to react to disturbances in front of them earlier by adjusting their speed. The variable $b_n(t)=on$ if car n has its brake lights on and $b_n(t)=off$ if they are off.

Several empirical observations suggest that drivers react in a temporal- rather than a spatial-horizon (George, 1961; Miller, 1961). For this reason the velocity-dependent temporal interaction horizon

$$t_n^s(t) := \min(v_n(t), h)$$

is introduced in the model. The constant h determines the temporal range of interaction with the brake light $b_m(t)$ of the car m ahead. Car n does only react to $b_m(t)$ if the time to reach the back of car m, assuming constant velocity (v_n = *const.*) and car m standing still, is less than $t_n^s(t)$, i.e.,

$$t_{n,m}^h(t) := \frac{d_{n,m}(t)}{v_n(t)} < t_n^s(t).$$

The estimations for h vary from 6s (George, 1961), 8s (Miller, 1961), 9s (Highway Capacity Manual, 1965) to 11s (Edie and Foote, 1958). Another estimation can be obtained from the analysis of the perception sight distance. In (Pfefer, 1976) velocity-dependent perception sight distances are presented that, for velocities up to 128km/h, are larger than 9s. Therefore h is set to 6s as a lower bound for the time headway (Knospe, 2002).

The third modification of the Nagel-Schreckenberg model implemented in the simulator is a velocity dependent randomization. In every time-step and for every car n with car m in front of it, the probability that car n brakes is calculated as:

$$p = p(v_n(t), b_m(t)) := \begin{cases} p_b, & \text{if } b_m(t) = on \text{ and } t_{n,m}^h(t) < t_n^s(t), \\ p_0, & \text{if } v_n(t) = 0, \\ p_d, & \text{default.} \end{cases}$$

The parameter p_0 tunes the upstream velocity of a wide moving jam and p_d

controls the strength of the fluctuations.

With this parameter set the model is calibrated to the empirical data. Leaving h fixed, the best agreement can be achieved for $d_s = 7$ cells, $h = 6$, $p_b = 0.96$, $p_0 = 0.5$ and $p_d = 0.1$. For a detailed analysis of the parameter set see (Knospe, 2002).

One more variable is needed so that lane changes can be done in parallel, $l_n \in$ {left, right, straight} notes if car n should change the lane during the current time-step or not.

For every time-step $t \to t+1$ the simulator performs the three following blocks sequentially. In every block the steps are executed for all cars in parallel. The definition of the gaps $d_{n,s}^{eff}(t)$ and $d_{r,n}^{eff}(t)$ in the lane-change-blocks is an obvious extension of the above definition, one simply inserts a copy of the car n on its left or right side.

Overtake on the lane to the left:

- Step 0: Initialization:

  For car n find the next car m in front on the same lane, the next car s in front on the lane left to car n, and the next car r behind car s. Set $l_n :=$ straight.

- Step 1: Check lane change:

  if $b_n(t) = off$ and $v_n(t) > d_{n,m}(t)$ and $d_{n,s}^{eff}(t) \geq v_n(t)$ and $d_{r,n}(t) \geq v_r(t)$, then set $l_n :=$ left.

- Step 2: Do lane change:

  if $l_n =$ left, then let car n change lane to the left.

Return to a lane on the right:

- Step 0: Initialization:

  For car n find the next car m in front on the same lane, the next car s in front on the lane right to car n, and the next car r behind car s. Set ln :=

straight.

- Step 1: Check lane change:

if $b_n(t) = off$ and $t^h_{n,s}(t) > 3$ and $(t^h_{n,m}(t) > 6$ or $v_n(t) > d_{n,m}(t))$ and $d_{r,n}(t) > v_r(t)$, then set $l_n :=$ right.

- Step 2: Change lane:

if $l_n =$ right, then let car *n* change lane to the right.

Move forward (drive):

- Step 0: Initialization:

For car *n* find the next car *m* in front. Set $p_n(t) := p(v_n(t), b_m(t))$ and $b_n(t+1) := off$.

- Step 1: Acceleration:

$$v_n(t+\frac{1}{3}) := \begin{cases} v_n(t), & \text{if } b_n(t) = on \text{ or } (b_m(t) = on \text{ and } t_n^h(t) < t_n^s(t)), \\ \min(v_n(t)+1, v_{\max}), & \text{default.} \end{cases}$$

- Step 2: Braking:

$$v_n(t+\frac{2}{3}) := \min(v_n(t+\frac{1}{3}), d^{eff}_{n.m}(t)).$$

Turn brake light on if appropriate:

$$\text{if } v_n(t+\frac{2}{3}) < v_n(t), \quad \text{then } b_n(t+1) := on.$$

- Step 3: Randomization with probability $p_n(t)$:

$$v_n(t+1) := \begin{cases} \max(v_n(t+\frac{2}{3}) - 1, 0), & \text{with probability } p_n(t), \\ v_n(t+\frac{2}{3}), & \text{default.} \end{cases}$$

Turn brake light on if appropriate:

$$\text{if } p = p_b \text{ and } v_n(t+1) < v_n(t+\frac{2}{3}), \quad \text{then } b_n(t+1) := on.$$

- Step 4: Move (drive):

  $$x_n(t+1) := x_n(t) + v_n(t+1).$$

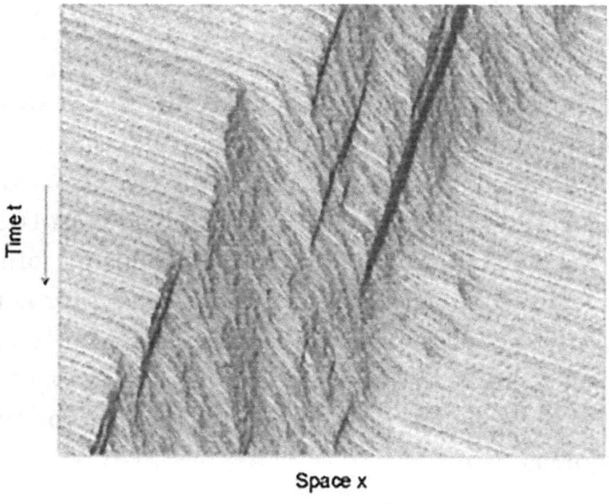

Figure 7. Space time plot of a periodic one-lane road simulated with the brake light model for a density of 27 veh/km. The cars are moving from left to right. Each dot represents one vehicle.

For a review of multi-lane traffic, i.e., different lane-changing rules, see, for example, (Nagel *et al.*, 1997). For a detailed discussion of the different cellular-automaton models see (Schreckenberg *et al.*, 1997; Chowdhury *et al.*, 1999; Helbing *et al.*, 2000) and the references therein.

In Figure 7 a space time plot of the model for a one lane road is shown. According to empirical results three qualitatively different microscopic traffic states can be observed (Knospe *et al.*, 2001).

Validation of the Model

A core requirement in the discussion of the simulation model is the detailed comparison with empirical data. Only if the model maps the real world sufficiently, it is capable to deal as the kernel of the online-simulation. Furthermore, one of the most puzzling points for any model is to reproduce significant empirical data on a macroscopic and microscopic level as well as the empirical observed coexistence of stable traffic states and especially the upstream propagation of wide moving jams through both free flow and synchronized traffic with constant velocity and without disturbing these states (Kerner, 2001).

Thus, for the comparison with empirical data, different aspects are observed to compare the simulation with the empirical results. First, a two-lane segment with an on-ramp is simulated. For the sake of simplicity, we used symmetric lane changing rules, i.e., the change to the right is modeled just like the one to the left. In addition, the lane changing rules on the on-ramp are formulated less stringent in the sense that incoming cars accept smaller gaps for lane changing and the incentive criterion is dropped, i.e., the cars always try to change their lane. Note, that the general results do not depend on the details of the applied rules. This lane change behavior at the on-ramp is also followed by the simulation of the whole network.

In analogy to the empirical setup, the simulation data are evaluated by a virtual loop detector, i.e., the number of cars passing a given link is measured as well as their velocity. This allows for the calculation of aggregated minute data of flow, speed and occupancy like for the empirical data.

The simulation run emulates a few hours of highway traffic, including the realistic variation of the number of cars that are fed into the system. Thereby, a large input rate leads to the emergence of synchronized flow, whereas at small rates small short-living jams evolve, as expected, in the vicinity of the on-ramp, because of the local perturbations (Helbing *et al.*, 1999). For the sake of simplicity, only one type of cars is modeled in the simulation. This leads to a smaller variance of the data points in the free flow region. So, the agreement between the results of the simulation and the empirical data is not only valid for the averaged results, but also for the statistical characteristics.

The simulation shows that the empirical results can quantitatively be recovered (see Figure 8). A detailed analysis of the two dimensional region of synchronized traffic in the fundamental diagram reveals a high correlation between the two lanes with respect to the velocity time-series of both lanes.

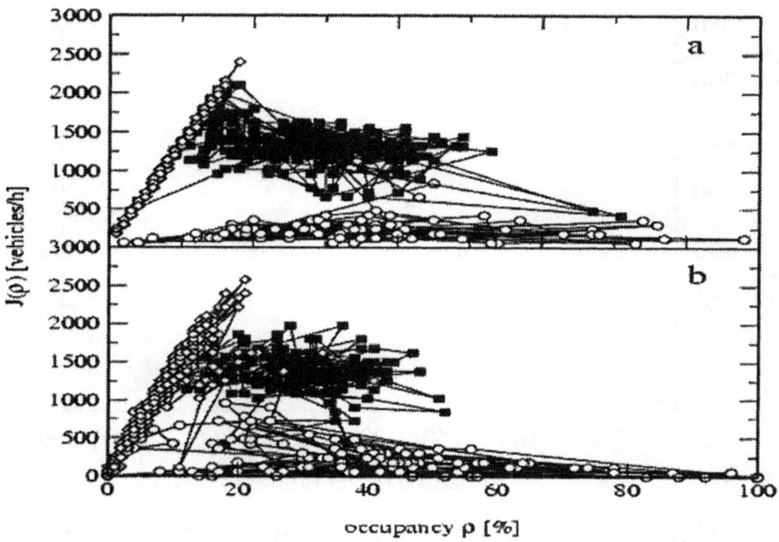

Figure 8. Comparison of the simulation results (a) with real traffic data (b). Diamonds correspond to free flow, squares to synchronized traffic, and circles to wide jams. Each point represents the average over an one-minute interval. The empirical data are from a detector on the A40 near Moers junction at 2000-12-12 (synchronized state) and near Bochum-Werne junction at 2001-02-14.

Furthermore, the upstream velocity of the downstream front of a wide jam, that is directly related to the deceleration probability p_0 is reproduced by the simulation model, i.e., by a calculation of the density autocorrelation function in the congested state of the system that was initialized by a mega jam. As a result one obtains an average upstream jam velocity of 2.36 cells/time-step (= 12.75 km/h) that is in accordance with empirical results (Kerner, 1996). For more detailed discussion of the parameter chosen see (Knospe, 2002).

Another observable useful for comparison is the travel time. Floating car data offer the opportunity to obtain a data set of measured travel times that is large and of sufficient quality. In Figure 9 an example is shown of 30 different travel times that are compared with simulated travel times up to 400s. The data is stemming from 32 vehicles that pass synchronized or congested traffic. The mean relative error for the depicted data set of measured and simulated travel times is 0.26.

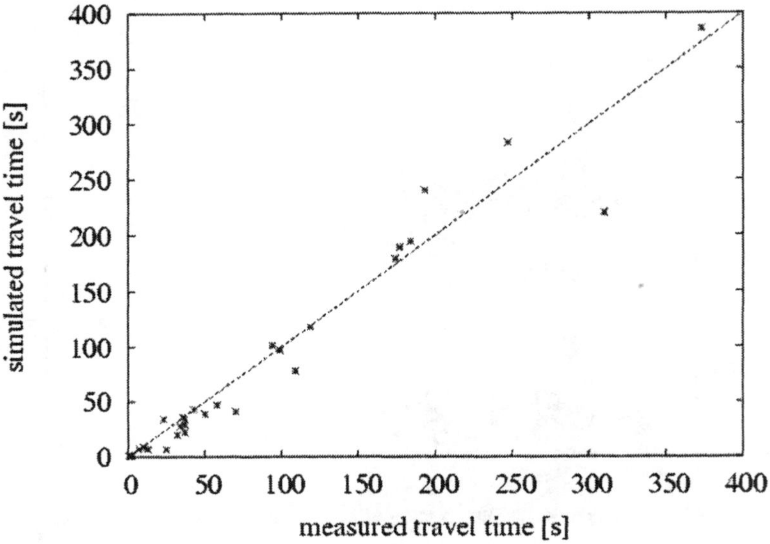

Figure 9. A comparison of simulated and measured travel times. Each cross indicates a floating car measurement. The x-value indicates the real, the y-value the simulated data, so that for exact simulated travel times the crosses have to lie on the bisecting line. The measured data are stemming from 30 vehicles passing synchronized or congested traffic on the BAB 40 of Monday 2002-03-11.

Moreover, the stability of the synchronized phase is described correctly. Even a compact jam, which moves through the region of synchronized traffic, cannot disturb it (see Figure 10). This superposition of the different traffic states is in compliance with empirical observations (Kerner, 2001).

A comparison of empirical and simulated data on a microscopic level further leads to good agreements between simulated and empirical data, i.e., beside others, the time-headways distribution is reproduced by the simulation model as well as the OV curves of the three traffic states are in excellent agreement with empirical findings. As a deeper analysis and comparison of the simulation results and empirical findings would exceed the scope of this article, we refer to the detailed discussion in (Knospe, 2002).

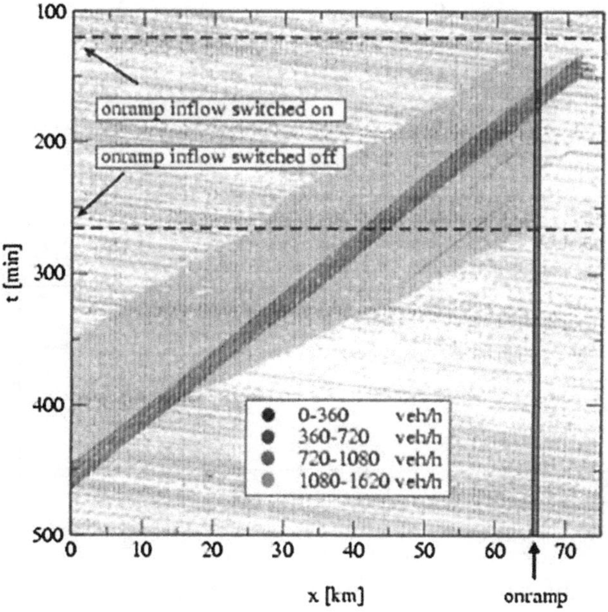

Figure 10. Wide moving jams in coexistence with synchronized traffic. Space/time evolution of the flow. The figure shows the stability of the synchronized flow in the presence of a wide moving jam crossing the region.

TOPOLOGY

An important point in the design of a simulator is the representation of the road network. Therefore, the network is divided into links. The main links connect the junctions and highway intersections representing the carriageway. Each junction and intersection consist of another link, like on/off-ramps or right/left-turn lanes. The attributes of each link are the length, the number of lanes, a possible speed limit, and the connecting links. In case of more than one connecting link, like at on-/off-ramps or highway intersections, there is also a turning probability for each direction. The turning probability is calculated by taking into account the measured traffic data. All these spatial and functional data was collected to build a digital image of the topology of the whole network.

Table 1. Design parameters of the North Rhine-Westphalian autobahn network

Area	34,000 km²
Inhabitants	18,000,000
On- and off-ramps	830
Intersections	67
Online loop detectors	4,200
Offline loop detectors	200
Number of links	3,698
Overall length	2,250 km

Another crucial information concern the positions of the installed loop detectors. They also have to be included in the digital map of the network. The positions in the simulation are called 'checkpoints', and at these checkpoints the simulation is adapted to the measured traffic flow of the loop detectors (see section Traffic Data). There are two types of detectors: online and offline. The online detectors are directly connected to the simulation and provide traffic data every minute. Since the offline detectors are not accessible by our computer network, only a sample data set can be used. Table 1 shows some design parameters of the network. North Rhine-Westphalia has more or less one fifth of whole Germany with respect to many numbers, e.g., number of cars, inhabitants, length of the autobahn network, etc..

ONLINE-SIMULATION

Using mathematical traffic models, designed to reproduce observed traffic phenomena on artificial and usually significantly simplified road networks (e.g., two lanes, no on-/off-ramps, periodic boundary conditions), in order to simulate real traffic in large and topologically complex road networks, makes some additional considerations necessary. Firstly, one has to introduce data-structures to model the topology of the network, which have to be flexible enough to model every possible topological appearance, but, at the same time, have to allow for an efficient algorithmic implementation of the traffic dynamics. This was considered in the last section. Secondly, one has to adopt some strategy to incorporate the measured data from the checkpoints into the simulation. We will discuss this point in the subsequent section.

Data Incorporation

As already mentioned, the simulation is provided with traffic data measured by loop detectors on the autobahn minute by minute. Clearly, the simulation can not provide accurate and reliable results on the current traffic state if these data are not assimilated in an appropriate way. When designing a strategy to incorporate the data, one has to have in mind, that the loop detectors are not impeccable, neither principal nor practical.

Sometimes a checkpoint delivers measurements that are not consistent with the measurements of neighboring checkpoints. Such data have to be eliminated or corrected by a filter, which examines the data before it is forwarded to the simulation. Another problem is, that the checkpoints are installed locally and therefore only can perform measurements of vehicles driving over them. This means that when the average velocity of the cars that drive over a checkpoint is high, one can deduce from this measurement information about the traffic state in a larger neighborhood of the checkpoint. However, if it is low, this is not necessarily the case.

A further example of the deficit of local measurements is that if the flow is high, the *cars/trucks-ratio* measured by a checkpoint can be expected to give a fair estimate of the real *cars/trucks-ratio* in a neighborhood of the checkpoint, but, if the flow is low, this ratio may fluctuate heavily.

However, knowing these limitations of the checkpoint data, it is possible to avoid over-interpretations of those measurements based on a weak fundament (to local or ensemble to small) and use the checkpoint data for a successful feedback control of the traffic.

Current Data

The simulation models the traffic dynamics in the network, but is lacking information about the boundaries. To map the current traffic state, the simulation is supplied with the traffic flow J_{meas} every minute. Thus, every minute the traffic state in the simulation is compared with real world measurements and, if required, adjusted. This is performed at the checkpoints. The driving force of the strategy is the difference between the simulated number of cars in an area of length Δl around the checkpoint i and the real traffic flow J^i_{meas} measured during one minute at the corresponding loop detector.

In general there are two possible situations: If there are too many vehicles in the simulation at the time step t in the area, i.e., $J^i_{sim,\Delta l}(t) > J^i_{meas}(t)$, the difference $J^i_{sim,\Delta l}(t) - J^i_{meas}(t)$ is removed from the area in the next minute. If the number of vehicles simulated is lower than the measured number ($J^i_{sim,\Delta l}(t) < J^i_{meas}(t)$), $J^i_{meas}(t) - J^i_{sim,\Delta l}(t)$ vehicles are added ($J^i_{sim0,\Delta l}(t)$ is the number of simulated vehicles before the adapting process):

$$J^i_{sim,\Delta l}(t+1) = J^i_{sim0,\Delta l}(t+1) + J^i_{meas}(t) - J^i_{sim,\Delta l}(t).$$

Especially, if vehicles are added to the simulation, it is important to add them "adiabatically", i.e., without destroying the current state of the system.

Network-Wide Traffic State

An obvious application of the acquired knowledge of the current traffic state is to provide it to the drivers by the use of Internet applications. In a descriptive and easy to understand web site every driver gets the possibility to check the current traffic state, especially for the roads he wishes to use. In a simple to navigate window the user can choose between an overview of the traffic state of the whole autobahn network of North Rhine-Westphalia and different parts of the network in a higher resolution. Each of these 1,200 tracks in the network is colored according to the current traffic state on it: light green corresponds to free flow, dark green to dense traffic, yellow to congested, and red represents jammed traffic.

As an additional information the current road works can be displayed. Each of these indicate by the color (red or green), whether one has to assume strong hindrances at these sites or not.

Figure 11 and Figure 12 give an impression of the design of the pages. Moreover, the current traffic state is online under *http://www.autobahn.nrw.de* since 19 September 2002. Rising hits of the site and a broad feedback of the user's show the benefit one can get by this first complete traffic state information tool.

Beside this the simulation offers the opportunity to get higher level information about the traffic situation, i.e., the estimation of travel times by the means of virtual floating cars. As every car drives in the simulation according to the physical rules discussed above a virtual floating car can be inserted at any position in the network and move along a given route. Following individual cars along their way one simply can measure their travel times. Finally an average over many trips is calculated.

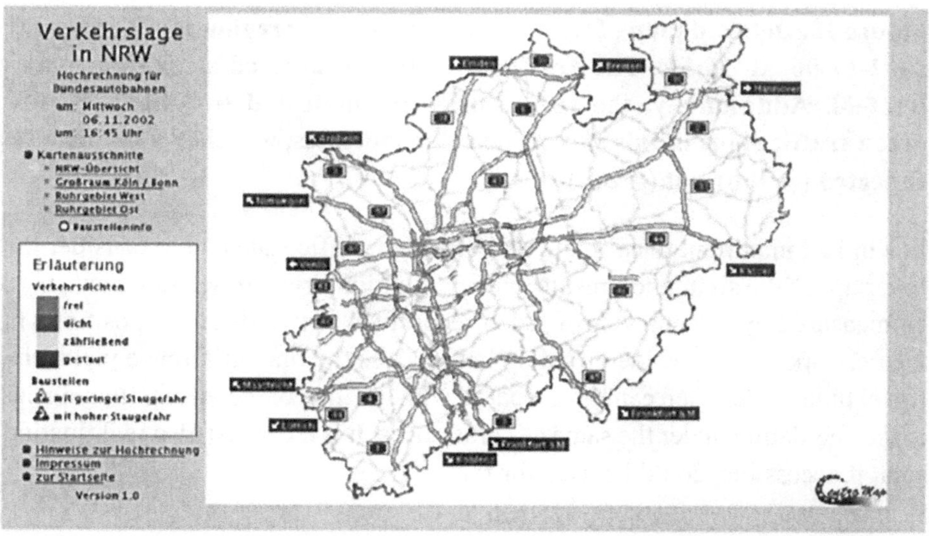

Figure 11. Visualization of the traffic state of the whole autobahn network in North Rhine-Westphalia for Wednesday 2002-11-06 at 16:45 just in the beginning of the afternoon rush hour. Especially in the region around cologne and the center of the Ruhr Region the traffic suffers from congestions.

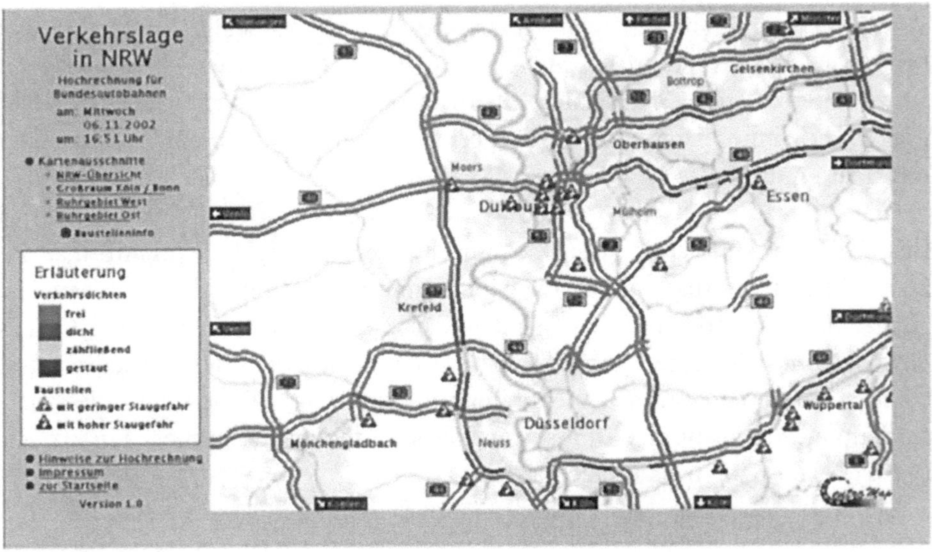

Figure 12. Detailed map of the western Rhine-Ruhr region for Wednesday 2002-11-06 at 16:51. The traffic is highly congested near Essen and Krefeld. Additionally, the road works are indicated by small red and green traffic signs in this view. The color indicates, whether a jam can be expected (red) or not (green).

It is in fact important to calibrate the estimated traffic state and to consider the 'overlap' of the simulation results with reality. Empirical travel times can, e.g., be measured by the analysis of video recordings at two different positions in the network. If a car is identified on both videos the time difference yields the travel time, which then can be compared with the time of the virtual floating car in the simulation under the same circumstances. It is clear that also real floating cars, if accessible, could be used for this.

Aside the benefits mentioned the control strategies of Variable Message Signs (VMS), which aim at an improvement of the capacity of the network, were studied on the basis of the online-simulation. The impact of different underlying control algorithms and constructional modifications, which are essential for the effectiveness of the VMS, were also a topic of this research project. For that VMS strategies were implemented in the microscopic traffic simulation. Using this technique the VMS and traffic data of the freeway A57 were analyzed in detail. One result was that an enhancement of the positive effect of the VMS could be achieved by a new algorithm based on historical data. Furthermore a optimal distance between the location of a detector device and the 'corresponding' VMS could be determined (Chrobok *et al.*, 2001).

SUMMARY AND OUTLOOK

In this article an online traffic simulator is presented that provides the traffic state of the autobahn network in North Rhine-Westphalia. Based on the traffic data of about 4,000 detection devices and a highly realistic microscopic simulation tool the traffic state of the whole autobahn system is generated. Using this information as the basis for a java applet every road user with Internet connection can get this information about the current traffic state easily. Congested areas are simply identified and the travel route can be adapted according to the traffic state.

This traffic information of the autobahn network is only the starting point of an embedded traffic state information system of the whole region in development. Implementing the most important secondary roads in the first step and enhancing this by the road network of the 11 cities and 4 districts (with 2.57 million automobiles and 127,000 trucks) in the Ruhr region shall lead to a main module of a complete traffic management system without any borders within the project "Ruhr Pilot". This regional traffic management system is not restricted to road traffic, but incorporates also train, plane, and ship transport. Traffic information and forecast will cover public transport, individual vehicular traffic as well as freight traffic

Beside the current traffic state, an anticipatory forecast of the traffic state is the main goal in the near future. The scenario is quite simple, but the impacts are highly correlated and a matter of present investigations: Using the current traffic state as the basis for a traffic forecast for the next thirty or sixty minutes will lead to an image of the traffic state at that time. But this is only the first part, as this directly leads to a very important question: How do the decisions of the road users change the traffic state after he becomes aware of it?

Therefore, the decisions of the drivers must be adapted by the forecast to give a valuable information of the coming traffic state. This question is the main issue of the actual research project "SURVIVE" (see SURVIVE for further information).

ACKNOWLEDGEMENT

The authors would like to thank the "Landesbetrieb Straßenbau NRW" for data support and the Ministry of Transport, Energy and Spatial Planning of North Rhine-Westphalia for financial support.

REFERENCES

Ahmed, Samir A. (1983). Stochastic processes in freeway traffic, *Traffic Engineering and Control* **24**, pp306–310.

Barceló, J. and Casas, J. (1999). The use of neural networks for short-term prediction of traffic demand, In *Proceedings of the 14th International Symposium on Transportation and Traffic Theory*, Ceder, A. (Ed.), Pergamon, Amsterdam, pp419–443.

Barlovic, R., Santen, L., Schadschneider, A., and Schreckenberg, M. (1998). Metastable states in cellular automata for traffic flow, *Eur. Phys. J.* **B 5**, pp793–800.

Barrett, C., Wolinsky, M., and Olesen, M.W. (2000). Emergent local control properties in particle hopping traffic simulations, In Traffic and Granular Flow, Wolf, D.E., Schreckenberg, M., and Bachem, A. (Eds.), *World Scientific*, Singapore, pp169–173.

Chowdhury, D., Santen, L., and Schadschneider, A. (2000). Statistical Physics of Vehicular Traffic and Some Related Systems, *Phys. Rep.* **329**, pp 199–329.

Chrobok, R., Grunewald, T., Pottmeier, A., and Schreckenberg, M. (2002). Analysis and validation of variable message signs using cellular automaton traffic simulations, In Proceedings of the 9th *Meeting of the EURO Working Group on Transportation*, DellOrco, M. and Ottomanelli, M. (Eds.), Bari, pp469–472.

Chrobok, R., Kaumann, O., Wahle, J., and Schreckenberg, M. (2000). Three categories of traffic data: historical, current, and predictive. In *Proceedings of the 9th IFAC Symposium Control in Transportation Systems 2000*, Schnieder, E. and Becker, U. (Eds.), IFAC, Braunschweig, pp 250–255.

Dia, H. (2001). An object-oriented neural network approach to short-term traffic forecasting, *Euro. J. Op. Res.* **131**, pp253–261.

Edie, L.C., and Foot, R.S. (1958). Traffic flow in tunnels, *Proc. HRB*, **37**, pp334-344.

Esser, J. and Schreckenberg, M. (1997). Microscopic simulation of urban traffic based on cellular automata, *Int. J. of Mod. Phys. C* **8,** pp1025–1036.

George, H. (1961). *Measurement and evaluation of traffic congestion*, Bureau of Highway Traffic, Yale University, pp43–68.

Helbing, D., Herrmann, H., Schreckenberg, M., and Wolf, D.E. (Eds.) (2000). *Traffic and Granular Flow '99*, Springer, Heidelberg.

Helbing, D. (1997). Empirical traffic data and their implications for traffic modeling, *Phys. Rev. E* **55**, R25–R28.

Helbing, D., Henneke, A., and Treiber, M. (1999). Phase diagram of traffic states in the presence of inhomogenities, *Phys. Rev. Lett.* **8**, pp4360–4363.

Highway Capacity Manual (1965). *HRB Spec. Rep. 87.* U.S. Department of Commerce, Bureau of Public Road, Washington, D.C.

Kaumann, O., Froese, K., Chrobok, R., Wahle, J., Neubert, L., and Schreckenberg, M. (2000). On-line simulation of the freeway network of North Rhine-Westphalia, In *Traffic and Granular Flow '99* (Helbing, D., Herrmann, H., Schreckenberg, M., and Wolf D.E.), Springer, Heidelberg, pp 351–356.

Kerner, B.S. and Rehborn, H. (1996). Experimental features and characteristics of traffic jams, *Phys. Rev. E.* **53**, pp R1297-1300.

Kerner, B.S. and Rehborn, H. (1997). Experimental properties of phase transitions in traffic flow, *Phys. Rev. Lett.* **79**, pp4030–4033.

Kerner, B.S., Rehborn, H., and Aleksic, M. (2000). Forecasting of traffic congestion, In *Traffic and Granular Flow '99* (Helbing, D.,. Herrmann, H.J., Schreckenberg, M., and Wolf, D.E.), Springer, Heidelberg, pp339–344.

Kerner, B.S. (2001). Complexity of synchronized flow and related problems for basic assumptions of traffic flow theories, *Network and Spatial Economics* **1**, pp35–76.

Knospe, W. (2002). Synchronized traffic: Microscopic modeling and empirical observations, Ph.D. Thesis, Gerhard-Mercator-University Duisburg, Germany (http://www.ub.uni-duisburg.de/ETD-db/theses/available/duett-08212002-212839/unrestricted/index.html).

Knospe, W., Santen, L., Schadschneider, A., and Schreckenberg, M. (2000). Towards a realistic microscopic description of highway traffic, *J. Phys. A*

33, L1–L6.

Knospe, W., Santen, L., Schadschneider, A., and Schreckenberg, M. (2001). Human behavior as origin of traffic phases, *Phys. Rev. E* **65**, pp015101-1–4.

Miller A. (1961). A queuing model for road traffic flow, *J. of the Royal Stat. Soc. B* **23**, pp64–75.

Nagel, K., Esser, J., and Rickert, M. (2000). Large-scale traffic simulations for transport planning, In *Ann. Rev. of Comp. Phys.* VII, Stauffer, D. (Ed.), World Scientific, Singapore, pp151–202.

Nagel, K. and Schreckenberg, M. (1992). A cellular automaton model for freeway traffic, *J. Physique I* 2, pp2221–2229.

Nagel, K., Wolf, D.E., Wagner, and P., Simon, P. (1998). Two-lane traffic rules for cellular automata: A systematic approach, *Phys. Rev. E* **58**, pp1425–1437.

Nair, A. S., Liu J.-C., Rilett, L, and Gupta, S. (2001). Non-linear analysis of traffic flow, In *Proc. of the 4th International IEEE Conference on Intelligent Transportation Systems, Stone* (B., Conroy, P., and Broggi, A.), IEEE, Oakland, pp683–687.

Pfefer, R.C. (1976). New safety and service guides for sight distances, *Transportation Engineering Journal of American Society of Civil Engineers* **102,** pp683-697.

Rickert M. and Wagner P. (1996). Parallel real-time implementation of large-scale, route-plan-driven traffic simulation, *Int. J. of Mod. Phys.* C **7**, pp133–153.

Schreckenberg, M., Neubert, L., and Wahle, J. (2001). Simulation of traffic in large road networks, *Future Generation Computer Systems* **17**, pp649–657.

Schreckenberg, M. and Wolf, D.E. (Eds.) (1998). *Traffic and Granular Flow '97*, Springer, Singapore.

Smith, B. L., Williams, B. M., and Oswald, R. K. (2002). Comparison of parametric and nonparametric models for traffic flow forecasting, *Transp. Res. C* **10, 7–8,** pp1133–1152.

Sun H., Liu H.X., and Ran B. (2003). Short term traffic forecasting using the local linear regression model, Accepted for *Proceedings of the 82nd Transportation Research Board Annual Meeting* and for Publication in the *Transportation Research Record.*

SURVIVE *http://www.traffic.uni-duisburg.de/survive/.*

Treiterer, J. (1975). Investigation of traffic dynamics by areal photogrammatic techniques. Tech. report, Ohio State University Tech. Rep. PB 246, Columbus, USA.

Van der Voort, M., Dougherty, M., and Watson, S. (1996). Combining Kohonen maps with ARIMA time series models to forecast traffic flow, *Transp. Res. C* **4**, 5, pp307–318.

Van Iseghem, S., and Danech-Pajouh, M. (1999). Forecasting traffic one or two days in advance - an intermodal approach, *Recherche Transports Securite,* **65**, pp79–97.

Wahle, J., Annen, O., Schuster, C., Neubert, L., and Schreckenberg, M. (2001). A dynamic route guidance system based on real traffic data, *Euro. J. Op. Res.* **131**, pp302–308.

Wild, D. (1997). Short-term forecasting based on a transformation and classification of traffic volume time series, *Int. J. of Forecasting,* **13**, pp63–72.

Williams, B. M. (2001). Multivariate vehicular traffic flow prediction: an evaluation of ARIMAX modelling. In *Proceedings of the 80th Annual Meeting of the Transportation Research Board,* Mira Digital Publishing , Washington D.C..

DATA AND PARKING SIMULATION MODELS

William Young
Department of Civil Engineering, P.O. Box 60, Monash University Victoria 3800, Australia
Bill.Young@eng.monash.edu.au

Tan Yan Weng
School of Civil and Environmental Engineering, Nanyang Technological University, 50 Nanyang Avenue, Singapore 639798, Singapore
cywtan@ntu.edu.sg

ABSTRACT

Discrete event parking simulation models aim to replicate travel and parking behaviour. Parking behaviour relies on driver behaviour and its interaction with the land-use, transport, vehicle and human systems. In order to understand the parking system it is necessary to understand how drivers, vehicles and the road environment interact. This system can never be fully understood nor modelled, however, as technology creates new methods of collecting and synthesising data and increases the power and speed of computer systems it is possible to increase the accuracy, reliability and applicability of simulation models. This paper outlines some of the present limitations in discrete event simulation models and data, and then explores the interaction between data collection, model accuracy and the validity of models. It highlights some of the recent developments in technology and indicates how these can be used to improve many aspects of *travel and parking simulation models.*

INTRODUCTION

Discrete event parking simulation models are potentially the most realistic replication of parking systems (Young 2000, 2001). Their ability to replicate decision making in discrete components of the system and their interaction with other elements of the parking system using very small time increments creates the opportunity to replicate travel and parking situations to a high level of detail.

The increasing availability of computationally powerful and fast computer systems allied with increased graphics and interactive capabilities are making discrete event traffic simulation models more available and useful to the traffic fraternity. Models like PARAMICS (Quadstone Ltd, 2000), VISSIM (PTV Planung Transport Verkehr AG, 2001), DRACULA (Liu et al, 1995), and AIMSUN (Transport Simulation Systems, 2001) are being used at an ever-increasing rate to analyse and convince clients of the affects of traffic infrastructure changes. These models will, however, always be simplifications of the real world situation and as such their accuracy will always be questioned (Young 1990). Their accuracy depends on the quality of the specification of the model and the accuracy with which data used to calibrate and validate them can be collected (Young et al, 1989). Figure 1 shows that as the complexity of models increase the specification error decreases while the measurement error increases. The more accurate model will be the one that minimises the sum of the specification and measurement error. Discrete event simulation models allow for unlimited increase in the specification of the model. If the quality of measurement improves then more complex models can be developed. The increasing accuracy that technology introduces into the collection of data allows the quality of data to improve and hence the model to increase its complexity.

Parking models rely on the level of understanding that model developers and users have on travel and parking behaviour. Travel and parking behaviour relies on driver behaviour and its interaction with the land-use, transport, vehicle and human systems. In order to understand the parking system it is necessary to understand how drivers, vehicles and the road environment interact. Integral to increasing this understanding of behaviour, and hence the quality of models, is an increase in the accuracy of the data used to develop them.

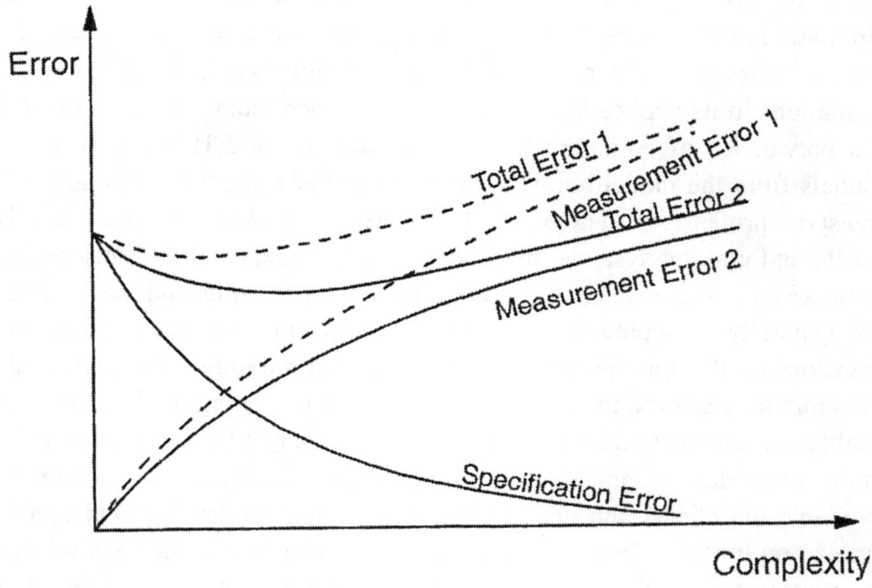

Figure 1 Model accuracy.

This paper explores the interaction between travel and parking simulation models, and the data used in developing them. It reviews the development of data collection techniques and the impact these developments are having on the models within the context of the model development process. In doing this it will relate developments in data collection to the development of the PARKSIM suite of parking simulation programs. PARKSIM was initially developed in 1986 (Young 1986; Young and Thompson 1987). It has been extended from a model that looks only at movement in parking lots to one that incorporates the traffic system around surface parking lots (SAPTSIM Le and Young 1998) and mixed-use multi-storey parking lots (EPSILON, Tan and Young 2002). This paper introduces the model development process outlining the areas where data influences model accuracy. Methods used to improve the quality of data are outlined within the context of model limitations.

THE USERS OF MODELS

In the context of the relationship between data and model reliability it is useful to explore the groups of people who use models. In days past traffic simulation models were the domain of researchers and model builders. These people were generally aware of the assumptions made in the model and limitations in its applications since they developed them. At the present time the prevalence of commercial traffic simulation models moves the use of models from the model builders and tends to make them the province of the transport profession. These people are generally skilled in the analysis of traffic and parking systems, but may not fully understand the intricacies and basic assumptions made in developing the model. Clients and the community are generally presented with the results of the models and they can make decisions on the appropriateness of the remedial measure. The increased use of computer graphics, to allow people to see what is going on, in these models enables the clients to have more confidence in the results. As models become more user-friendly and the data describing the transport system more available the clients and community may be able to develop and apply the simulation models. In many cases this group will lack a good knowledge of the transport system and an understanding of the model assumptions. In this case the validity of models and their ability to be applied to particular situations becomes critical, since these groups are more interested in the outputs of models than whether the assumptions underpinning its development are appropriate to the situation studied. Clearly, the validation of the simulation models for the particular applications clearly becomes more important. The availability of commercial software and their application by the broad community places considerable emphasis on the processes used to calibrate and validate the models.

THE MODEL DEVELOPMENT PROCESS

Figure 2 presents a view of the model development process and the role of information in each of the steps (Young, 1990). It starts with the determination of the problem to be addressed (Step 1). Consequent on the determination of the problem is the clarification of objectives (Step 2) to be achieved and the criteria (Step 3) to be used to measure the effectiveness of achieving these objectives. Technological improvements have created a situation where simulation models are being used to address increasingly complex issues.

This in turn is requiring a greater understanding of the systems being replicated. Improving the quality and diversity of the data used in creating these models is required.

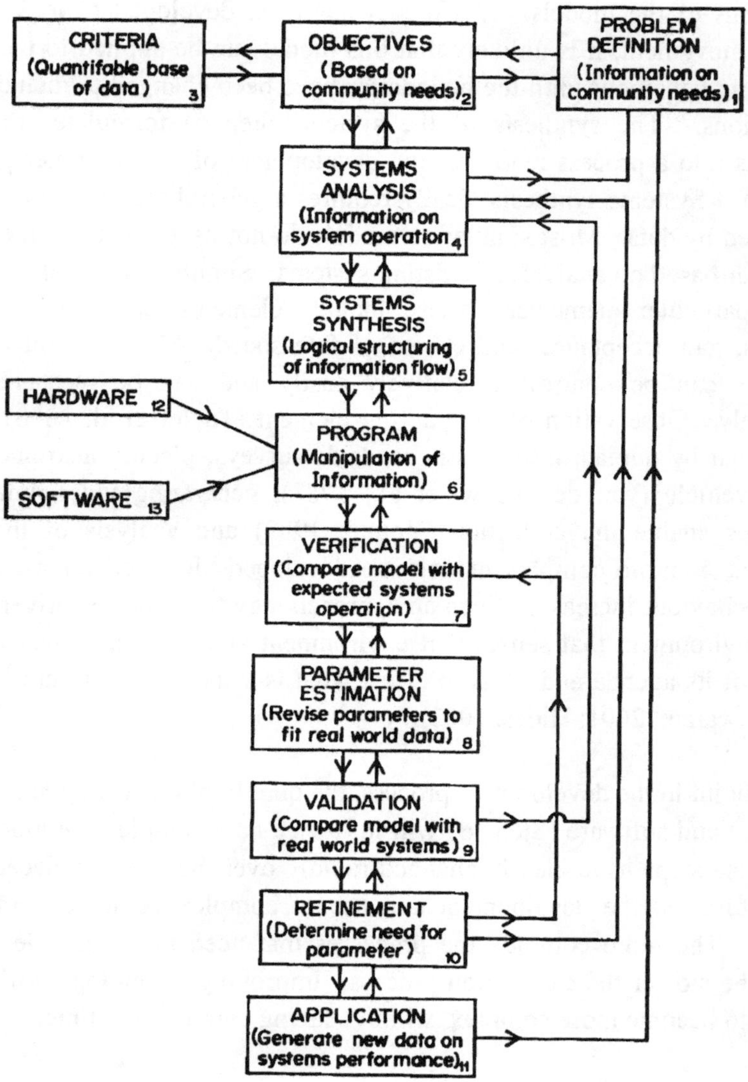

Figure 2: The model development process (Young et al 1989).

Once the aim of the model has been determined, then the boundary conditions (space, time and human) can be specified and the system can be analysed (Step 4) in order to determine the interactions to be represented. The determination of the boundaries to be set in the model determines the limitations of the models. If a model has been developed to look only at vehicle movement, it is unlikely that this model can be applied to pedestrian or parking behaviour until the boundaries have been changed to include these interactions. The synthesis of the system (Step 5) formulates the main elements into a process ready for the development of the computer program (Step 6). Systems synthesis ideally requires a general theory, which can be supported by data. Most simulation models do not have a general theory but are rather based on analysis of existing systems. Simulation models are made up of particular elements. These include elements like acceleration of vehicles, gap acceptance, and choice of free speed. Measurement of these elements can be controlled relatively easily and can be measured very accurately. Observation of the parking element (Taylor et al, 2000) can be carried out by human observers in roadside surveys, placing instrumentation in the vehicle (van der Touw, et al., 1983), generating video images of processes enable the collection (Young, 1989) and analysis of this data. However, as more complex models are developed, the need for a theory of driver behaviour increases. Such an approach may focus on the driver as part of an environment that senses that environment and acts on it over time, in pursuit of its agenda and so as to effect what is senses in the future (Au and Parameswaran, 2001; Hidas, 2002),

At this point in the development process, the quality of the available hardware (Step 12) and software (Step 13) will influence how complex the model is to be. These steps have clearly changed rapidly over the last few decades and have facilitated the development and use of complex computer simulation models. The more complex the processes that need to be included in the model the slower the calculation process. Improving technology will enable models to become more complex, while reducing computation time.

Once the model has been represented on the computer, it is necessary to verify (compare the model with the expected system: Step 7), estimate parameters (Step 8), and validate the model with real word situations (Step 9). Parameter estimation (Step 8) requires calibration at both the microscopic and macroscopic levels. The individual parameters can be estimated by collecting specific data relating to the components. The driver behaviour in different countries and even cities may vary, hence calibration at this microscopic level is important. At the macroscopic level, overall parameter estimates can be derived to provide particular overall performance measures. This may include consideration of temporal and spatial route choice considerations and origin-destination setting. The calibrated and validated model can be refined to make it perform more efficiently (Step 10) for specific applications.

Initial applications of the model (Step 11) are usually restricted to the definition of the model (Step 4), however, the development of new problems and situations that can be analysed will extend the use of the model and should result in improvements to the model to more accurately replicate these situations. The proliferation of commercial traffic simulation models has meant that many people now have access to powerful analysis tools. In order to use these tools, the analyst must be aware of the assumptions made in the model and the appropriateness of these assumptions to the situation being analysed. If the model is not appropriate for particular situations then the results could be misleading. Models may be applied to situations that are not in existence and hence there may be a need to develop data collection systems that mimic the new situations so driver's reactions to these situations can be tested.

A major concern with discrete event parking simulation models has been the time it takes to obtain a result. Given that simulation models replicate real world behaviour, they can be very slow in achieving finality. Further, since they are commonly stochastic in nature the models must be run a number of times in order to obtain an indication of the variation in output variable, and hence a measure of confidence in the result. The rapid increase in use of simulation models clearly indicates that the threshold of time taken to use these models has been overcome.

Integral to all of the model development steps (Step 1 to 11) is the quantity and quality of data. During the initial stages of model development (Steps 1 to 3) the data is broad and often based on subjective assessment. A closer investigation of the system (Steps 4 to 6) requires more detailed study of the processes involved, while, a broad view of the model and data needs is being developed. Graphics and simplified models are useful in verification. The determination of the accuracy of the model (Steps 7 to 10) requires more focused data collection. Validation of the model requires it to be compared with actual parking systems. Accurate data is needed to ensure the model works correctly. Data are required to validate the model's components, the overall drive cycle, search patterns and the macroscopic outputs from the model. The application of the model (Step 11) often assumes the model is correct and appropriate for the particular situation. Each application of the model should improve its performance. This may not in fact be the case and data should be used to determine if the application is appropriate. This is particularly important when the model is applied to situations for which it was not originally been developed.

The question of the reliability of the model and the type of data needed relies on the problem being analysed (Step 1) and the objectives to be achieved (Step 2). Models are developed at a number of levels. Each individual action can be studied and modelled independently. The individual elements need to be combined into algorithms, which present the model flow. This flow relates to the movement from the origin to destination along links and through the network. Each of these processes will be discussed below.

REPLICATION OF PARKING AND TRAFFIC

Simulation models replicate real world situations. As such they have generally been built using data of observed driver behaviour, this is often termed revealed preference data in the transport literature. Revealed preference data is relatively easy to collect and essential in providing the framework for developing the simulation model. This section of the paper will review some of the basic elements in parking simulation models and illustrate how revealed data collection techniques are assisting in improving the replication of the parking system.

Speed, acceleration and braking

An integral part of a travel and parking system is the determination of free speed, acceleration and deceleration of vehicles. These elements can be measured directly, through direct observation, video or using instrumented vehicles (Taylor et al., 2000).

Video offers an increase in accuracy over manual methods of data collect due to its ability to repeat readings to check accuracy. Automatic extraction of data using visual imaging can also increase the speed at which data can be collected (Young 1989). Variations in speed measured using video recording and analysis along the links in parking systems are shown in Figure 3 and 4. It can be see there is quite a variation and that in fact the estimation of a free speed in short aisles in parking systems may not be regular.

Figure 3 indicates that at least one vehicle exceeds the general speed limit 60 km/hr in parking lots. Further, the design of roads in the vicinity of parking systems generally assumes that the 85-percentile traffic flow will be associated with the design speed. This indicates there will be 15 percent of vehicles travelling at higher speeds than the design speed. McLean (1978) in his development of relationships between speeds and the geometric characteristics of rural roads supported this view. Clearly, assumptions that vehicles travel at or below the design speed in travel and parking simulation models will result in an underestimate of the performance of the traffic system.

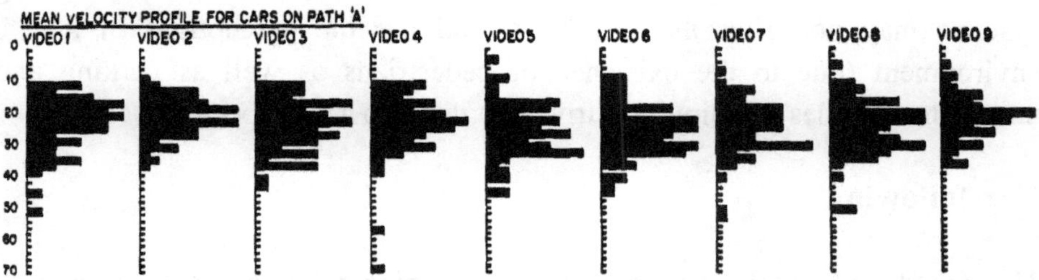

Figure 3: Speed profiles in parking lot aisles.

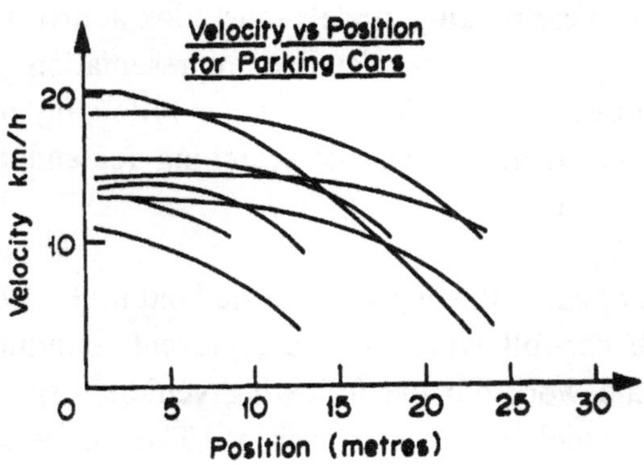

Figure 4: Parking lot speeds along lanes.

Recent development in Global Positioning Systems and data loggers (Greaves and De Gruyter, 2002) has enabled the measurement of the acceleration, deceleration and speed characteristics in real time in many vehicles with many drivers. The Global Positioning System (GPS) provides the potential to passively collect latitude, longitude, altitude, bearing, and speed to high levels of precision. As such, GPS provides the capability of collecting previously unavailable en-route operational characteristics (speeds, idling, cruise, acceleration, deceleration) that can be used to validate the driving behaviour profiles (drive-cycles) underlying model-based estimates of fuel consumption and vehicle emissions. The development of drive cycles and the investigation of variations in these is integral information in developing travel and parking simulation models. Figure 5 shows such a drive cycle. It can be see that the variations in travel are considerable and relate closely to the design of the parking system.

Figure 5 raises the question about the appropriateness of a free-flow speed for vehicles in parking lots. This concept has been used in many traffic simulation models to represent the steady state motion of vehicles. Parking systems may not offer the length of road nor the perceptions of a safe environment (due to the existence of pedestrians as well as parking and unparking vehicles) required for drivers to develop a free speed.

Car-following

The speed and acceleration parameters combine with the interaction with other vehicles to result in traffic flow. One component of this interaction is car-following. Since car parking models must look at driver behaviour in and around the parking system, a detailed representation of car-following behaviour is required. In particular, the car-following behaviour of cars following other cars in the process of searching for and finding a parking place must be represented.

The initial studies of car-following were carried out in the 1950's (Chandler et al, 1958). These car-following models represent longitudinal interactions among vehicles and work out the following vehicle's speed in response to stimulus from the vehicle or vehicles in front. The stimulus is represented in terms of space and speed differences.

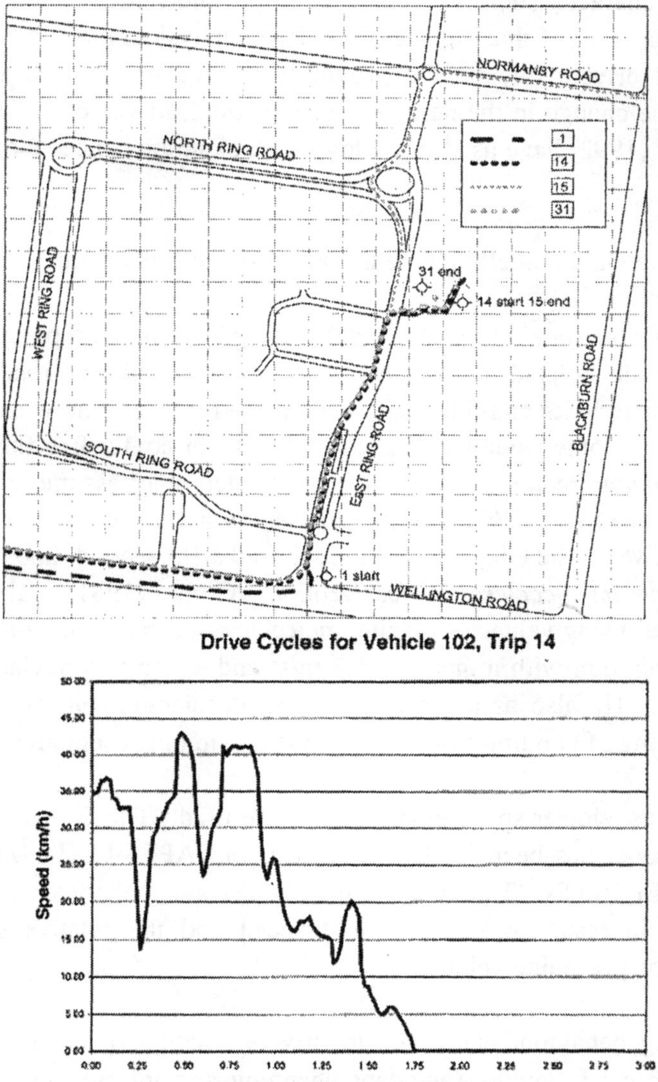

Figure 5: Drive cycle in a parking lot.

There are many approaches to developing car-following models. These have been reviewed by Brackstone and McDonald (1999). Chandler et al (1958) assume that the driver's acceleration is proportional to the speed difference between the two interactive vehicles. Gipps (1981) considers a safe following distance below which collisions are unavoidable. Helly (1959) relates the acceleration characteristics of a vehicle to that of the vehicle in front of it and the vehicle in front of that vehicle also. Psychophysical models develop the concept that drivers rely on the proximity change to the lead vehicle by looking at the change in the perceived size of the lead vehicle. Kikuchi and Chakroborty (1992) have used fuzzy logic to look at the concepts of too close or too far.

In general, the main modelling parameters for car following are:

- *Free-flow speed:* Free-flow speeds are generally input parameters and may be related to the character of the link or road. These characteristics could include pedestrian crossings, link width and link length. Some models use speed limits as a proxy for free flow speeds.
- *Minimum headway:* Car-following models may assume a minimum safe headway. This may be represented by either a time or distance headway. The Gipps (1981) model used a 1-2 second time headway.
- *Maximum acceleration/deceleration:* Gipps (1981) validated his model using values of maximum acceleration sampled from a normal distribution with a mean of 1.7 m/s^2 and a standard deviation of 0.3 m/s^2. He also assumed a maximum deceleration under emergency braking of two times the value of the maximum acceleration.

In parking lots, slower speed environments are used. The Gipps (1981) car-following model has been implemented in both SAPTSIM (Le and Young, 1998) and EPSILON (Tan and Young, 2002) and calculates a vehicle's acceleration in response to its desired speed and the relative speed and distance of the preceding vehicle.

Car-following behaviour can relate to drivers reaction time. Hence a key dimension in car-following (and lane-changing and intersection models) is reaction time (Brackstone and McDonald, 2001). The driver's ability to react to situations and make particular decisions takes time. This involves a risk and a chance that drivers may make a mistake and be involved in a traffic accident. Some simulations models (e.g. PARAMICS) introduce the concept of aggression in drivers. Hence some drivers may develop more aggressive behaviour while moving through the system. This aggression relates more to levels of gap acceptance and car-following. Risk taking and the level of attention of drivers that could result in a road accident are rarely included in traffic simulation models. Hence, drivers will be unable to react to an unsafe situation and hence will not be involved in a road accident.

Taken at its most simple level, the reaction time assumed in the determination of stopping sight distance in highway design in Australia is 2.5 seconds. If all vehicles were travelling at the same speed, then a vehicle that immediately stops would require the following vehicle to be travelling at headway of 2.5 seconds. This separation headway would result in traffic flows of 1,440 vehicles per hour. However, research has shown that freeway traffic moves at much lower headways. Research places maximum capacities of roadways at around 2,000 vehicles per hour per lane. This is generally associated with speeds of approximately 60 kph. The classic speed-flow relationship indicates higher speeds will result in lower traffic flows. Headways on freeways during peak times have been recorded as low as 0.5 seconds (Thompson, Young and Taylor, 1989). The use of half-second headways and ignoring minimum safe headways would result in traffic flows of 7,200 vehicles per hour. However, the adoption of half-second headways would clearly assume unsafe behaviour since no vehicle could stop if the vehicle in front suddenly stopped at this speed.

Gipps (1981) recognised that headways lower than the 2.5 seconds existed on arterial roads and freeways, and changed his simulation update interval to 2/3 seconds so that low headways could be included. The safety rational used was that drivers were looking at what was occurring several vehicles in front of them in order to adopt such small headways. Another rational is that drivers during peak periods accept due to experience that they can travel at shorter headways. Clearly, the negative side of this occurs when a driver for some reason does something unexpected. This would result in a number of vehicles involved in a multiple accident. Such situations are not uncommon on freeways.

Safe headways are incorporated into many simulation models. Drivers may in fact use headways that are considerably lower than acceptable levels. The adoption of safe headways in the traffic stream may result in an underestimate of traffic flows and hence incorrect estimates of the performance of the traffic system. Further in parking situations the headways and speeds are likely to be such that lower flows and reaction times are present.

Car-following models have been developed through observation. However, the overall performance of the models in terms of typical traffic flow situations has been used as a base for justifying the car-following procedures.

The previous discussion indicates that the present car-following models appear to provide appropriate overall estimates of traffic flow performance in the overall transport system. There is clearly a need to measure the parameters used in car-following models. Improved measurement of the relative positions of vehicles and their interaction can be carried out using in-vehicle detection equipment. Brackstone and McDonald (2001) have used an instrumented car and automotive radar to measure the interaction between the instrumented car and the following vehicle. Data collected at this level will provide detailed estimates of many of the parameters required for the improvement of car-following models. Video techniques could also be used to investigate the psychophysical models and driver's interaction to vehicles in front of them.

Lane changing

Since car parking models must look at behaviour in and around the parking system, a detailed representation of lane changing is required. Lane-changing models look at traffic streams and consider individual drivers' ability and propensity to change lanes. There are a variety of reasons or desires for lane changing. The following are a list of but a few: moving past a parking vehicle, passng a vehicle searching for a parking space, bus stopping at bus stops; bus moving into reserved bus lane, and other vehicles move off a bus lane, or move away from a lane that is leading to a reserved bus-lane section; avoiding an incident (parked vehicle, road works, accidents); making junction turning movements; and overtaking a slower moving vehicle.

All travel and parking simulation models should consider lane-changing objectives and the complex decision-making behaviour. More specifically, the models need to represent lane-changing decisions such as (Gipps, 1986): Is it possible to change lanes? Is it necessary to change lanes? and Is it desirable to change lanes? To answer these questions, the models need to work out: whether it is physically possible and safe to change lanes, whether it is for avoiding obstructions, transit lanes or for junction turning, and whether it is to gain speed.

The various lane-changing desires can be generally classified into two types: mandatory where a lane-changing has to be carried out by a certain position on the current link, or discretionary where a lane-changing will be carried out if traffic condition in the target lane is better. For example, a vehicle would only change lanes to gain speed if the speed offered by the adjacent lane is higher by a pre-defined factor.

Many of the microsimulation packages available commercially do not consider explicitly the vehicles' route, ie vehicles make random decisions as to which turn to make at individual intersection. Thus, in these models, there would not be a reason for lane-changing in order to be in the correct lane for junction turning movements, or at least it won't be a mandatory type of lane-changing. Both SAPTSIM and EPSILON consider route choice and lane changing within the broader context of route choice.

Once a lane-changing desire (whether mandatory or discretionary) is triggered, a gap-acceptance model is used to find the gaps in the target lane acceptable to the subject vehicle to merge. The parameters considered here are front gap and rear gap (lag) in the traffic stream of the target lane, and the critical gap acceptable by the driver as safe to merge. The front gap is measured between the rear of the vehicle just in front of the subject vehicle in the target lane; and the rear lag between the rear of the subject vehicle and the one immediately behind it in the target lane. The front and rear gaps are derived from the positions (and speeds) of the vehicles involved; so they are deterministic once the positions and speeds of the vehicles are determined. The critical gap acceptable by drivers, on the other hand, is a behaviour parameter, which is represented differently by different microsimulation models.

In most of microsimulation models, as with the traditional gap-acceptance model, the acceptable gaps used in lane-changing models are either fixed for all drivers or fixed for each individual driver. They do not represent variability in driver's gap-acceptance behaviour with changes of traffic conditions and with "urgency" of a driver's lane-changing desire. Further many models do not consider that common situation of "forced" lane-changing and the safety implications of such actions (Hidas, 2002).

Considerable effort is required in understanding lane changing. New data collection techniques using vehicle simulators and analysing drivers choice are required to ensure that lane-changing models are correct and accurate.

Overtaking

Since car parking models must look at behaviour in and around the parking system another situation that involves lane changing and gap acceptance is overtaking. Overtaking and lane changing cannot be observed easily and hence do not have the same level of data and understanding as does driver behaviour at intersections (Troutbeck, 1990).

Some overtaking may take place on links that forbid it. The movement of vehicles on rural roads involves the need to overtake slower vehicles. These manoeuvres may be taken with some risk. The introduction of overtaking lanes on rural roads to break up the flow of vehicles and reduce the frustration of drivers aims to reduce the amount of overtaking in such instances. Clearly, the overall improvement consequent on the introduction of overtaking lanes must be compared to the performance of the system with some risk taking if it is to be a true comparison.

Another dimension of overtaking relates to the road rule that overtaking should not take place on the near side. Recently, the overtaking of vehicles on the near side has been controlled in Australia. It is thought that the faster vehicles should move to the central lane and slower vehicles move to the outside lanes. Lane changing would be required to ensure vehicles can get to the outside lanes to exit freeways. The introduction of such behaviour cannot be analysed in models which allow vehicles in the outside lanes to overtake vehicles in the inside lanes. Clearly, the behaviour of drivers in Australia is not consistent with this action and the capacity of the roads is higher as a consequence.

Gaps and gap acceptance in traffic

The interaction between vehicles is often represented through the gap, time or distance, between vehicles. A gap in a traffic stream offers the opportunity for another vehicle to move into it. The interaction between different streams of traffic relates to the opportunity and willingness of drivers to accept gaps in traffic. Time gaps are collected using observation from a fixed point in the transport system, while distance gaps are measured along a length of road. Considerable research has been devoted to this fundamental descriptor of traffic flow (Blunden 1971, Brilon et al. 1999). Most of the basic studies in this area have been carried out by human observers but automatic detection using induction loops and video monitors have provided considerable data on time gaps in traffic. The gaps in traffic are critical to understanding the interaction between vehicles on different legs of an intersection, overtaking and lane changing situations.

Measuring gap acceptance is the process of collecting information of the critical gap drivers will accept when moving from one position in the traffic stream to another. The gap-acceptance models deal with problems of finding an acceptable gap in traffic stream(s) for a vehicle to cross or to merge. Such gaps cannot be easily measured. The earliest method of determination of an acceptable gap involved looking at the gaps accepted and the gaps rejected (Raft and Hart, 1950). The critical gap was defined as where these two curves intersect. Gaps are usually represented in time (seconds). Average acceptable gaps of between 3 and 3.5seconds are often observed. The model parameters used may include a critical gap and a variation in acceptable gaps.

At intersections vehicles start to react to gaps in traffic and traffic controls (signals or giveway signs) at a downstream intersection when they are within some distance of the stopline. Only the lead vehicle in each lane reacts to intersection control; the following vehicles follow the preceding ones according to car-following rules until they become the lead vehicle. There may be a number of intersection types that need consideration. These may include signalised, giveway and roundabout intersections. Signalised intersections will be discussed in the next section. Traveling towards a giveway intersection, vehicles may aim to stop just before the stopline, and only when they are a few meters away from the stopline where they can see the situation on the major road will they start to look for gaps to join in or to cross the major flows. The acceptable gap is individual based and can vary with the length of time the individual has waited at the giveway sign. Vehicles approach a roundabout as though approaching a priority junction and give way to circulating traffic on the roundabout.

Gap acceptance models involved risk. Driver behaviour can be a function of the situation, for instance, to represent the fact that a driver may choose a smaller gap that he/she would normally take after having waited a long while (Taylor et al. 2000). Some times the choice of inappropriate gaps could result in an accident and the consequent increase in delay to traffic. However, even though risk is present the acceptance of a small gap is unlikely to result in a road accident and hence the true level of risk taking or attention is not represented in the traffic simulations.

Most gap acceptance situations are carried out with vehicles facing the traffic stream they are entering. In parking situations, some cars may reverse out into the traffic stream. Unparking manoeuvres are also gap acceptance situations. These situations require vehicles to move into a stream of traffic while moving backwards.

Signalised intersection behaviour

Signalised intersections present a situation where drivers react to information provided from traffic management systems. At a signalised intersection, when the signal has just changed to green, the head of the queue checks whether its path is clear before moving off. During the remaining green period, vehicles move across the intersection at a speed determined by the car-following rules: the lead vehicle follows the last vehicle in the outbound lane it turns into. At the instant the signal changes to amber, the vehicle nearest to the stopline will consider whether to stop. If it is too close to the stopline, it will either go ahead if it can pass the stopline within the amber period with its current speed, or alternatively make a random decision whether to carry on moving or to stop. If the decision is to stop, it applies its maximum deceleration if necessary; similarly, if it decides to go on, it may accelerate at its maximum acceleration. This decision is then maintained throughout the remaining amber period. A vehicle is allowed to move across at the start of a red signal only if it cannot stop at the stopline with its maximum deceleration.

One change that is occurring in traffic behaviour as the result of the widespread use of signalised intersections is vehicles not stopping at red lights. Increasingly, traffic signals are being used as safety devices and to manage the flow of traffic. These lights have a series of movement each controlled by red and green lights. An all-red period is used to provide a safety mechanism. The general assumption in traffic simulation models would be that drivers stop at traffic signals when the lights are red. Such an assumption is unlikely to be correct and the number of accidents occurring when vehicles go through red lights supports this. Further, Pretty (1974) found that traffic signals performed better than police control in a capacity sense as a consequence of drivers using the amber and all-red period. Assumptions of stopping on red lights may not result in true estimates of the traffic flow at traffic lights. The development of speed cameras at intersections allows both the signal timing to be monitored and the acceleration and deceleration characteristics of vehicles to be monitored.

At intersections, vehicles and often prohibited from turning right (left in Australia) during the red phase. This is sometimes relaxed through the use of signs. If vehicles disobey the rules then the capacity of the intersection may be increased but the potential safety decreased. Further, pedestrians often cross the road when there is no traffic light or when the pedestrian light is showing red. This involves a risk and this risk is rarely incorporated into the pedestrian model. Data on red light running, illegal turn behaviour and pedestrian movement should be included in models if a realistic representation of the system is required.

Parking and unparking

The times required to park and unpark (Figure 6) is crucial elements in parking simulation models. They influence the size of the acceptable unparking gap in traffic flow; the delay to vehicles moving past a space a vehicle is parking in and the efficiency of the parking layout. Measurement of parking and unparking times can be difficult because of the short time, complexity of the manoeuvres and in the case of multi-storey facilities, the lack of suitable vantage points and obstructed view.

Human observers patrolling car parks have a problem with being in the correct location at the correct time. They can locate themselves where vehicles will park and wait or they can roam around trying to follow vehicles. Parallax and problems with defining when the parking manoeuvre starts create variability in the results. However, this approach does provide a base for assessing the accuracy of other methods. Parking and unparking manoeuvres using instrumented vehicles will also provide a base for looking at parking behaviour. This approach may be a little contrived and therefore should be compared to data collected in real life parking situations. Determining parking and unparking times, unlike many situations in transport, does lend itself to controlled experimentation. Experiments with vehicles moving into and out of parking places with predefined parking layouts allow numerous parking configurations to be analysed. An advantage of this approach is that parking situations not used in many parking lots can be studied. Validation of data collected using this approach requires comparison with field data.

Video taping (Young, 1989) and analysing the tapes in the office provides a powerful method of measuring parking and unparking times in surface parking lots. It overcomes the sampling problem associated with real world observation. However, parallax problems due to the lack of high positions to mount the video make it difficult to measure changes in velocity.

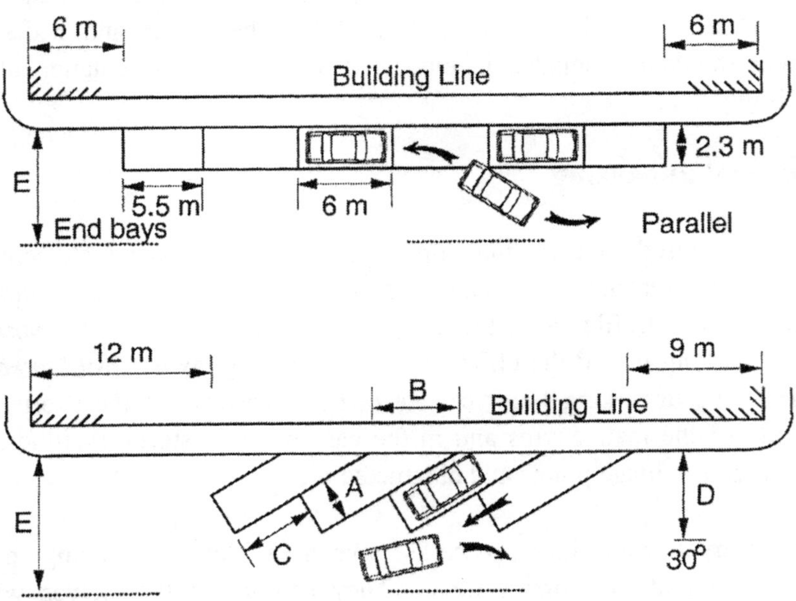

Figure 6: Parking and unparking procedures.

Data loggers with global positioning systems (see Figure 5) provide a rich data base and the ability to define particular parking situations accurately. Data loggers, like instrumented vehicles, allow the change in speed and location of the vehicle to be ascertained. Hence a very accurate estimate of parking and unparking behaviour can be made.

In summary, the measurement of parking and unparking times can be carried out in numerous ways. Each of these provides a useful estimate. Taken in total they provide a very flexible and accurate method of measuring parking behaviour.

Drive Cycles

The individual elements outlined above must be combined in order to represent the entire traffic system. These elements are usually combined to take into account the direct interaction and activities that people undertake. However, do drivers drive in this way? Do they prepare themselves for particular situations? Do they slow down when they see a red light or do they drive slower while searching for a parking place or looking for a road sign?

Measurement of these combinations requires more complex data collection techniques. Drivers combine elements into a driving cycle (Figure 5). This cycle varies depending on conditions and the relationship to other drivers. Drive cycles involve the development of acceleration, speed monitoring, car-following and deceleration. Each of these elements are derived independently using the above techniques but are combined into the discrete simulation model depending on the environment that the vehicle is in.

Determination of the drive cycle of vehicles can be carried out using instrumented vehicles. These vehicles can be driven around the road system or can follow other vehicles to determine how vehicles move through the system. Recent technological developments have allowed the data collected on drive cycles to be broadened. Data loggers that incorporate a Global Positioning System can be placed in any vehicle with any driver. The data collected (see Figure 5) can be used to investigate variations across drivers, vehicle, land use characteristics or combinations of these underlying variants. Such data can be used in the systems analysis (step 4), parameter estimation (Step 8) and validation (Step 9) of the model development process.

REPLICATION OF DRIVERS DECISION MAKING PROCESSES

The previous discussion has focused on what drivers do. It has discussed the collection of observed (or revealed) data. In order to meet the demands of understanding the traffic system in the future, it is necessary to start to address the question of why drivers do what they do? Hidas (2002) points to the consideration of drivers as agents who react to their environment. Such a view presents a significant step toward the incorporate of driver behaviour into simulation models. Data that will enable study of drivers reason for doing things required a variety of approaches. One approach is the use of virtual experiments to collect stated preference data. This approach combined with revealed data provides a strong base for interpreting drivers reactions in many existing and future situations. This section of the paper will begin to address this issue in the context of travel and parking simulation models.

Interaction between road users

Most existing discrete event traffic simulation models look at movement along lanes and the interaction (car-following, lane-changing etc) between vehicles in lanes. They consider the size (length, width and height) of the vehicles in terms of the space used on the roads but do not generally consider interactions broader that those related to lane movement. This limits simulation models applications and prejudges many model outcomes.

The movement in transport policy towards the allocation of road space between vehicles, bicycles, pedestrians and public transport relies on the interaction between vehicles in separate lanes. Further, at many intersections narrower lanes are used to increase capacity throughput. The change in speed characteristics relates to lane narrowing or the proximity of more vulnerable drivers. Models that take into account the spacing between vehicles need to be developed to ensure the correct traffic regime occurs.

An important dimension of studying parking systems is the spatial interaction between pedestrians and drivers (Yue and Young, 1998). Pedestrians, bicyclists and even some vehicles may not use the lanes marked in the parking lot. The driver's reaction to this and its influence on speed characteristics need to be studied. Virtual experiments may provide some insights into these situations.

Route choice

The application of discrete event simulation models and the complexity of the traffic network they consider has progressively expanded as computing power has increased and computation time decreased. Initial models focused on intersection movement (eg INSECT Nairn and Partners, 1986), link movement was replicated next (eg MULTSIM Gipps, 1982), with more complex networks (eg PARKSIM (Young 1986); DRACULA, (Liu et al., 1995); PARAMICS (Quadstone Ltd., 2000)) being progressively introduced.

The theoretical challenge with networks is the extension of the route choice algorithms in the model. Route choice algorithms are based on comparing the cost associated with taking alternate routes between an origin and a destination pair. In simulation models this cost is usually measured in time, however, as tolling and road pricing become more common monetary cost will be incorporated in many models.

There are many approaches to determining route choice. Each drive can be allocated a destination or route on entry to the simulated space (Static route choice) or the route choice decisions can be developed progressively as the driver moves through the system (Dynamic route choice). The factors changing the route choice may be observed queue length, incidents and/or expected delay. The decision on which route to take can be based on minimum path and probabilistic route choice. Further, some simulation models attempt to model drivers knowledge of the transport system (unfamiliar or familiar drivers) by varying the travel cost along routes drivers may be unfamiliar with. These route choice models have been validated using observed data. Route choice is an area where direct observation and questionnaires have provided base data for modelling route choice decisions.

A particularly complex route choice situation relates to the need to search for and the choice of a parking space near the destination zone. In this case there is no definite destination, just a general area where the driver would like to end up. The parking search process has been replicated using observed data and questionnaire studies (Le and Young, 1998). However, the data collected in these situations was collected either after the trip had been made or prior to the trip being made. In both cases there can be inaccuracies. The data collected prior to making the trip represents what the driver wishes to do. The data collected after the trip represented the completed trip and many of the decisions made during the trip could be forgotten or not properly measured.

An approach that is gaining interest for collecting parking search data relates to stated preference data collection techniques (Tan and Young, 2002; Bonsall, Palmer and Balmforth, 1998; Bonsall and Palmer, 2001). Stated preference (SP) approaches place the respondent in particular situations and investigate their reaction to these situations. This approach to data collection has been used in discrete choice modelling for many years. Its application to traffic situations has been a little slower but with the advent of driving simulators and interactive computer graphics their application is set to grow.

SP traffic and parking data can be generated through experiments set up on computer screens (Figure 7). Drivers can be subjected to particular traffic situations and their reactions to these monitored. It is possible to investigate many variations in traffic situations using this approach. In the case of parking search the driver can be subjected to variations in the travel time to the parking lot, location of and utilisation of parking stations, time to destination, and different levels of information. Drivers' reactions to the traffic situations are monitored at particular points in the system. It is therefore possible to collect in depth information on the drivers' decision points.

Figure 7: Graphics representation of parking system (Bonsall and Palmer, 2001).

Bonsall et al (1998) have applied the SP approach to search for a parking station, investigating the time to parking location, utilisation, prices and other parameters. The decision process was modelled using hierarchical logit models. They found the approach robust and useful. Tan and Young (2002) are applying the approach to study movement to and within a parking station. A four level decision support system is being investigated. The application of the resultant models to real world parking situations will provide support for validating the models.

Modelling of the provision of information through parking guidance system (Young, 1987; Polak et al. 1990) is an application of simulation models that requires a deep understanding of the search patterns of drivers and their reaction to information. This requires data collection procedures that are flexible and able to test drivers' reactions to new situations. Bonsall et al. (1998) have provided some initial investigation in this area.

One major concern about stated preference approaches is, do they replicate reality? Clearly, this is a question that must be addressed. Various techniques (Loviere, Hensher and Swait, 2001) have been used in the discrete choice literature. These range from comparison of stated and revealed preference data and the use of correction factors to the integration of the two data sets and the derivation of combined models. Notwithstanding these concerns, the overall outcome of the combination of stated preference models and revealed data to validate the approach is likely to provide a more robust search model which can be subjected to verification using observed data.

Total trip considerations

Route choice models in most traffic simulation models tend to focus on the end of the vehicle trip. This is used to determine parking accumulation and use. This may involve a parking choice or may be just a traffic zone. Drivers are unlikely to consider this the end of the trip and would consider the shopping activity they wish to undertake or the business meeting as a more significant end. The incorporation of the total trip into simulation models will provide a more accurate estimate of the decisions made at each point in the trip and a better understanding of what people require of the traffic and parking system.

Driver risk

Models make assumptions (e.g. with respect to gap-acceptance, car-following, or adherence to speed limits) about the safety-related behaviour of drivers (Young, Liu and Bonsall, 2001). Some of these assumptions (eg reaction time, safe headways etc.) have been introduced in previous sections of this paper. These assumptions are often based on observation of drivers. These assumptions may not replicate the real behaviour of those drivers who adopt seemingly unsafe behaviour (eg. running red lights at signalised intersections, following vehicles too closely on freeways or accepting smaller reaction times on links) in order, for example, to reduce their journey time. These actions result in the performance of the system that we observe but may also result in conflict and some times in accidents.

An approach used in some simulation models is to describe drivers in terms of aggression or awareness. These characteristics are used in some simulation models (eg PARAMICS) to determine desired speed, lane selection, gap acceptance and yielding behaviour (Abdulhai et al. 1999). Hidas (2002) proposed a modelling approach that incorporates "forced" and "co-operative" lane changing procedures. To this point in time they have not been used to describe risk-taking behaviour.

Good design will seek to maximise safety and may design systems assuming safe behaviour. However, is the safety of a design necessarily enhanced by making unrealistic assumptions about behaviour in simulation models? If safety is to be introduced into simulation models, then there is a need to quantify risk. Risk represents the probability of an accident occurring. Drivers accept risk when they drive but simulation models assume that an accident will not occur and there is no risk. Most simulation model will not allow an accident to occur and will assume that the driver is able to avoid all accidents. Clearly, data related to risk is required in order to make simulation models more reliable.

How can an understanding of the risk taking characteristics of driver's be understood? Clearly, it requires drivers to be subjected to situations where risk is involved and their reactions measured. Simulated vehicle behaviour provides one approach to this problem.

Improved understanding of aspects like gap acceptance can be obtained by setting up virtual experiments where drivers are subjected to various gaps, speed characteristics etc. These virtual experiments may assist in determining driver's risk taking and improving model performance.

OUTCOMES

The previous discussion has related to simulation models and their structure. Clearly, there is a need to improve these models to enable them to make more accurate predictions. Simulation models also provide the base information on which to make estimates of the external influences of traffic systems. Some of these outputs are fuel consumption, emissions, noise and community impact. The need to improve the quality of the environment in and around parking systems makes it essential to gain a more detailed understanding of how parking systems and the environment they are located in, interact.

Energy consumption

Measurement of energy consumption has been carried out for several decades. Simulation models lend themselves to providing very accurate measures of fuel consumption since they can provide measures of speed, acceleration and braking over very small time increments.

Models of fuel consumption have been developed for various situations (Bowyer et al. 1984). The lowest level model, the instantaneous fuel consumption model (Biggs and Akcelik, 1986), can be incorporated into simulation models. These models have been developed using instrumented vehicles. Fittings attached to the fuel pump of vehicles have allowed this information to be collected easily. The models have however only been collected for typical vehicle types. Broader data collection of the vehicle fleet is required to ensure the accuracy of the simulation model predictions. This could be facilitated in the near future since the vehicle diagnostic procedures built into many recent vehicles can be harnessed to provide data useful for monitoring energy consumption and travel behaviour. Hand held computers could be used to download the information on speed, fuel consumption and engine operating condition from the on-board vehicle diagnostic system.

Emissions

Like energy consumption the instantaneous emissions from vehicles can be estimated using simulation models. Information on emissions is not as easily collected as that for fuels consumption. A common method is to develop drive cycles and engine maps from instrumented vehicles. These drive cycles and engine maps can be transferred to laboratory situations and the emissions measured in the laboratory using a Dynamometer. Another approach is to use remote sensing. Remote sensing uses infrared and, in some cases, ultraviolet spectroscopy to measure the concentration of pollutants in exhaust emissions as a vehicle passes a sensor on the roadway. Both these approaches are limited in their application. On-board emission measurement is widely recognised as a desirable approach for quantifying emissions from vehicles since it allows data to be collected under actual conditions on the roadway. Frey et al (2001) have utilised this approach to study tailpipe emissions of CO, NO and Hydrocarbons. The approach offers considerable opportunities for testing a vast variety of vehicles emissions and their interaction with the road system due to the portability of the measurement system.

Emissions from the vehicle are only part of the overall issue. The movement of emissions around closed facilities like parking lots is important to ensure the environment drivers require. The combination of vehicle emission measuring and data on emissions in closed facilities is required to understand these situations.

Noise levels

The noise created by vehicles is a combination of the vehicle noise, tyre on road noise and road surface characteristics. Clearly, these are difficult to measure, however, there is a need to quantify this aspect and incorporate detailed algorithms into the simulation models.

Diffusion of sound in small area requires the measurement of noise levels in complex shapes and its movement and reflection with regard to the different materials.

Community impacts

The easiest transport groups to model are vehicles moving in predetermined lanes. It is not surprising, therefore, that most simulations in the transport area look at vehicle movement on roads. These impacts are interpreted by trained professionals and the road system designed to enable the impacts to be minimised.

Transport, however, provides access to many people wishing to make trips. These people may be pedestrians, bicyclists, and people with disabilities. The usage of transport systems by a broad range of people will ensure an efficient allocation of resources. Simulation models by their character can be used to indicate the impacts of traffic systems for a variety of users. More complex models that include the linking of trips (eg pedestrian movement, parking, and bicycle movement) are required. The development of these models requires a considerable increase in understanding of the behaviour of these groups. This requires detailed data collection of people behaviour and travel patterns.

CONCLUSIONS

The increasing availability of highly developed discrete event traffic simulation models has resulted in their increased use in the investigation and prediction of traffic impacts. Alongside this increase in application is an increasing sophistication of the traffic scenarios being investigated. In order to apply these models confidently to these situations, their specification and development must be consistent with the application. This requires the comparison and validation of the models with appropriate data.

Alongside the increased use of simulation models is the increasing sophistication of data collection technology. This paper has investigated the link between data and the accuracy of discrete event travel and parking simulation models. The quality of the data used in the development and validation of simulation models and its ability to replicate the real world situation has a considerable impact on the accuracy of the models. The above discussion has highlighted some of the issues associated with the design of simulation models and assumptions that underlie them.

Discrete event travel and parking simulation models make assumptions about behaviour. Some of these are directly observable, others cannot be observed and are therefore implicit in the models.

Developments in global positioning systems, vehicle detection and video systems have enabled observation of traffic behaviour to be improved and replicated with greater confidence. These techniques provide new methods of validating and analysing the driver's behaviour in travel and parking simulation models.

Given the increasing sophistication of the applications of simulation models, the models must move beyond the observation of behaviour and try to understand why this behaviour is taking place. For instance the search and risk taking characteristics of drivers need to be explored and incorporated into models. The use of interactive computer graphics and stated preference procedures allow model developers to explore some of the areas that have not been explored using conventional observation techniques.

REFERENCES

Abdulhai, B. et al (1999). Simulation of ITS on the Irvine FOT Area using Paramics 1.5" Scalable Microscopic Traffic Simulator": Phase 1: Model Calibration and Validation, Institute of Transportation Studies. University of California, Berkley, California.

Au, S. and Parameswaran, N. (2001). Attitudes for agents in dynamic worlds, *FLAIR 2000*, Key West, Florida.

Blunden, W.R. (1971). *The Land Use / Transport System*, Pergamon Press: Oxford

Brilon, W. et al (1999). Useful estimation procedures for critical gaps, *Transportation Research Part A*, **33(3/4)**, pp161-186.

Bonsall, P.W. et al (1998). PARKIT – A simulated world for parking choice research, Presented *at 9th World Conference on Transport Research*, Antwerp.

Bonsall, P. and Palmer I. (2001). Modelling driver's car parking behaviour using data from a travel choice simulator, Presented *at 9th World Conference on Transport Research*, Seoul.

Bowyer, D.P. et al (1984). Guide to fuel consumption analyses for urban traffic management, *AIR* 390-9, Australian Road Research Board, Vermont South.

Biggs, D.C. and Akcelik, R. (1986). An energy-related model of instantaneous fuel consumption, *Traffic Engineering and Control* , **27**, pp320-5.

Brackstone, M. and McDonald, M. (1999). Car-following: a historical review, *Transportation Research,* **Part F 2**, pp181-196.

Brackstone, M. and McDonald, M. (2001). Driver behaviour studies in the motorway operations platform grant, International Driving Symposium on Human Factors in Driver Assessment, *Training and Vehicle Design*, Aspen, Collarado.

Chandler, R.E. et al (1958). Traffic dynamics: studies in car following, *Operations Research*, **6,** pp165-184.

Frey, C.H. et al (2001). Measurement of on-road tailpipe CO, NO, and Hydrocarbon emissions using a portable instrument, *Proceedings of the Air and Water Management Association*, Orland, Florida.

Gipps P.G. (1981). A behavioural car-following model for computer simulation, *Transportation Research*, **15B**, pp105-111.

Gipps, P.G. (1982). Computer user manual: MULTSIM: A program package for multi-lane traffic simulation, Commonwealth Scientific and Industrial Research Organisation. Division of Building Research, Melbourne, Australia.

Gipps P.G. (1986). A model of the structure of lane-changing decisions, *Transportation Research*,**20B**, pp403-414.

Greaves, S. and De Gruyter, C. (2002). Profiling Real-World Driving Behaviour using Passive Global Positioning System (GPS) Technology, Institute of Transport Engineers Conference, Melbourne.

Helly, W. (1959). Simulation of bottlenecks in single lane traffic flow, Presentation at the Symposium on Theory of Traffic Flow. Research laboratories, General Motors, New York, pp207-238.

Hidas, P. (2002). Modelling lane changing and merging in microscopic traffic simulation, *Transportation Research*, **Part C** (In press)

Kikuchi, C. and Chabroboty, P. (1992). Car-following model based on fuzzy inference system, *Transportation Research Record*, **1365**, pp82-91

Le, H. and Young, W. (1998). Modelling shopping centre traffic movement (1): Model validation, *Transport Planning and Technology*, **21 (3)**, pp203-33.

Liu, R. et al (1995). DRACULA: dynamic route assignment combining user learning and microsimulation, *Proceedings of 23rd PTRC European Transport Forum*, Seminar E, pp143-152.

Louviere, J.J, Hensher, D.A., and Swait, J.D. (2001). Stated choice methods: Analysis and application, Cambridge University Press: Cambridge, England

Mclean, J.R. (1978). Observed speed distributions and rural road traffic operations, *Proc 9th Australian Road Research Board Conference* **9(5)**, pp235-244.

Nairn, R.J. and Partners (1986). Intersection simulation model (INSECT), Department of Main Roads, NSW.

Polak, J. et al (1990). Parking guidance systems, *Traffic Engineering and Control*, **31(10)**, pp519-25.

Pretty, R. (1974). Police control of traffic at intersections, *Proc 7th Conference of Australian Road Research Board* **7(4)**, pp83-95.

PTV Planung Transport Verkehr AG (2001). VISSIM 3.50 User Manual, PTV Planung Transport Verkehr AG, Karlsruhe, Germany.

Quadstone Ltd (2000). Modeller v3.0: User Guide, Quadstone Limited, Edinburgh, Scotland.

Raff, M.S. and Hart, J.W. (1950). A volume warrant for urban stop signs, Eno foundation for highway traffic control, Saugatuck, Connecticut.

Tan, Y.W. and Young, W. (2002). Modelling traffic and parking in multi-storey parking systems, *International Conference on Seamless & Sustainable Transport,* Singapore.

Thompson, R.G. et al (1989). Microcomputer analysis of video-based speed and headway data, *Proc. of International Conference on Applications of Advanced Technology in Transport Engineering*, San Diego, California, pp462-7.

Transport Simulation Systems (2001). AIMSUN: Version 4.0 User Manual, Transport Simulation Systems

Taylor, M.A.P. et al (2000). Understanding traffic systems, Ashgate: London, England

Van der Touw, J.W. et al (1983). Fuel consumption on urban traffic – a guide to planning experiments, *Transportation Research*, **17A**, pp219-231.

Troutbeck, R.J. (1990). Traffic considerations in the appraisal of routes for road trains in rural areas, *Proc. 15th Australian Road Research Board Conference* **15(4)**, pp85-118.

Young, W. (1986). PARKSIM/1 (1): a simulation model of driver behaviour in parking lots, *Traffic Engineering and Control*, **27(12)**, pp606-13.

Young, W. and Thompson, R. (1987a). PARKSIM/1 (2): a computer graphics approaches for parking lot layouts, *Traffic Engineering and Control*, **28(3)**, pp120-3.

Young, W. (1987b). Parking guidance systems, *Australian Road Research Journal*, **17(1)**, pp40-7.

Young, W. (1989a). The application of VADAS to complex environments, *Journal of American Society of Civil Engineers*, Transportation **115(5)**, pp521-36.

Young, W. et al (1989b). Microcomputers in Traffic Engineering, Research Studies Press: Taunton, England.

Young, W. (1990). Parking policy, design and data, Department of Civil Engineering: Monash University, Melbourne, Australia.

Yue, W.L. and Young, W. (1998). Using Parksim2 to measure safety in parking lots, *J. of the Institute of Engineers Singapore*, **38(5)**, pp10-20.

Young, W. (2000). Modeling Parking,. *Handbook of Transport Modelling* (In Hensher, D.A. and Button, K.J.), Pergamon Press: Oxford

Young, W. (2001a). "Parking systems" In *Transport Handbook* **Volume 3** (Hensher, D.A. and Button, J.)

Young, W. et al (2001b). Traffic simulation, risk and modellers liability, Presented at *CAITRE 2000 Annual Conference*, CSIRO Canberra, 6-8 December 2000.

SAGA OF TRAFFIC SIMULATION MODELS IN JAPAN

Hirokazu Akahane
Faculty of Engineering, Chiba Institute of Technology, 2-17-1 Tsudanuma, JAPAN 275-0016
akahane@ce.it-chiba.ac.jp

Takashi Oguchi and Hiroyuki Oneyama
Faculty of Engineering, Tokyo Metropolitan University, 1-1 Minamiosawa, JAPAN 192-0397
(oguchi-takashi, oneyama-hiroyuki)@c.metro-u.ac.jp

ABSTRACT

This paper narrates a history of developments in the traffic simulation models in Japan. Starting from 1971, basically two kinds of logic existed for reproducing dynamic traffic flow. These were the *Block Density Method* and *the Input-Output Method.* These methods were compared to the calculation engine of the network traffic simulation model. Subsequently, these methods evolved, were modified, and became more advanced as found in the AVENUE and SOUND simulation models. This development catered better to the changes in needs required by newer traffic simulation models and resulted in part from developments in computer technology. The first models do not include drivers' route choice behaviour, but the later do. Initially these methods were applied to urban expressways; however, they later expanded to include surface street networks.

INTRODUCTION

Figure 1 illustrates a classification of the network of traffic simulation models developed and applied inside and outside of Japan. The decision for application is dependent upon the degree of details of the traffic flow model and the size of the road networks. In this figure it can be seen that the Japanese models (written in bold letters) together with famous models developed in Europe and the United States are distributed over a wide range. It is interesting to note that the simulation engines of the Japanese models were based on the same logic as the foreign ones, despite being independent developments.

Table 1 demonstrates a history of the development of some traffic simulation models together with the authors who partially contributed to or applied those particular models. The development of subsequent models originated from a comparison study between the *block density method* (BDM) and the *Input-Output Method* (IOM). A study was proposed in order to select an on-line traffic simulation model that was to be installed in the traffic control system at the Metropolitan Expressway Public Corporation (MEPC) in Tokyo. Its objective was to carry out short-term predictions of traffic conditions.

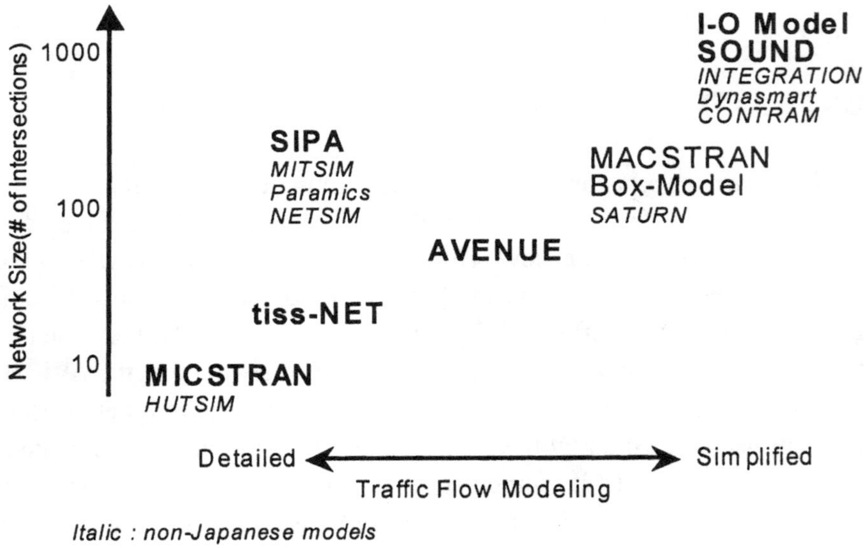

Figure 1. Network simulation models in Japan

The IOM with almost all of its original core components was installed in the traffic control system at the Metropolitan Expressway in Tokyo (the total length today is 270.4 kilometres as shown in Figure 2) for on-line real-time predictions of traffic conditions. As shown in Table 2, the IOM is a kind of fluid model that utilises the relationships between the traffic density and the flow rate (basic characteristics of traffic flow). It basically consists of the *upstream* and *downstream operations*. The '*upstream operation*' is started from the upstream ends of a road network, in which traffic demand goes down to reachable links. Then, a '*downstream operation*' is conducted from the downstream ends of the road network, during which it is checked as to set on the nodes of the loops. No route choice model has been built into the IOM.

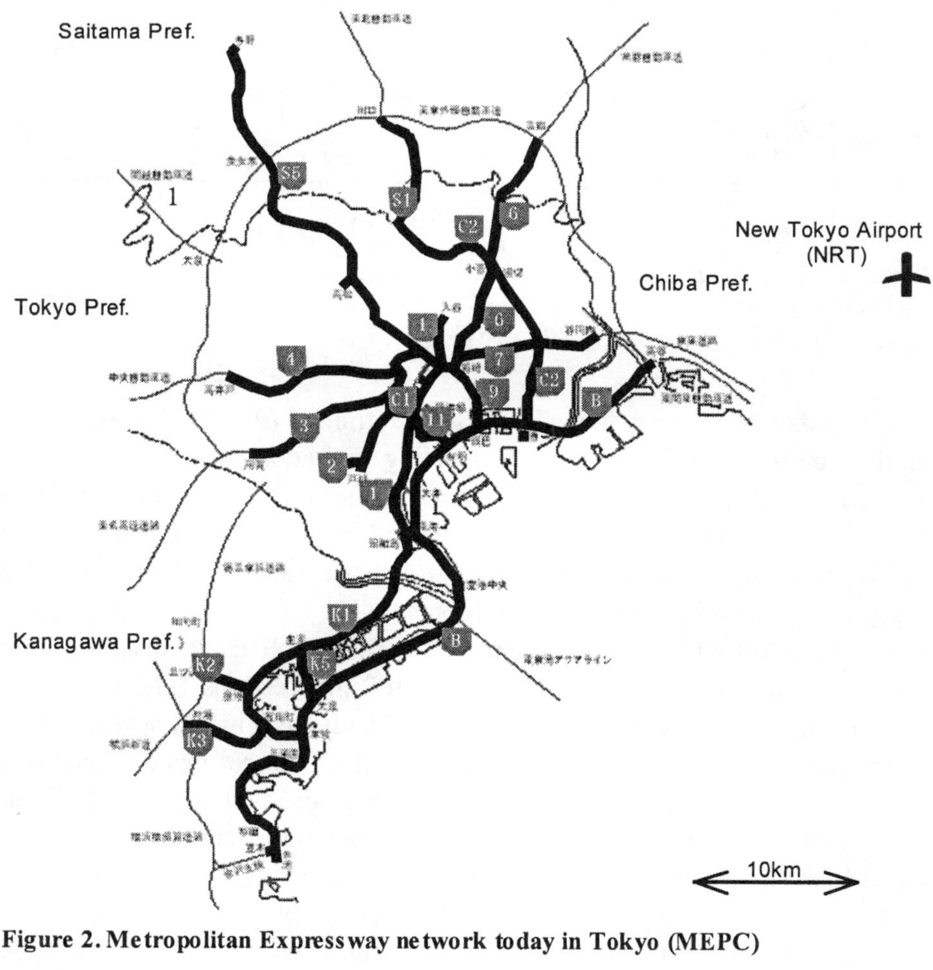

Figure 2. Metropolitan Expressway network today in Tokyo (MEPC)

Table 1. A History of the development of simulation models related to authors in Japan.

Year	AVENUE (for streets)	SOUND (for expressways)	Input Output Method (for expressways)
1970	Block Density Method(BDM)		Input Output Method (IOM)
1989	DESC[a] (Simplified-BDM)		
1993	AVENUE[b] (Hybrid -BDM)	BDM with route choice model[c]	
1995		SOUND[d] (Packet & Density-Speed Relationship)	
1996	AVENUE Ver.2[e] (+Multi-Scanning)	SOUND Arterial [f] (+Point Queue: for streets)	←
2001	AVENUE Ver.3 (Commercial Package)	SOUND/Express SOUND/A-21	

a) Ozaki(1989), b) Horiguchi et al.(1993), c) Kuwahara et al. (1993), d) Yoshii et al. (1995), e) Horiguchi, et al. (1996), f) Okamura et al.(1996)

On the other hand, even though the BDM refrained from implementing the traffic control system due to the computing limitations of those days, two streams of traffic simulation models were produced. One of them starts from DESC (Ozaki, 1989) developed for streets with the hybrid composition of the BDM (the simplified-BDM to be more precise) and a micro-model for representing the behaviours of gap acceptance. The DESC was developed into the present AVENUE (Horiguchi et al., 1993). AVENUE can display traffic conditions of streets that consist of about 100 intersections with a dynamic route choice model (Horiguchi et al., 1996). Another is a model developed for expressways, which includes a dynamic route choice model, and was originally developed based on the BDM (Kuwahara et al., 1993). Then SOUND (Yoshii et al., 1995), with a dynamic route choice model, was developed and introduced the vehicle-packets moving model according to relationships between traffic density and speed. SOUND introduced the *Point-Queue* theory in order to represent traffic conditions on streets in addition to on expressways (Okamura et al., 1996). In addition to conventional uses, these functional features were developed to evaluate the effects of the Intelligent Transport Systems (ITS) and the Transport Demand Management (TDM).

This paper details a history of the development of traffic simulation models related to authors in Japan, in relation to the changes in the needs of traffic simulation models and the performance of computers at that time.

Table 2. Outlines of the Three Traffic Simulation Models

	Input Output Method	SOUND	AVENUE
Traffic Flow	Fluid model based on relationship between density and flow rate.	Vehicle-packets move based on relationship between speed and flow rate.	Individual vehicle moves based on Hybrid Block Density Method.
Route choice	(N.A.)	Dynamically chosen based on travel information.	Dynamically chosen based on travel information.
Segmentation of drivers	(N.A.)	Classified into conservative and sensitive for travel information.	Classified into multiple segments.
Network classes	Applied to expressways.	Applied to streets and expressways.	Applied to streets.
Network size	Large (Metropolitan Expressways, 263.4 kilometres in total)	Large (100 - 1000 square kilometres in area)	Small but detailed (with 1 - 100 intersections)
Use	Included in the traffic control system of Tokyo Metropolitan Expressways - Short-term prediction of traffic conditions.	Evaluates effects of urban traffic management schemes - Provision of predicted travel times - Road pricing.	Evaluates impacts of management schemes and urban developments - Road construction, signal control, lane use - Shopping malls, amusement parks, bus terminals.

COMPARISON OF THE BLOCK DENSITY METHOD AND THE INPUT-OUTPUT METHOD

Purpose of Traffic Simulation Development

In 1971 the *Japan Society of Traffic Engineers* (JSTE) considered developing a traffic simulation model. It was the first time a technique was considered for predicting the future of unsteady traffic conditions; such as the expansion and reduction of traffic congestion (queuing vehicles), and the evaluation of the effects of new traffic control measures on the Metropolitan Expressway network in Tokyo (MEPC)(JSTE, 1971).

At that time, two kinds of calculation methods, the *Block Density Method* (BDM) and the *Input-Output Method* (IOM), were researched as potential traffic simulation models. Both of these two algorithms were coded in FORTRAN in a tentative way. The calculated speed of computers, the amount of the memory used etc. were examined in the tentatively created computer programs, using the computer FUJITSU FACOM 270-30.

The Fundamental Idea of the BDM

In the BDM a road network is expressed with links and nodes, and each link is divided into still finer blocks. Block length dL should be at least equal to or more than $V_f\,dt$, in order for vehicles not to jump over a block at the time of the calculation step (scanning interval dt), even in the case of free speed V_f. If dL is longer than $V_f\,dt$, vehicles can reach only part of the way through the block with free speed in one scanning interval, therefore the state of the block is difficult to be expressed as one value. In conclusion, one of the most important fundamental features of the BDM is to set up the dL equal to $V_f\,dt$.

In view of the fact that the length of a link takes various values, it is necessary for the application to express link length as a multiple of dL, dL must be a common divisor of the entire link length that exists in the road network. The longer dL becomes, the less the number of blocks there are and also the longer dt becomes. As a result calculation time can be saved.

The relationship between the flow rate Q_i and the traffic density K_i (Q-K relationship) is pre-determined for each block i.

$$Q_i = f_i(K_i) \tag{1}$$

The f_i function must be able to describe two states in the real traffic flow: uncongested flow condition and congested flow condition. The function form of this is arbitrary when K_i can have two values for each value of Q_i.

When the fundamental characteristics of traffic flow are taken into consideration, the Q-K relationship must include the following boundary conditions: the maximum flow rate Qc_i must exist in the flow rate Q_i; the density corresponding to Qc_i is set to Kc_i (critical density); the minimum value

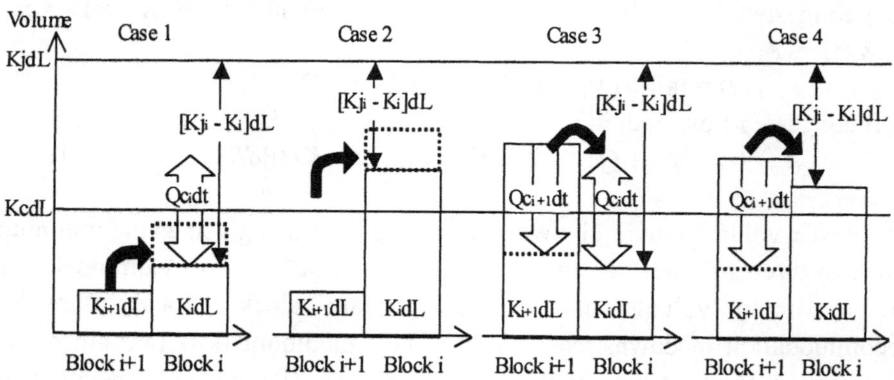

Figure 3. **Relationship between the amount of vehicles existing within each block and the amount of movements.**

of traffic flow Q_i is zero; and the traffic density corresponding to this flow condition includes two values in uncongested and congested flow conditions, the former being $K = 0$ and the latter being $K = Kj_i$ (jam density, the maximum of K).

The function shown in equation (2) (parabola curve, when $Kj_i = 2Kc_i$) was used in the 1971model.

$$f_i(K_i) = Qc_i - Qc_i\,((Kc_i - 2\,K_i) / Kc_i)^2 \tag{2}$$

The other type of function representing the Q-K relationship sets the constant speed in uncongested flow conditions as depicted in the following equation. It is drawn as two straight lines that cross at the point of critical density.

$$f_i(K_i) = V_f K_i \quad \text{(when } K_i < Kc_i\text{)},$$
$$f_i(K_i) = Qc_i\,(Kj_i - K_i) / (Kj_i - Kc_i) \quad \text{(otherwise)} \tag{3}$$

In the BDM, the flow rate corresponding to the traffic density of a block is computed using the Q-K relationship for every scan, and based on this flow rate vehicles are moved through the adjoining two. The fundamental logic is shown in equation (4) (Kuwahara et al., 1997). Case 1, 2, 3, and 4 in Figure 3 illustrate the cases of 1), 2), 3) and 4) in equation (4), respectively.

$A_{i+1,i}(t) = Q_{i+1,i}(t)\ dt$

1) if uncongested flow both in block $i+1$ and i $[K_{i+1}(t) \leq Kc_{i+1}, K_i(t) \leq Kc_i]$,

$$A_{i+1,i}(t) = \min\{ f_{i+1}(K_{i+1}(t))dt,\ Qc_i\ dt,\ [Kj_i - K_i(t)]dL \}.$$

2) if uncongested flow in block $i+1$, congested flow in block i $[K_{i+1}(t) \leq Kc_{i+1}, K_i(t) > Kc_i]$,

$$A_{i+1,i}(t) = \min\{ f_{i+1}(K_{i+1}(t))dt, f_i(K_i(t))dt,\ [Kj_i - K_i(t)]dL \}.$$

3) if congested flow in block $i+1$, uncongested flow in block i $[K_{i+1}(t) > Kc_{i+1}, K_i(t) \leq Kc_i]$,

$$A_{i+1,i}(t) = \min\{ Qc_{i+1}\ dt,\ Qc_i\ dt,\ [Kj_i - K_i(t)]dL \}.$$

4) if congested flow both in block $i+1$ and i $[K_{i+1}(t) > Kc_{i+1}, K_i(t) > Kc_i]$,

$$A_{i+1,i}(t) = \min\{ Qc_{i+1}\ dt, f_i(K_i(t))dt,\ [Kj_i - K_i(t)]dL \}. \qquad (4)$$

The traffic volume moving between adjoining blocks is given as the minimum value of three volumes; the maximum volume out of the upstream block $i+1$, the maximum volume into the downstream block i, and room for accommodation in downstream block i. The maximum flow rate out of the upstream block $i+1$ becomes the flow rate $f_{i+1}(K_{i+1}(t))$, which corresponds to the density of the block in uncongested flow condition, or becomes the maximum flow rate Qc_{i+1} of the block in congested flow condition. The maximum flow rate into downstream block i becomes the maximum flow rate Qc_i of the block in uncongested flow condition, or becomes the flow rate $f_i(K_i(t))$ corresponding to the density of the block in congested flow condition. The product of the flow rates and dt equals the traffic volume which can be moved out or in. The amount of accommodation available in downstream block i is the product of the block length dL and the difference between the jam density Kj_i and the density $K_i(t)$ at that time. In addition, when a Q-K relationship is expressed in equation (3), since $f(K(t))dt = V_f\ K\ dt = K\ dL$ becomes an identity in uncongested flow condition, the maximum volume which can be moved out of the upstream block is equal to the volume which exists in this block.

After calculating the volume that moved between adjoining blocks in equation (4), the traffic density $K_i(t+dt)$ of each block after one scanning interval dt is given as the equation (5) based on '*the formula of continuation*' (preservation law of traffic volume).

$$K_i(t+dt) = K_i(t) + (A_{i+1,i}(t) - A_{i,i-1}(t))/dL \quad (5)$$

The reason why the number order of blocks is given here is contrary to the traffic flow direction, from the downstream to the upstream, is merely for calculating convenience at merging sections. One of the worthy features of the BDM is the fact that the simulation result is identically independent from the calculation order for updating the traffic density of blocks. This fact was proved by the *Cell Transmission Model* (Daganzo, 1994), which has almost the same fundamental logic of as the BDM.

Based on the BDM, the traffic density of each block is updated one by one in every scan, and a dynamic change in time and the space of traffic flow conditions is reproduced. These characteristics of dynamic change can quite well reproduce qualitatively and quantitatively the movement of the discontinuity front of the traffic density (propagation of a shock wave). This is expected in the theoretical fluid model of traffic flow, and in the backward propagation of a vehicle queue (generation of a queue) caused by capacity restrictions.

In addition the first BDM developed in 1971 instructed that at a diverging section the traffic flow diverges into two directions based on a pre-determined diverging ratio without a route choice function. At a merging section traffic flow merges with the ratio of two inflow traffic volumes (traffic density of the blocks).

The Fundamental Idea of the IOM

The IOM expresses a road network with links and nodes, and the method of moving traffic volume through a node between the upstream link and the downstream link connected to the node is modelled as below (JSTE, 1971).

Suppose the scanning interval is *dt* and the traffic capacity (maximum flow rate) of node *k* is Qc_k, the inflow volume $I_k(t)$ into node *k* (traffic demand), outflow volume $O_k(t)$ from node *k* (actual flow), and the number of vehicles $W_k(t)$ queuing at node *k* are calculated in equation (6). This represents the basic idea of the Point-Queue Model.

$$O_k(t) = I_k(t), \qquad W_k(t) = 0 \qquad (\text{when} \quad I_k(t) \le Qc_k\, dt)$$
$$O_k(t) = Qc_k\, dt, \quad W_k(t) = I_k(t) - O_k(t) \quad (\text{when} \quad I_k(t) > Qc_k\, dt) \qquad (6)$$

Although the outflow is equal to the inflow of a node when the inflow traffic to the node does not exceed the traffic capacity resulting when a queue is not being generated, the outflow is also identical to the traffic capacity when the inflow traffic to the node exceeds the capacity and the excess volume becomes the number of vehicles in a queue.

The number of the adjoining upstream node of node k through a link is set to $k+1$. Presuming that the link travel time from a node $k+1$ to k is set to $T_{k+1,k}$ and this is given from outside of the simulation system based on free speed, the relationship between $O_k(t)$ and $I_k(t)$ are given below in equation (7).

$$I_k(t) = O_{k+1}(t - T_{k+1,\,k}) \qquad (7)$$

Using equation (6) and (7), the calculation is started from the upstream ends of a road network and repeated sequentially from upstream to downstream ('*upstream operation*'), and each $I_k(t)$ and $O_k(t)$ are set once. When however, a queue in a certain node k is accumulated and extended to the upstream node $k+1$ adjoining from the node k, the calculation method mentioned above cannot reach the right solution, therefore a re-calculation which revises $I_k(t)$ and $O_k(t)$ is performed as stated below.

The link, which connects node k and the adjoining upstream node $k+1$, is set to $i_{k+1,\,k}$, and the Q-K relationship of the link $i_{k+1,\,k}$ is pre-determined in equation (1), like in the BDM. The traffic density corresponding to the rate of the outflow of a node $k+1$ (time differentiation of $O_{k+1}(t)$) can generally take two values; $K_{k+1,\,1}$ and $K_{k+1,\,2}$ ($K_{k+1,\,1} \le K_{k+1,\,2}$). The maximum number of vehicles in a queue which can be accommodated in a link $i_{k+1,\,k}$ will be given in equation (8), when the link length is set to $L_{k+1,\,k}$.

$$N_{k+1,\,k} = (K_{k+1,\,2} - K_{k+1,\,1})\, L_{k+1,\,k} \qquad (8)$$

When the number of vehicles $W_k(t)$ given in the '*upstream operation*' is larger than $N_{k+1,\,k}$, the excess number of vehicles is set to $E_{k+1}(t)$, $O_{k+1}(t - T_{k+1,\,k})$ and $W_k(t)$ are revised as equation (9).

$$E_{k+1}(t) = W_k(t) - N_{k,\,k-1}$$
$$W_k(t) = W_k(t) - E_{k+1}(t)$$
$$O_{k+1}(t - T_{k+1,\,k}) = O_{k+1}(t - T_{k+1,\,k}) - E_{k+1}(t)$$

$$\text{(when } W_k(t) > N_{k+1,\,k}\text{)}$$
$$\text{no need for the revision of } O_{k+1}(t - T_{k+1,\,k}) \text{ and } W_k(t)$$
$$\text{(when } W_k(t) \leq N_{k+1,\,k}\text{)} \quad (9)$$

Accordingly $O_{k+1}(t)$, $I_{k+1}(t - T_{k+1,\,k})$ and $W_{k+1}(t - T_{k+1,\,k})$ are also revised in equation (10).

$$I_{k+1}(t - T_{k+1,\,k}) = O_{k+1}(t - T_{k+1,\,k}),$$
$$W_{k+1}(t - T_{k+1,\,k}) = W_{k+1}(t - T_{k+1,\,k}) + E_{k+1}(t) \quad (10)$$

The traffic flow condition is fixed by the '*downstream operation*', which is started from downstream ends of a road network. Dynamic characteristics of traffic flow conditions of a road network can be simulated by repeatedly calculating each *dt*.

In the IOM the scanning interval *dt* can be taken arbitrarily, as long as it is in the range of the time resolution needed. A comparatively long scanning interval is then sufficient enough. It is not necessary to divide a link into two or more blocks like the BDM. In addition, at diverging sections, $O_k(t)$ is calculated according to the diverging ratio pre-determined from its origin-destination volume ratio (without route choice function). At merging sections, when the sum of $I_k(t)$ of the merging links does not exceed the maximum flow of node *k*, $O_k(t)$ can be given by the sum. When the sum exceeds this maximum, $O_k(t)$ is equal to the maximum and is divided in proportion to $I_k(t)$ of each merging link.

Adoption as the traffic control system of the MEPC in Tokyo

Both the BDM and the IOM were developed for an on-line real time traffic simulation system, which reproduces the dynamic traffic conditions of a road network based on the computer technology of those days. Since the spacing between merging and diverging sections is close in the Metropolitan Expressway, the block length *dL* in the case of the BDM becomes about 100m at a maximum. Supposing also that the average travel speed of blocks being 60 km/h, the scanning interval *dt* would become 6 seconds in this case. When *dL* becomes shorter, the calculation load and memory consumption must be increased. Therefore, *dt* is set to 6-7 seconds in the case of the BDM. On the other hand, in the case of the IOM, there are such no restrictions in *dt*, *dt* is set to about 2 minutes, which is the time resolution needed for predicting dynamic traffic conditions.

The technique based on the IOM was adopted as the traffic simulation model to be used for the traffic control system of the MEPC, because of the importance of the on-line real-time nature. In addition the computer system used for the application of the traffic control system of those days was TOSBAC 7000-25: it had a main memory of 130KB, a sub memory of 1,536KB, and a clock frequency of 1MHz (MEPC, 1977).

THE ERA OF THE INPUT-OUTPUT METHOD

Although the IOM won the battle with the BDM and acquired the status as the on-line model for the traffic control system of the MEPC, there were many problems that needed improvement in order for it to be applied to actually predict traffic conditions. Therefore, various improvements have been continuously performed by the MEPC.

In an earlier improvement the '*Flow rate vector*' was introduced to consider the change in traffic conditions in a shorter time than the scanning interval *dt* (JSCE, 1984). A subsequent improvement was based mainly on the simulation logic itself and the preparation of input data and parameters settings (JSCE, 1996). The logic of the IOM contained a flaw in a backup of the inflow traffic to the main line and on-ramp where the traffic flows excessively into a diverging section. As for the parameter settings, topics discussed included how to set the Q-K pattern and traffic capacity. It is preferable to update parameters automatically with an on-line model using the data observed and accumulated every day. Therefore, the automatic updating techniques (such as the 'Kalman filtering method' and the 'exponential smoothing method') which can include the bias rectifying method with traffic sensor data and give consideration to heavy vehicles, were developed and introduced (Akahane et al., 1987).

Improvement of the IOM

Introduction of the 'Flow Rate Vector'

The basic logic in the calculation of traffic flow in the IOM consists of mainly two operations: the '*upstream operation*' and the '*downstream operation*'. With the IOM, basically traffic demand is set and excess over a traffic capacity is judged at every *dt* scanning interval. Therefore, if *dt* is set to a large value, to some extent, excessive traffic may flow according to the change in traffic demand because a constant traffic volume is assumed regardless of the demand change.

To prevent this problem of excess traffic flow, the '*Flow Rate Vector*' method was developed and examined for the '*upstream operation*'. A scanning interval *dt* is divided into *n* sections and the change in the input and output traffic volume within *dt/n* is expressed by a vector formation. The 'Inflow rate vector' and 'Outflow rate vector' are defined as follows:

$FI_{k,\, t(i)}$ (Inflow rate vector): Traffic volume at time $t(i)$ which is going to pass through point *k*,

$FO_{k,\, t(i)}$ (Outflow rate vector): Traffic volume at time $t(i)$ which pass through point *k* after the check of traffic capacity,

i : suffix denoting the order of the flow rate vector, $i = 1, \ldots, n$.

At first, speed $V_{k-1,k}(t)$ and travel time $T_{k-1,k}(t)$ are calculated in the link $i_{k,k-1}$ at the time t using the outflow rate $\Delta O_k(t-dt)$ at the node k and density $K_{k-1,k}(t-dt)$ in the link $i_{k,k-1}$, in the previous scan t-dt, and length of section $L_{k-1,k}$ as shown below:

$$V_{k-1,k}(t) = \Delta O_k(t-dt) / K_{k-1,k}(t-dt) \tag{11}$$

$$T_{k-1,k}(t) = L_{k-1,k} / V_{k-1,k}(t) \tag{12}$$

Here, travel time $T_{k-1,k}(t)$ in the link $i_{k,k-1}$, is the time taken by the vehicle at the node k-1 at the time t-dt to arrive at node k, and all the vehicles in the link $i_{k,k-1}$ at the time t-dt can reach the node k.

Travel time $T_{k-1,k}(t)$ can be converted into J [unit time], expressed by a unit of *dt/n*.

$$J = \frac{T_{k-1,k}(t)}{dt/n}. \tag{13}$$

Each element *1, 2, …, J* of the inflow rate vector $FI_{k,t(i)}$ contains a number of existing vehicles. If the density of the small section is assumed to be uniform, the elements of the inflow rate vector are expressed in equation (14):

$$FI_{k,t(i)} = \frac{E_{k-1,k}(t-dt)}{J}, \qquad (i=1,\ldots,J) \qquad (14)$$

where $E_{k-1,k}(t-dt)$ denotes the number of vehicles in the link $i_{k,\,k-1}$ at the time *t-dt*. Each element *J*+1, *J*+2, …, *n* of the inflow rate vector are equal to each element 1, 2, …, *n-J* of the outflow rate vector in an upstream node.

$$FI_{k,t(i)} = FO_{k-1,t(i-J)}, \qquad (i=J+1,\ldots,n) \qquad (15)$$

It is then checked as to whether the traffic capacity C_k in the time unit *dt/n* has been exceeded or not for each element of the inflow rate vector at node *k*. The outflow rate vector is calculated in equation (16).

$$\begin{aligned} &FO_{k,t(i)} = FI_{k,t(i)}, && \text{for } FI_{k,t(i)} > C_k/n, \\ &FO_{k,t(i)} = C_k/n, \quad FI_{k,t(i+1)} = FI_{k,t(i+1)} + \left(FI_{k,t(i)} - C_k/n\right), && \\ & && \text{for } FI_{k,t(i)} \le C_k/n \end{aligned} \qquad (16)$$

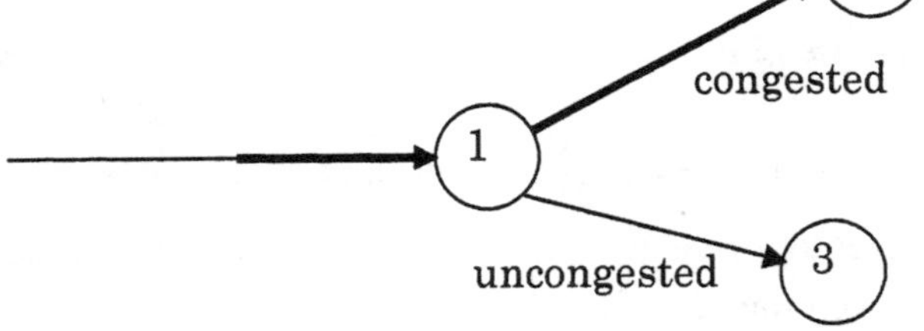

Figure 4. Diverging section (with excess flow).

The input and output traffic volume of node *k* at a scanning interval *dt* are calculated as the summation of all elements of the flow rate vector. Using the 'flow rate vector' method, the change in traffic demand is taken into consideration, the reproducibility of traffic volume was verified. There were however, no obvious improvements with this method. Therefore, this method, which required a complicated calculation process, was not adopted as the on-line traffic control system.

Prevention of Excess Flow at Diverging Sections

In the IOM, a pre-determined diverging ratio is adopted at diverging sections without dealing with any directional flow (origin-destination volume). In the '*upstream operation*' outflow traffic from the diverging node is assigned by the diverging ratio to the downstream link and in the '*downstream operation*'. The excess flow is then pushed back upstream. The traffic volume pushed back upstream is processed in the following scanning interval. At that time the traffic volume is assigned by a pre-determined diverging ratio regardless of the real direction of the traffic volume. This results because the direction from which the traffic is pushed back is not memorized. Therefore, in the case that one of the diverging links is congested and the other is not congested, the traffic volume, which originally should be assigned only to the congested direction, is assigned to both the congested and the uncongested directions. This results in the excess flow going to the uncongested link.

Following is an explanation regarding the diverging section as drawn in figure 4. Here the directional diverging ratio is denoted as B_{12} and B_{13}, respectively ($B_{12}+B_{13}=1$). Link 12 is congested and link 13 is uncongested. It is then assumed that the backup volume $B_2(t)$ will be pushed back from only the congested link 12 at the time t.

Outflow $O_1(t+dt)$ at node 1 at the following scanning interval $t+dt$ is calculated in the '*upstream operation*'. This outflow $O_1(t+dt)$ is then divided into the backup flow of the previous scanning interval and outflow $O'_1(t+dt)$, which passes originally through the scanning interval $t+dt$, which is shown in equation (17).

$$O_1(t+dt) = O'_1(t+dt) + B_2(t) \tag{17}$$

The directional outflow can be distributed by the diverging ratio, as illustrated in equation (18).

$$\begin{aligned} O_{12}(t+dt) &= O_1(t+dt)\times B_{12} = \{O'_1(t+dt)+B_2(t)\}\ B_{12} \\ O_{13}(t+dt) &= O_1(t+dt)\times B_{13} = \{O'_1(t+dt)+B_2(t)\}\ B_{13} \end{aligned} \tag{18}$$

Suppose also that the outflow from congested link 12 is restricted to the fixed value O'_{12}, which is determined by the capacity of node 2 and the maximum number of vehicles able to exist in link 12, in the '*downstream operation*', traffic volume $B_2(t+dt)$ must be pushed back upstream as in equation (19).

$$\begin{aligned} B_2(t+dt) &= O_{12}(t+dt) - O'_{12} \\ &= \{O'_1(t+dt) + B_2(t)\}B_{12} - O'_{12} \end{aligned} \tag{19}$$

The outflow of node 1 can be given in equation (20).

$$\begin{aligned} O_1(t+dt) &= O_1(t+dt) - B_2(t+dt) \\ &= O_{13}(t+dt) + O'_{12} = \{O'_1(t+dt) + B_2(t)\}\, B_{13} + O'_{12} \end{aligned} \tag{20}$$

However, since backup volume $B_2(t)$ should be assigned only to link 12, the revised outflow for each direction $O^*_{12}\ (t+dt)$ and $O^*_{13}\ (t+dt)$, and backup volume $B^*_2(t+dt)$, should be given in equation (21).

$$\begin{aligned} O^*_{12}(t+dt) &= O'_1(t+dt)\, B_{12} + B_2(t) \\ O^*_{13}(t+dt) &= O'_1(t+dt) \quad B_{13} \\ B^*_2(t+dt) &= O^*_{12}(t+dt) - O'_{12} = O'_1(t+dt)\, B_{12} + B_2(t) - O'_{12} \end{aligned} \tag{21}$$

Therefore, equation (22) denotes the outflow $O^*_1(t+dt)$ of node 1 after revision.

$$\begin{aligned} O^*_1(t+dt) &= O^*_1(t+dt) - B^*_2(t+dt) \\ &= O^*_{13}(t+dt) + O'_{12} = O'_1(t+dt)\square_{13} + O'_{12} \end{aligned} \tag{22}$$

Since the outflow of diverging node 1, shown in Figure 4, is increased by $O^*_1(t+dt) - O_1(t+dt) = B_2(t)\, B_3$, there should be less congestion than conditions without the holding diverging direction.

The logic in calculating traffic volume is improved as the traffic volume, pushed back to the upstream, holds the original direction. It allows the volume to be put into the direction of the following scan. Consequently the reproducibility of the amount of traffic congestion was improved drastically (JSTE, 1996).

Limitations of the IOM

As the network of the Metropolitan Expressway extended, alternative route choice options increased, as a result a route choice model was especially needed to consider the effect of traffic information or to detour trips in the case of an incident. To introduce the route choice model into the IOM, the destination of each vehicle should be identified with a label. However, it is difficult to identify the destination since the '*upstream* and *downstream operations*' are repeated alternately in the IOM. If the identification of destinations were introduced, it would conceivably be complicated and difficult to handle. As a result route choice behaviour cannot be included in the IOM.

The calculation order of the nodes must be clearly prescribed and be strictly adhered to in the IOM. This is because the calculation result depends on this order. In the case of a ring road, the cut point, which becomes the upstream and downstream end, must be set. To avoid prediction errors, the '*cut-points*' should be set in an uncongested section. However, sometimes the whole section of the inner ring road of Metropolitan Expressway in Tokyo may get congested. In this case prediction errors may become larger.

THE REVIVAL OF THE BLOCK DENSITY METHOD

A Simulation Model for the Impact Analysis of Traffic Signal Controls

Although the IOM was put into practical use in the network of the Metropolitan Expressway in Tokyo, a traffic simulation model called DESC (*Dynamic Evaluator of Signal Control*) was developed in the middle of 1980s to evaluate and research the signal control system at the *Japan Traffic Management Technology Association* (TMT) (Ozaki, 1989). The DESC is a systemized model of the earlier models developed by TMT (1986) and Mukai et al. (1986). The earlier models were based on the BDM to evaluate the impact of *area traffic control* and traffic signal coordination.

The fundamental logic of the BDM at first (JSTE, 1971) contained some problems with reproducing shock wave propagation of the discontinuity front with traffic density when the capacity fall caused by an incident in the downstream dissolved (the case of 3) as in equation (4). The discontinuity front will distribute spatially as a shock wave propagates upstream, and density changes gradually over many blocks. The reason why a shock wave cannot arise in one block is that the traffic density in the block must be taken as one value. This problem can be reduced by making both the scanning interval dt and the block length dL shorter. The bi-linear relationship shown in equation (3) is used for the Q-K relationship in the DESC. In order to solve the problem mentioned above of reproducing shock wave propagation, scanning interval dt must be shortened to one second. Also if the free speed of urban streets is considered to be about 30 to 50 km/h, the block length dL should be set to 10 to 15 m.

In the DESC, the amount of memory consumed by a computer increases because of the shortened scanning interval dt. As a result the calculation time becomes lengthy. To save on the computation load, the logic of the BDM defined in equation (4) is simplified in the DESC. When N_i is the maximum number of vehicles accommodated in block i, Qc_i is the maximum flow rate into block i, and $n_i(t)$ is the existing number of vehicles in block i at time t; the traffic volume $A_{i+1,\,i}(t)$, which moves in the time dt (from t to $t+dt$) from the upstream block $i+1$ to the adjoining downstream block i, is given in equation (23).

$$\begin{aligned} A_{i+1,\,i}(t) &= \min\{\, n_{i+1}(t),\, Qc_{i+1}\,dt,\, Qc_i\,dt,\, [N_i - n_i(t)] \,\} \\ &= \min\{\, K_{i+1}(t)\,dL,\, Qc_{i+1}\,dt,\, Qc_i\,dt,\, [Kj_i - K_i(t)]dL \,\} \end{aligned} \tag{23}$$

The traffic volume which can flow out of block $i+1$ cannot exceed the maximum volume $Qc_{i+1}\,dt$ nor the existing number of vehicles $n_{i+1}(t)$ (= $K_{i+1}(t)dL$) in the block. The traffic volume which can flow into the adjoining downstream block i cannot exceed the maximum volume Qc_i nor acceptable number of vehicles which is given by subtracting the maximum number of vehicles Qc_i (= $Kj_i\,dL$) accommodated in the block by the existing number of vehicles $n_i(t)$ (= $K_i(t)dL$) at time t. Since the logic adopted by the DESC is simplified in the Q-K relationship of traffic congestion, it can be called the *Simplified Block Density Method* (S-BDM)).

In the S-BDM, a congested traffic condition is always treated as the condition of jam density (Kj). Therefore, when a definite lower traffic density than jam density in the congested flow condition (queue) forms upstream of a bottleneck, the reproducibility of the expansion and the reduction of a waiting queue is not expressed correctly. When however aimed at a surface street network controlled by traffic signals, flow conditions in queue caused by traffic signals can be estimated as jam density.

Although the directional traffic volume at an intersection in the DESC is given by a pre-determined diverging rate, the processing of a right-turning vehicle sets up a gap acceptance function that: treats each vehicle dispersedly, detects the gap of an opposing through traffic and reproduces the probability of right-turning behaviour. It tries to reproduce the traffic capacity fall phenomenon of the intersection by the existence of right-turning vehicles by blocking a through lane without an exclusive right-turning lane, or influencing the overflow of queuing vehicles on exclusive right-turning lane composed of particular blocks. Two kinds of functions are prepared for the gap acceptance function depending on the sizes of the gap accepted: in the case that one vehicle turns to the right independently; and in the other case that following vehicles can also turn to the right with the preceding car in the same gap. It also has the function of simulating the vehicle detectors with traffic counts and time occupancy, and the function of reproducing platoon dispersion in order to express the randomness of each vehicle.

Although the *Cell Transmission Model*, proposed by Daganzo (1994), has a strong fundamental resemblance to the S-BDM in the DESC it does not however, consider the restricting condition of the maximum volume Qc_i into the downstream block i as shown in equation (23). Daganzo (1994) discovered that the assumption of the third term of equation (23) being set to N_i - $n_i(t)$ (= $[Kj_i - K_i(t)]dL$) is equivalent to the assumption of the back propagation speed of the queue (shock wave speed) being set to be V_f. This is not realistic; therefore he proposed a modified model in which the back propagation speed of the queue is set to w ($< V_f$) in general form. Equation (24) incorporates the modification of calculating the moving volume between adjoining blocks.

$$A_{i+1,i}(t) = \min\{ n_{i+1}(t), Qc_{i+1}\, dt, \alpha^*[N_i - n_i(t)] \}$$

where $\alpha = 1$, if $n_{i+1}(t) \leq Qc_{i+1}(t)\, dt$

$\alpha = w / V_f$, if $n_{i+1}(t) > Qc_{i+1}(t)\, dt$ (24)

Inclusion of a route choice function into the BDM

In the 1990s, twenty years had passed since the development of the first BDM, and the road network of the Metropolitan Expressway had expanded in leaps and bounds. Consequently, the possible options of route choices to drivers increased. Drivers thus became able to choose a route according to traffic information, regarding traffic congestion or travel time, which was shown on a VMS (Variable Message Sign). Traffic management strategies and detour guidance in the case of an incident also became more necessary.

Although the several traffic simulation models which could deal with route choice behaviour were already developed such as: SATURN (Van Vliet and Hall, 1991), COMTRAM (Leonard et al., 1989), DYTAN (Kido et al., 1978), and the Box-Model (Iida et al., 1991, 1996), these models were aimed at being applied to surface streets and were insufficient to reproduce the propagation of traffic density occurred in the Metropolitan Expressway in Tokyo. The propagation of traffic density causes the phenomenon where the extension of traffic congestion originating from a downstream bottleneck blockades upstream diverging and merging points. The phenomenon causes traffic capacity declines to occur frequently, because there are so many diverging and merging sections and junctions in the Metropolitan Expressway. In order to reproduce such a phenomenon, it is necessary to simulate the extension of traffic congestion when considering the propagation of traffic density more precisely. It is necessary that the computational load and memory capacity don't become extremely large when applying the simulation model to large expressway networks such as the Metropolitan Expressway in Tokyo. Thus, the requirements for a newly developed model are: 1) route choice model inclusion, 2) high reproducibility of the propagation of traffic density, and 3) less computational load and memory capacity.

In the early 1990's, a traffic simulation model based on the BDM including route choice function was developed by Kuwahara et al. (1993) for a road network of the Metropolitan Expressway in Tokyo. It was developed within the guidelines expressed above. Although the fundamental logic of this model is the same as the original BDM, destination information is given to each individual vehicle to incorporate a route choice function. As a result, the traffic density in a block is classified into destinations.

$$K_i(t) = \sum_s K_{is}(t) \tag{25}$$

Where, s is the label of a destination, $K_{is}(t)$ is the density at time t in block i where the traffic volume has the destination label of s. Thus the traffic volume in a block becomes possible to be classified into each route by giving the label s to the destination.

The route of each OD pair is re-determined as the shortest time path according to the travel time of the links derived by the traffic simulation model for every fixed time *DT* (*DT* is longer than the scanning interval *dt*, a multiple of *dt*). The direction to destinations according to the shortest time path calculated for every *DT* is recorded in just the upstream block of a diverging point (diverging block). Therefore, the traffic existing on the road network at the time of the every iteration of finding the shortest time path, will progress according to the reformed new route of the next scanning interval. In a diverging block, traffic volume is moved to the downstream block in the direction based on the chosen route. Therefore in this model, the diverging ratio at a diverging point changes according to the result of searching for the shortest time path.

If you intend to implement a route choice function to the IOM, management of a destination label meets difficulties since calculations must be done not only from the upstream side but also done from the downstream side in turns. Therefore, the BDM was adopted as the traffic simulation model including the route choice function. There are difficulties however, in increasing the amount of memory. This is necessary in the management of the traffic density in each block according to the destination label *s*. The calculation time is also extended to calculate the density of blocks separately according to the destinations in each scan.

A general structure of a simulation model developed for such purposes including a route choice model and a vehicle movement model is shown in Figure 5. Each vehicle moves on a network within the vehicle movement model, and the direction of it at a diverging section is determined in the route choice model based on traffic condition such as travel time. The authors only focused on the vehicle movement model and outlined the development circumstances and model structure.

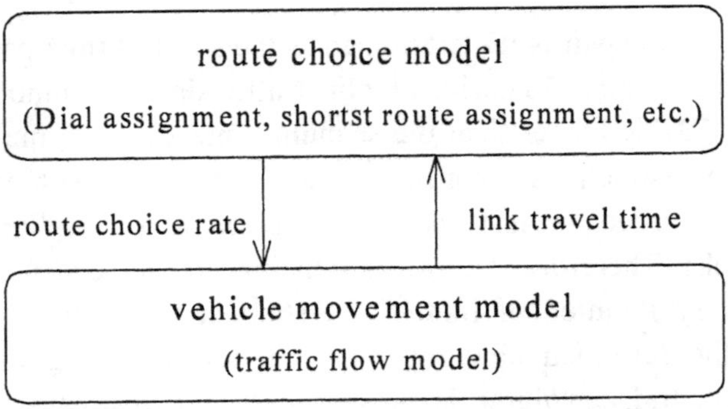

Figure 5. Structure of a simulation model

Development of the AVENUE

The AVENUE (an Advanced & Visual Evaluator for road Networks in Urban arEas) was developed by Horiguchi et al. (1993, 1996) in order to evaluate the various measures of traffic operations. It was developed not only for evaluating the improvements of intersections or signal control measures such as those taken into consideration with the DESC, but also for considering traffic restrictions, route guidance, and planning for parking areas on an area-wide level.

In the AVENUE, a traffic flow is expressed with logic based on equation (4), and without the simplification of the BDM. Therefore, the AVENUE can handle several types of bottlenecks at ordinary sections due to road works and street parking. In the AVENUE, vehicles are basically moved on the logic not of a car-following behaviour, but of the BDM. The main reason for adopting the BDM was due to the following four points. Firstly, observations and calibrations of parameters concerning car-following behaviour seemed to have difficulties and, in this case, the traffic capacity cannot be handled explicitly. Secondly, although there are a lot of vehicles not following the leader in low traffic volume conditions, such a phenomenon cannot be treated by any existing car-following models. Thirdly, in order to strictly treat the delay, which is one of the most important indices for evaluating the performance of a traffic simulation, it is important to reproduce the traffic capacity precisely. Lastly, it is comparatively easier to observe the bottleneck capacity of traffic flows than to observe the parameters of a car-following model.

The scanning interval of the AVENUE is set to 1 second like the DESC and the Q-K relationship is described as a bi-linear function (equation (3)). The traffic volume moving between adjoining blocks are described as movements of vehicles, which have attributes such as destinations, types of vehicles, etc., for modelling route choice and lane change behaviour. The AVENUE is a kind of unified and generalized model of the DESC. It includes a route choice function, and is able to be applied to the network of the Metropolitan Expressway. The logic of the AVENUE (called as Hybrid Block Density Method (H-BDM)) is comprehensive, as shown below.

In the H-BDM, since a link is divided into blocks for each lane, the blocks are connected to each other in the shape of a cascade. The flow calculation is based on the H-BDM and is applied not only to the going-straight direction of the same lane but also to the lane changing direction. Regulating traffic's moving directions, vehicle types, etc. at an intersection can be done in each block at the downstream end of an intersection approach, and all the downstream exiting blocks of an intersection will then only accept what fills this regulation. With the combination of several regulations of these blocks, various lane operations, such as an exclusive right and left turning lane and a bus lane etc., can be reproduced. The logic of H-BDM has already been verified based on a verification process (Horigucih et al., 2001), the result is exhibited at: http://www.i-transportlab.jp/products/avenue/verification/.

Moreover, since it was thought not to be necessary to strictly reproduce a propagation of shockwave far upstream of an intersection, a method for allowing longer block length (not a fixed length of a block) was developed in order to save on computational load by Horiguchi et al. (1996). It was named the Multi-Scan Hybrid Block Density Method (MSH-BDM).

In the AVENUE, the Object-Oriented Modelling method is adopted. It's therefore easier to add other functions, such as a signal control function and a vehicle detector function, and so on. It is also possible to apply some car-following models of each vehicle discretely, if needed, in particular at some roadway sections (for example, weaving sections etc.) in which the traffic phenomena cannot fully be reproduced with the logic of the BDM solely. It is also possible to apply two or more models of route choice functions. For example, it is possible to reproduce the traffic flow including buses with fixed routes and the other general vehicles choosing a route probably in accordance with the Dial Algorithm.

The AVENUE is designed as a package software, which is preferred by general users. The user interface, which was not properly designed to the traffic simulation model previously, is well-designed in the AVENUE. It also has the advantage that the input operation of network creations, of the OD traffic, etc. can be done simply with a mouse on a computer screen (or given by text files) and the results of simulation calculations can be shown graphically.

ALTERNATIVES TO THE BLOCK DENSITY METHOD

Limitations of Improvements to the BDM

Kuwahara et al. (1993) developed a traffic simulation model including a route choice behaviour based on the BDM. However, this method had the problem that the number of blocks and destination labels increased enormously as a network became larger. For example, suppose the block length becomes 60/3.6[m/s] × 3[sec] =50m when the scanning interval is 3 seconds and the free flow velocity is 60 km/h. Then the number of blocks in the link of 1km is 1000 / 50= 20, at the same time the number of blocks classified by the destination label becomes 20×100=2,000 when the number of the destination is 100. On the other hand, the possible number of vehicles existing in the link of 1km is at most 160, if the jam density is 160 vehicles/km/lane. Obviously the number of memories for a route choice actually used is at most for 160 vehicles per lane, other memories (more than 1,800) are not used for any vehicles. Therefore the use of the BDM for the development of a large-scale network simulation model including a route choice model had come to an end. As a result a new vehicle movement model needed to be introduced.

The Development of Vehicle Movement Logic Based on the Q-K Relationship

In order to solve the problem mentioned above, a new large scale network simulation model with a vehicle movement model based on the Q-K relationship was developed by Yoshii et al. (1995), named the 'SOUND'

(Simulation On Urban road Network with Dynamic route guidance)'. In this model the relationship between traffic flow Q and traffic density K is set up for every link, and each vehicle or packet, which aggregates several number of vehicles into one, is moved using this Q-K relationship.

Figure 6 shows the vehicle movement logic of the SOUND. Two vehicles A and B exist at time t on a link, and only the movement of the former vehicle A has been completed at the following time $t+dt$. L denotes a moving distance of vehicle B from t to $t+dt$, and S denotes spacing between vehicle A and B at the time $t+dt$ after the movement of vehicle B is completed. The moving distance L of vehicle B can be determined in order to match the pre-determined Q-K relationship. When the Q-K relationship is denoted as $Q = f(K)$, the relationship between velocity V and spacing S (S-V relationship) can be obtained as in equation (26) since $S=1/K$, and $Q=KV$.

$$V = S \times f\left(\frac{1}{S}\right) \tag{26}$$

Here, given that $L=Vdt$, the moving distance L of the vehicles in the scanning interval dt is described in equation (27).

$$f\left(\frac{1}{S}\right) \times S = \frac{L}{dt} \tag{27}$$

In equation (27), both L and S can be determined since scanning interval dt and Q-K function f are pre-determined and $S+L$ can be obtained by the movement of the former vehicle. This calculation process of vehicle movements is performed from downstream.

At a merging section, vehicles entering from merging links move to a downstream link independent from the traffic volume of the other link and each other, when there is no congestion in both of the merging links. On the other hand, where one or both of two merging links are congested, the merging ratio of the congested link is equal to or exceeds the pre-determined merging ratio. At a diverging section, vehicles diverge in accordance with the diverging ratio obtained by the route choice model as shown in figure 5.

To save the computational load and memory consumption, instead of calculating the total of the vehicles movements, a group of vehicles, which has the same destination and same route choice characteristics, is substituted as a '*packet*'. All the vehicles in a packet move together in the same way just like the movement of a packet.

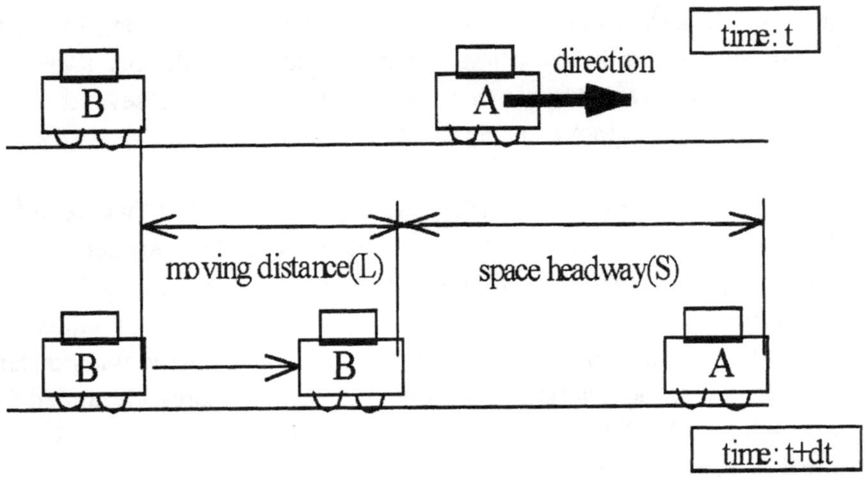

Figure 6. Method of vehicle movement in the SOUND.

Since the attributes, such as a destination and route choice characteristics, can be given to each vehicle, the model can be also applied to analyse the effect of traffic information on route choice behaviour. As for the route choice characteristics, the model can contain several route choice layers. For example the '*route choice layer*' which always chooses a route based on the newest travel time, and the '*fixed route layer*' which always runs the same fixed route.

THE UNION OF THE BDM AND THE IOM

Needs for the 'SOUND Arterial'

In recent years the need for simulation models that can reproduce dynamic traffic changes as well as be applied to the large areas arose. Also, the model needed to evaluate the traffic management measures applied to a wide area, such as evaluating road-pricing measures in the Tokyo Metropolitan Area. In addition it became very important that the surface streets network was also included to evaluate traffic controls on urban expressways such as a flexible toll system like a multiple-entry-free toll. Evaluating travel time predictions in the case of an incident, which causes a route choice among expressways and surface streets, had also become imperative. These social and technical requirements strongly motivated the development of the surface street version of the SOUND, named the 'SOUND Arterial'. This was achieved by improving the vehicle movement logic of the original SOUND (Okamura et al., 1996).

Although it is possible to apply vehicle movement logic using the Q-K relationship in the street version of the SOUND, there was however, a problem with the computational load and memory consumption in the case of an application to a large area network. Therefore in the SOUND Arterial, the *'Point-Queue' Model*, which was simpler than the Q-K model and was already used in the IOM, was adopted again to avoid too high a computational load. As was mentioned before, it was necessary for the models to express the propagation of traffic density more accurately, such as in the Q-K model and the BDM so as to reproduce traffic congestion in the urban expressways. However, it was not necessary to reproduce propagation of density so accurately on surface streets since main bottlenecks are signalised intersections. It can also be approximately assumed that vehicles confronting a red signal are queuing in jam density and that reproducibility of density propagation is not so important. This is the reason that the Point-Queue model was adopted in the SOUND Arterial. It might be a reasonable and appropriate choice to separate the vehicle movement logic by road type, the Q-K model for expressways and the Point Queue model for surface streets in order to satisfy both needs for model accuracy and a reduction in the computational load at the same time.

The Vehicle Movement Logic of the 'SOUND Arterial' Based on the Point Queue Model

Vehicles moving in a packet are grouped according to similar attributes, destination, etc. Also each link is divided into two *vehicle lists*, a list of moving vehicles and a list of dischargeable vehicles, as shown in Figure 7. Packets in the list of moving vehicles obey the FIFO (First In First Out) principle – that is packets exit the list in the same order of entrance. In addition, each packet is switched over to the list of dischargeable vehicles after the free travel time T_F passes. The list of dischargeable vehicles shows queues in front of intersections and the packet in the list moves to the next link in accordance with the approach capacity of the intersection. Also, right and left turning vehicles can move to the next link depending on the extent of the number of vehicles existing in the exclusive right and left turn lane in spite of the FIFO principle even in the case that vehicles moving through are blocked at the intersection.

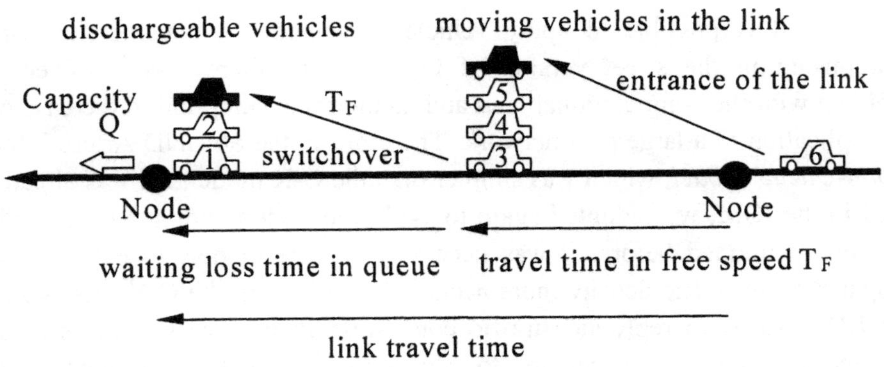

Figure 7. Vehicle moving logic of the'SOUND Arterial'.

In an earlier version of the SOUND Arterial (Okamura et al., 1996), reproducing the traffic congestion phenomena - caused by excess demand to a bottleneck capacity - was emphasized as the objective. The directional capacity of signalised intersections was therefore given as the saturation flow rate multiplied by a directional split. The traffic signal model was however, not implemented. The newest version of the SOUND was developed in 2001 as a integrated package model. It was composed of the model for expressways based on the Q-K model (SOUND/Express) and the surface street model based on the Point Queue model (SOUND/A-21, see Table 2). In SOUND/A-21, the traffic signal lightning was modeled and it is able to evaluate the travel times even taking into consideration unsaturated delays caused by traffic signal controls. The model has already been applied to several case studies on urban-scale large networks. For example, it was applied to evaluate the effect of the temporal demand shift in the Tokyo Metropolitan Area (Oneyama et al., 2001).

CONCLUDING REMARKS

The SOUND for streets is being improved by changing the algorithm for representing traffic congestion from the *Point-Queue* to the *Physical-Queue* in order to simulate more widely and precisely the influences of traffic congestion on road networks. However, more efforts are recently being made to verify and validate traffic simulation models such as the activities of Hanabusa et al. (2001). The objective of such activities was to set and adjust the parameters of models, to generate the input data, to evaluate and present the output of models, and to add components for estimating volumes of exhaust emissions; more so than to develop the models themselves. This means that the field of traffic simulation has evolved from a '*researching stage*' to a '*practical stage*'.

In Japan, the revised '*Road Traffic Law*', under which private sectors as well as public ones may collect, edit, process, and provide road traffic information, has been enforced since June, 2002. This has presented a situation where predicted information about traffic conditions has become widely utilized by private enterprises and end-users for scheduling trips and logistics. In such a situation, the prediction of traffic conditions by means of traffic simulation is greatly expected. To achieve the fruitful '*practical stage*', activities of academic societies in Japan (formerly in the *Japan Society of Civil Engineers* (JSCE), now in the JSTE) are performed in a *Clearing House* (http://www.jste.or.jp/sim/). A draft version of a '*standard verification process for traffic simulation model* ' (Horiguchi et al., 2001) and standard 'data-sets' for validation of reproducibility of real-world traffic phenomena have now been disclosed.

There still exists however, some research topics such as the estimation of OD traffic volumes, modelling and setting the values of parameters of route choice behaviours, and the construction of systems which can collect information about influential factors on traffic conditions. Two-way communication especially, brought by ITS technologies and/or general-purpose technologies (like the *Internet*), needs to be fully utilized for collecting and transmitting the necessary information.

This paper detailed a history of developments in traffic simulation models in Japan in relation with changes in needs of traffic simulation models and the performance of computers. The authors expect that this article may contribute to prospects for the future development of traffic simulation models.

Finally, the authors would like to acknowledge Prof. Masahiko KATAKURA

(Tokyo Metropolitan University), Prof. Masao KUWAHARA (University of Tokyo) and Dr. Ryota HORIGUCHI (i-Transport Lab) for their useful advice, and to every other related personnel or organizations for their precious information.

REFERENCES

Akahane, H. and Koshi, M. (1987). Updating of volume-density relationships for an urban expressway control system, *Proceedings of 10th ISTTT*, Boston.

Daganzo C. F. (1994). The cell transmission model: a dynamic representation of highway traffic consistent with the hydrodynamic theory. *Transportation Research B*, **28B**-4, pp269-287.

Hanabusa, H., Yoshii, T., Horiguchi, R., Akahane, H., Katakura, M., Kuwahara, M., Ozaki, H., Oguchi, T. and Nishikawa, I. (2001). Construction of a data set for validation of traffic simulations, *Journal of Infrastructure Planning and Management* (JSCE), **No.688/IV-53**, pp115-123 (in Japanese).

Horiguchi, R., Katakura, M. and Kuwahara, M. (1993). Development of a traffic simulator of urban street network - AVENUE -, Proceedings of *13th Symposium on Traffic Engineering*, pp33-36 (in Japanese).

Horiguchi, R. Kuwahara, M., Akahane, H. and Ozaki, H. (1996). A Network simulation model for impact studies of traffic management 'AVENUE Ver.2', *Proceedings of 3rd World Congress on Intelligent Transport Systems*, CD-ROM, ITS America, Orlando.

Horiguchi, R., Kuwahara, M. and Yoshii, T. (2001). A Manual of verification process for road network simulation models - an examination in Japan, *Proceedings of 7th World Congress on Intelligent Transportation Systems*, CD-ROM, ERTICO, Turin.

Iida, Y., Uchida, T., Fujii, S. and Takao, K. (1991). A dynamic traffic simulation model considering a phenomenon of traffic jam lengthening, *Proceedings of Infrastructure Planning* (JSCE), **No.14(1)**, pp301-308, (in Japanese).

Iida, Y., Fujii, S. and Uchida, T. (1996). A dynamic traffic simulation model considering a route choice behavior on road network,. *Journal of Infrastructure Planning and Management* (JSCE), **No.536/4-31**, pp37-47 (in Japanese).

Japan Society of Traffic Engineers (JSTE) (1971). Research on the traffic situation prediction technique in traffic control. Research Report (in Japanese).

Japan Society of Traffic Engineers (JSTE) (1984). Research on future traffic management System in the Metropolitan Expressway, No.3. Research Report (in Japanese).

Japan Society of Traffic Engineers (JSTE) (1996). Research on future traffic management System. Research Report (in Japanese).

Japan Traffic Management Technology Association (TMT) (1986). Research on the investment effects of traffic safety devices (in Japanese).

Kido, T., Ikenoue, K. and Saito, T. (1978). Model of O-D routes and traffic Assignment in street network (DYTAM-I), Reports of the National Research Institute of Police Science, Research on Traffic Safety and Regulation, **Vol.19, No.1**, pp1-10 (in Japanese).

Kuwahara, M., Ueda, I., Akahane, H. and Morita, H. (1993). A study on a traffic simulation model incorporating vehicle route choice for urban expressways, *Traffic Engineering*, **28**-4, pp11-20 (in Japanese).

Kuwahara M., Yoshii T. and Horiguchi, R. (1997). Reproduction of traffic flow with block density method, *Traffic Engineering*, **32**-4, pp39-44 (in Japanese).

Leonard, D. R., Gower, P. and Taylor, N. B. (1989). CONTRAM: structure of the model, TRRL Research Report RR178.

Metropolitan Expressway Public Corporation (MEPC). (1977). Outline of traffic control, Traffic control of Metropolitan Expressway, pp.14-28.

Mukai, S., Katakura, M. and Sakurada, Y. (1986). Evaluation of signal control by simulation experiments, *Proceedings of Infrastructure Planning* (JSCE), **No.9**, pp495-502 (in Japanese).

Okamura, H., Kuwahara, M., Yoshii, T. and Nishikawa, I. (1996). Development and validation of surface street network simulation model, *Proceedings of 16th Symposium on Traffic Engneering*, pp.93-96 (in Japanese).

Oneyama, H., Iryo, M. and Kuwahara, M. (2001). Simulation analysis of traffic congestion alleviation and environment improvement by demand spreading over time in Tokyo Metropolitan Area, *Proceedings of Infrastructure Planning* (JSCE), **No.24**, CD-ROM (in Japanese).

Ozaki H. (1989). A simulation model for evaluating area signal control (DESC), *Traffic Engineering*, **24-6**, pp31-37 (in Japanese).

Van Vliet, D. and Hall, M. D. (1991). SATURN 8,3-a user's manual-universal version, Institute of Transport Studies, University of Leeds.

Yoshii, T., Kuwahara, M. and Morita, H. (1995). A network simulation model for oversaturated flow on urban expressways, *Traffic Engineering*, **30**-1, pp33-41 (in Japanese).

A STUDY ON FEASIBILITY OF INTEGRATING PROBE VEHICLE DATA INTO A TRAFFIC STATE ESTIMATION PROBLEM USING SIMULATED DATA

Chumchoke Nanthawichit and Takashi Nakatsuji
Hokkaido University, Kita 13, Nishi 8, Kita-ku, Sapporo, Japan
Chumchoke@yahoo.com
naka@eng.hokudai.ac.jp

Hironori SUZUKI
Japan Automobile Research Institute, Karima, Tsukuba, Japan
hsuzuki@jari.or.jp

ABSTRACT

Probe vehicle data has great potential for improving the estimation accuracy of traffic situations. The primary objective of this study is to present a systematic method of estimating traffic states on an expressway by combining probe vehicle data with conventional fixed detector data, using a Kalman filter technique. This paper discusses the parameter of identification of simulation models used prior to the introduction of the new method. Then, the method to estimate traffic states using integrated data from probe vehicles and detectors is described. Experimental results showed that the estimation accuracy could be improved by the proposed method.

INTRODUCTION

Probe vehicle data has great future potential for traffic surveillance systems. It could lead to revolutionary innovations for the advancement of traffic control systems. In order to assess what probe data can do for Intelligent Transportation Systems (ITS) and how it works, a variety of traffic data must be examined for extensive analytical purposes. However, the availability of probe data is still limited. At present, as the number of vehicles equipped with monitoring and communicating devices is still small, it is almost impossible to obtain probe data that covers extensive traffic situations. Under such circumstances, probe data must be generated using a traffic flow simulation technique. Simulation programs offer the opportunity to provide a variety of traffic data under various scenarios. Many researchers have used simulation data in their studies. It is important not only to have a good comprehension of simulation modeling, but also the model parameters that are carefully calibrated.

The primary objective of this study is to present a systematic method of estimating traffic states on an expressway in real time and to combine probe vehicle data with conventional fixed detector data. The method is based on a feedback approach; in which a macroscopic traffic simulation model first estimates traffic states, then they are adjusted according to observation information by using Kalman filtering technique (KFT). In this study, a variety of probe and detector data that covered extensive traffic situations were generated using a simulation program with INTEGRATION software (Rakha, 2001); and a trip-based microscopic traffic simulation to generate the hypothetical traffic data.

Prior to the introduction of the new estimation method, the identification of model parameters of both the macroscopic model and INTEGRATION will be discussed. The parameters were calibrated so as to minimize the difference between the simulated and actually measured data. Two methods are examined to identify the macroscopic parameters; one is based on a descent approach and the other is based on a random searching algorithm.

Many projects such as: the ADVANCE (Ivan et al., 1993, Sethi *et al.*, 1995, and Sen *et al.*, 1997) project in Chicago, the PATH (Sanwal and Walrand, 1995) program in California, the RTA project in the U.K, (Brackstone et al., 2001) the EURO-SCOUT (Oberstein, 1997) program in Germany, and the Internet ITS project in Japan, were launched to investigate the feasibility of using probe vehicles in obtaining real-time traffic data. Various kinds of information can be collected with this technique, for example: position, speed, time, lane used, link travel time, congested time/distance etc.; however so far, the applications for probe vehicle data are still limited. Most studies, such as Hellinga and Fu (2000, 2002), Chen and Chien (2001), and Choi and Chung (2001), have concentrated on travel time detection, while some authors, such as Ivan *et al.* (1993), Sethi *et al.* (1995), and Sanwal and Walrand (1995), have also used probe data for other purposes, including estimating O-D flows and detecting incidents. In the Nagoya area of Japan, more than 1500 taxis are taking part as probe vehicles to collect traffic data and weather conditions, which will be used for enhancing the taxi service as well as providing real-time information.

So far, very few studies have dealt with speed data from probe vehicles for estimating traffic states. In addition, apart from average speed, no interest has been paid to the estimation of the other fundamental traffic flow variables, namely traffic density and traffic volume. Moreover, an algorithm for combining probe data with conventional detector data to estimate traffic states is still lacking.

A summary of the theoretical methods used in this paper is found in the following section. These include, the macroscopic model, the methods for identifying the model parameters, and the new method for estimating traffic states (i.e. density, space mean speed and traffic volume). This new method uses a macroscopic traffic flow model with probe data, in which the KFT is applied to update the state variables estimated by a macroscopic model. Next the traffic data used in this study is discussed. The numerical experiment section will firstly examine the identification of model parameters by using different techniques. The later part will be devoted to the evaluation of a method for integrating probe vehicle data into fixed detector data for estimating traffic states on a freeway. Finally, we made some conclusions from the experimental analysis and made some suggestions as to how to improve the traffic state estimation.

THEORETICAL BACKGROUNDS

From amongst the various mathematical models for traffic flow, the macroscopic model which describes the traffic states in an aggregate manner, is one of the best tools for real-time applications due to its simplicity of traffic flow description and its computational efficiency. Although every macroscopic model has its own deficiencies, for example as discussed in, Daganzo (1995), Michalopoulos *et al.*(1993), and Lebacque and Lesort(1999), several techniques were introduced in order to compensate for these deficiencies. One such technique, the KFT, has been integrated into macroscopic models for the real-time estimation of traffic states. Cremer (1979), Payne *et al.* (1987), Pourmoallem *et al.* (1997), and Suzuki and Nakatsuji (2002) applied the KFT as a feedback tool to estimate traffic states on a freeway. In these studies, observed traffic data was taken from fixed vehicle detectors. However, with long separations between successive detectors, estimation results would probably deteriorate. In order to estimate traffic states more accurately, traffic information from additional sources is required. Probe vehicle technique has great potential in this respect due to its ability to cover an entire road network.

The Macroscopic Traffic Flow Model

In the macroscopic model, discretization in both time and space was adopted in the numerical processes to reduce the mathematical complexities. Figure 1 shows the space-time discretization of a freeway section on the assumption that traffic state is homogeneous within each segment. Each segment is Δx_j long. Traffic variables in the macroscopic model were defined as follows:

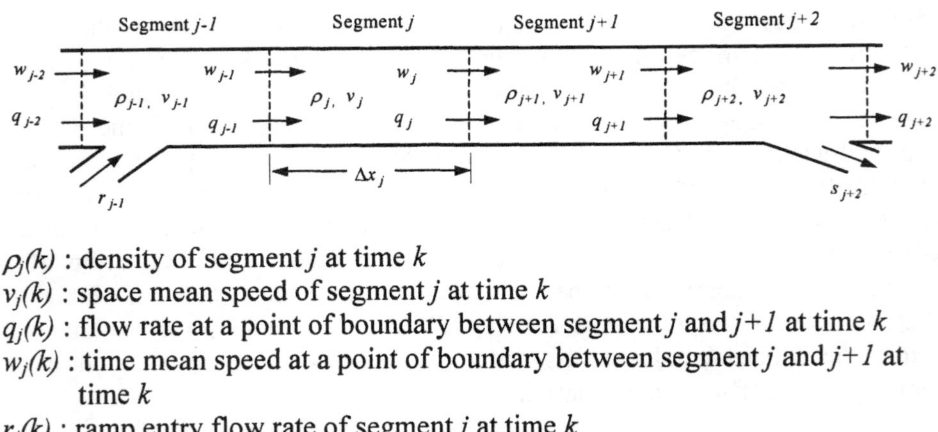

$\rho_j(k)$: density of segment j at time k
$v_j(k)$: space mean speed of segment j at time k
$q_j(k)$: flow rate at a point of boundary between segment j and $j+1$ at time k
$w_j(k)$: time mean speed at a point of boundary between segment j and $j+1$ at time k
$r_j(k)$: ramp entry flow rate of segment j at time k
$s_j(k)$: ramp exit flow rate of segment j at time k

Figure 1. Discretization of a Road Section.

Payne's macroscopic traffic flow model (Payne et al., 1971), which is one of the so-called higher order continuum models, was adopted in this study due to its simplicity to integrate with the KFT. Eq. (1) to Eq. (3) describe the discretized form of the model:

$$\rho_j(k+1)=\rho_j(k)+\frac{\Delta t}{\Delta L_j}\left(q_{j-1}-q_j+r_j-s_j\right)_{(k)} \tag{1}$$

$$v_j(k+1)=v_j(k)+\frac{\Delta t}{\tau}\left\{v_e\left[\rho_j(k)\right]-v_j(k)\right\}+\frac{\Delta t}{\Delta L_j}v_j(k)\left[v_{j-1}(k)-v_j(k)\right]$$
$$-\frac{\nu\cdot\Delta t}{\tau\Delta L_j}\frac{\rho_{j+1}(k)\lambda_j/\lambda_{j+1}-\rho_j(k)}{\rho_j(k)+\kappa} \tag{2}$$

$$q_j(t)=\alpha\left(v_j(k)\rho_j(k)\right)+(1-\alpha)\left(v_{j+1}(k)\rho_{j+1}(k)\right) \tag{3}$$

where τ is relaxation coefficient; ν is anticipation coefficient; and κ is density coefficient. k indicates time step, whereas x represents position; Δt is the time increment; λ_j is the number of lanes for segment j; α is a weighting parameter ranging from 0 to 1. Eq. (3) reflects the fact that the states of both neighboring segments may affect the volumes determined at the edge of each segment, as suggested by Cremer (1979). v_e is the speed at equilibrium state, which can be obtained from the density-speed relationship;

$$v_e(\rho)=v_f\left[1-\left(\frac{\rho}{\rho_{jam}}\right)^a\right]^b \tag{4}$$

where ρ_{jam} is the jam density, v_f is free-flow speed and a, b are sensitivity factors which are identified separately from the macroscopic model parameter, in advance. The macroscopic model parameters to be identified in this study are τ, ν, κ, and α.

Parameter Estimation Techniques

Since the parameters in the macroscopic model have significant effects on the performance of the simulation, they have to be identified carefully. The first step of the estimation process is the calibration of those parameters. In general, this procedure is formulated as an optimization problem, which can be solved based on an iterative comparison of model estimates with real traffic variables. The objective function, J, was set as the error between observed variables and model outputs:

$$J=\sum_{i=1}^{n}\left(\gamma_q\cdot(q_i-\hat{q}_i)^2+\gamma_w\cdot(w_i-\hat{w}_i)^2\right) \tag{5}$$

where γ_q and γ_w are the weighting factors of both volume and the speed errors. In this study, the reciprocals, $1/\sigma_q^2$ and $1/\sigma_w^2$, were used. For example, if the volume deviates in the range of 300 vph and the speed deviates in the range of 10 kph, the values of γ_q and γ_w would be 10^{-5} and 10^{-2}, respectively.

Nonlinear least square technique (NLT)

The typical approach to identify model parameters is the least square method. The parameters of ν, τ, κ, and α are estimated so as to minimize the objective function. The objective function, which is a nonlinear equation with respect to the unknown parameters, has to be transformed to linear form using the Taylor expansion technique. Once the equation becomes linear, the least square estimation technique can be applied as:

$$\frac{\partial J}{\partial \beta} = \left.\frac{\partial J}{\partial \hat{\beta}}\right|_{\hat{\beta}_0} + \frac{\partial^2 J}{\partial \hat{\beta}_1^2}\left(\hat{\beta}_1 - \hat{\beta}_{10}\right) + ... = 0$$

$$\hat{\beta} = \hat{\beta}_0 - \left[\nabla^2 \hat{\beta}_0\right]^{-1} \nabla \hat{\beta}_0 \quad \text{or} \tag{6}$$

$$\begin{bmatrix} \beta_1 \\ \beta_2 \\ \beta_3 \\ \vdots \\ \beta_p \end{bmatrix} = \begin{bmatrix} \beta_{10} \\ \beta_{20} \\ \beta_{30} \\ \vdots \\ \beta_{p0} \end{bmatrix} - \begin{bmatrix} S_{11} & S_{12} & \cdots & S_{1p} \\ S_{21} & S_{22} & & \\ \vdots & & \ddots & \\ \vdots & & & \\ S_{p1} & \cdots & \cdots & S_{pp} \end{bmatrix}^{-1} * \begin{bmatrix} S_{1y} \\ S_{2y} \\ \vdots \\ \vdots \\ S_{py} \end{bmatrix}$$

where $$s_{my} = \frac{\partial J}{\partial \hat{\beta}_m} = \sum_{i=1}^{n}\left(-2\gamma_q \cdot (q_i - \hat{q}_i)\cdot \frac{\partial q_i}{\partial \beta_m} - 2\gamma_w \cdot (w_i - \hat{w}_i)\cdot \frac{\partial w_i}{\partial \beta_m}\right) \tag{7}$$

$$S_{ml} = \frac{\partial^2 J}{\partial \hat{\beta}_m \partial \beta_l} = \sum_{i=1}^{n}\begin{pmatrix} 2\gamma_q \cdot \left(\dfrac{\partial q_i}{\partial \beta_m}\cdot\dfrac{\partial q_i}{\partial \beta_l} - (q_i - \hat{q}_i)\cdot \dfrac{\partial^2 q_i}{\partial \beta_m \partial \beta_l}\right) \\ + 2\gamma_w \cdot \left(\dfrac{\partial w_i}{\partial \beta_m}\cdot\dfrac{\partial w_i}{\partial \beta_l} - (w_i - \hat{w}_i)\cdot \dfrac{\partial^2 w_i}{\partial \beta_m \partial \beta_l}\right)\end{pmatrix} \tag{8}$$

β_m represents the model parameter of ν, τ, κ, and α. i denotes the observation data at each time step and *m*, *l* indicate the individual unknown parameters. Iterations are repeated until the changes of the unknown parameters are small enough, or no more improvement is required in the correlation between model variables.

Box's Complex technique (BCT)

This method has been used for parameter identification in the field of traffic engineering in several research studies (Cremer and Papageorgiou, 1981, and Papageorgiou *et al.*, 1989). It is a kind of random search technique, which has proven effective in solving the problem of the nonlinear objective function subject to nonlinear inequality constraints. The procedure should tend to find the global maximum due to the fact that the initial set of points is randomly scattered throughout the feasible region (Kuester and Mize, 1973). Unlike NLT and BCT, algorithm does not require any derivatives. First, a number of complex points, which are the sets of model parameters, were generated randomly. The points must satisfy both explicit and implicit constraints. In this study, the explicit constraints are the ranges of the model parameters, while the maximum and minimum bounds of traffic speed and volume are treated as the implicit constraints. If the explicit constraints are violated, the point is moved a small distance inside the violated limit. If an implicit constraint is violated, the point is moved one half of the distance to the centroid of the remaining points. After that, the objective function, J, is evaluated at each point. The point having the highest function value is replaced by a point reflected through the centroid of remaining points as seen in the following equation.

$$X_i(\text{new}) = \delta(\overline{X} - X_i(\text{old})) + \overline{X} \tag{9}$$

A value of $\delta = 1.3$ is recommended by Box (1965). A point that repeatedly gives the highest function value on consecutive trials is moved one half the distance to the centroid of the remaining points. Iteration repeats until the objective function values of each point are nearly equal. To decide whether the global optimum is reached and to investigate the effect of the initial value on this technique, the initial values were changed and the procedure was repeated.

The Kalman Filtering Technique (KFT)

The traffic estimation method adopted here is based on the feedback discipline using the Kalman filter. The KFT, a method for adjusting the state variables by the available measurement data, has been widely used in various engineering fields and has also been applied to traffic flow problems in several studies (Cremer, 1979, Payne *et al.*, 1987, Pourmoallem *et al.*, 1997, and Suzuki and Nakatsuji, 2002). Traffic states are described by a macroscopic model and then adjusted according to the KFT algorithm; the basic concept of this technique is the adjustment of state variables in proportion to the difference between the measured and estimated values of the observation variables.

The state variables, x(k), are determined by the system equations, while the observation variables, y(k), are measured in the system process. In a traffic state an estimation problem where measurement data can be obtained from fixed detectors only, traffic density and space mean speed are treated as the state variables, $x(t)=(\rho,v)_{(k)}$, whereas traffic volumes, q, and spot speeds, w, are treated as the observation variables, $y(t)=(q,w)_{(k)}$. The continuity equation, Eq. (1), and the momentum equation, Eq.(2), are treated as state equations, while observation equations consist of a relationship between traffic volume and state variables, as in Eq.(3), and a relationship between spot speed and state variables which might be set as the following equation:

$$w_j(k)=\alpha v_j(k)+(1-\alpha)v_{j+1}(k) \tag{10}$$

where α is the same weighting parameter as used in Eq.(3). To formulate the KFT, white noise errors were induced in both state and observation equations as follows:

$$x(k+1)=f[x(k)]+\xi(k) \tag{11}$$

$$y(k)=g[x(k)]+\varsigma(k) \tag{12}$$

where $\xi(k)$ and $\varsigma(k)$ are noises representing the modeling errors and measurement errors respectively. Next, the state and observation equations are linearized around the nominal solution $\tilde{x}(t)$, using Taylor's expansion. The system equations become:

$$\tilde{x}(k+1)\cong f[\hat{x}(k)]+\frac{\partial f}{\partial x}(x(k)-\hat{x}(k))+\xi(k)\cong A(k)\cdot x(k)+b(k)+\xi(k) \tag{13}$$

$$\tilde{y}(k)\cong g[x(k)]+\frac{\partial g}{\partial x}(x(k)-\tilde{x}(k))+\zeta(k)\cong C(k)\cdot x(k)+d(k)+\zeta(k) \tag{14}$$

where $A(k)=\frac{\partial f}{\partial x}$, $b(k)=f[\hat{x}(k)]-\frac{\partial f}{\partial x}\hat{x}(k)$

$$C(k)=\frac{\partial g}{\partial x},\ d(k)=g[\tilde{x}(k)]-\frac{\partial g}{\partial x}\tilde{x}(k) \tag{15}$$

The state vector, $\hat{x}(k)$, after obtaining the actual measurement data, y(k), is given as

$$\hat{x}(k) = \tilde{x}(k) + K(k)[y(k) - \tilde{y}(k)] \tag{16}$$

where K(k) is termed as Kalman gain which prescribes the reliability of the observed data. $\tilde{x}(k)$ and $\tilde{y}(k)$ are estimated by Eq. (11) and (12), respectively. Before processing the KFT, the initial values of the state variables and error covariance matrices have to be set.

COMBINATION OF THE PROBE DATA INTO THE OBSERVATION VARIABLE

Probe Data

We assumed that the probe data was reported at each specified time regardless of their positions, which could be obtained from a GPS-based system or a beacon-based system. There are several types of information on the current status of the probe vehicle that could be available on a real-time basis, including position and speed. The probe information that we attempted to apply as the observation variable for the KFT is the probe speed, since speed is explicitly included in the macroscopic model.

Combining the Probe Data into the KFT

Following is a method presentation of how to apply both the fixed detector data and the probe vehicle data into the KFT, for a traffic state estimation problem. This method was proposed in our previous work (Nanthawichit *et al.*, 2003). Some assumptions were set up to make the method application convenient. We modeled a road section as shown in Figure 2. The road section was discretized so that fixed detectors, if any, were located at approximately the middle of a certain segment to avoid influence from neighboring segments in the calculation of the observation variables. This method differs from the conventional approach mentioned above in that; if a fixed detector is located in the middle zone of a segment instead of the segment boundary, $w_j(t)$ in Eq.(10) would reduce to $v_j(t)$. Therefore, we may assume that the observed speed from the fixed detector, $w_{d,j}$, is the observation variable of space mean speed, and the observed volume from the fixed detector, $q_{d,j}$, is the observation variable of the segment flow. It is more advantageous if the harmonic mean of speed at detector is used instead of the arithmetic mean; this enables the assumption to be closer to the theoretical concept.

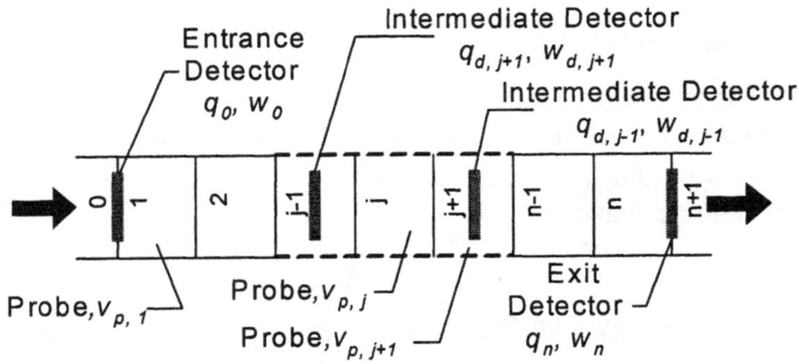

Figure 2. Road Section with Probe Vehicles and Fixed Detectors.

It was assumed that the fixed detectors must be available, at least, at both the entrance and exit of the road section as well as at every on/off-ramp. On- and off-ramp volumes are used as the source and sink terms in the continuity equation. The traffic volume and spot speed at the entrance and exit are used as observation variables for the KFT, and as boundary conditions for the macroscopic traffic flow model.

At every time step, the macroscopic model estimates the traffic states using input conditions at the boundaries, while the probe data are sorted by their location- which is determined by the segment the data came from. The probe data is then averaged, in a required aggregated data format at each segment. Together with the fixed detector data, the probe data is used as observation variables in the KFT to update the estimated states. In order to combine the data from different sources, a data fusion technique is required when both probe data and fixed detector data are available in the same segment. Choi and Chung (2001) may be referred to for data fusion techniques.

The observation equations for the entrance and exit detectors used for all patterns can be assumed as:

$$q_0 = \rho_1 \cdot v_1 \tag{17}$$

$$w_0 = v_1 \tag{18}$$

$$q_n = \rho_n \cdot v_n \tag{19}$$

$$w_n = v_n \tag{20}$$

Additional fixed detectors (or supplementary detectors) may be available between the entrance and exit detectors. Traffic volume and speed data from those detectors can be used as additional observation variables. As detectors are approximately located in the middle of certain segments, we might assume that the average speed measured at the detectors is equal to the space mean speed of the segments, $w_{d,j} = v_{d,j}$. The observation equations for the supplementary detectors are:

$$q_{d,j} = \rho_j \cdot v_j \tag{21}$$

$$v_{d,j} = v_j \tag{22}$$

where j identifies the number of segments with a fixed detector.

The average speed of the probe vehicles in each segment is considered to be an observation variable for the space mean speed in each simulation step. For the segments where both detector and probe data are available, the speed data from the probe vehicles should be integrated with the speed data from the fixed detectors (i.e. the observation speed, $v_{d,j}$, in Eq.(22) should be replaced with the speed integrated from both sources). In this study, for simplicity's sake, a weighted average of speed from both data sources was used to combine the data from different sources. Since the data used in this study is simulation data, the reliability of data from both sources may be considered the same. The same weight was assigned for both probe data and detector data. Other data fusion techniques could be used for this purpose, applying to real traffic data.

Generally, probe data may not be available for all segments for every time step. As a result, the number of observation variables from probes varies with time. The observation equation for the segments where only probe data is available at time step t reduces to:

$$v_{p,h} = v_h \tag{23}$$

where $v_{p,h}$ is the observed speed from the probe vehicles, and h identifies the segment that has only probe data.

TRAFFIC DATA

Simulation Field

In this study, a 5,550m road section was modeled emulating the road section of the Yokohane Line of the Metropolitan Expressway, in Tokyo Japan, in an inbound direction between the Namamugi Junction and the Taishi Ramp. The road section is divided into nine segments ranging from 400m to 800m with two on-ramps and one off-ramp. The number of lanes varies from two to three. Figure 3 depicts the geometry of the study road section.

Calibration Data

The macroscopic model parameters were calibrated using the traffic data measured at the simulation site. The data was collected using traffic detectors, which record data at one-minute intervals. Two typical traffic conditions were considered:

Case 1: low density traffic and quite smooth traffic collecting on Feb 21st (Wed.), 1996 during 16:00 to 16:30 hrs.

Case 2: traffic density changed rapidly from off-peak to morning peak on Feb 23rd (Fri.), 1996 during 06:30 to 07:00 hrs.

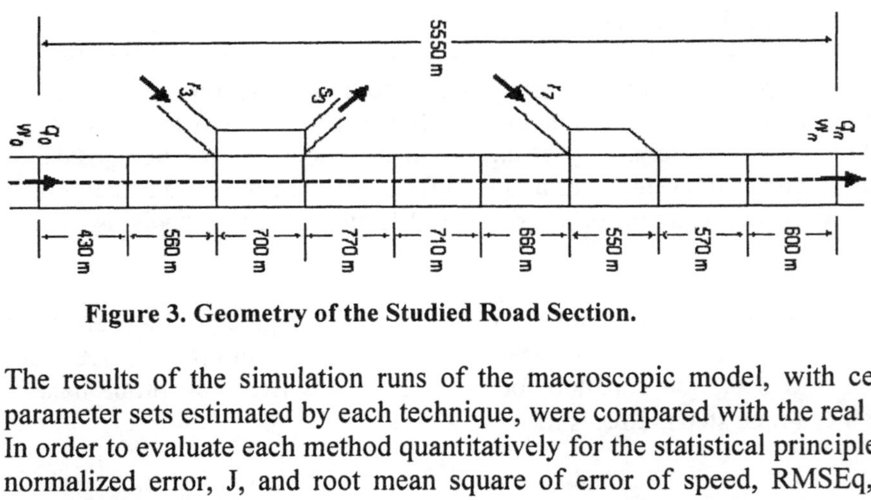

Figure 3. Geometry of the Studied Road Section.

The results of the simulation runs of the macroscopic model, with certain parameter sets estimated by each technique, were compared with the real data. In order to evaluate each method quantitatively for the statistical principle, the normalized error, J, and root mean square of error of speed, RMSEq, and volume, RMSEw, were calculated. In this study, all data points (the number of segments multiplied by the number of time intervals) were used for the statistical evaluation.

Simulation Data

INTEGRATION

INTEGRATION software (Rakha, 2001), a trip-based microscopic traffic simulation, was used to generate traffic data. It has various features suitable for Intelligent Transportation Systems applications, including the capability to generate fixed detector data and probe vehicle data, and has been used in several studies (Hellinga and Fu, 1999, 2002, and Hellinga and Gudapati,

2000) to generate hypothetical data. To present the real traffic situation properly before using, INTEGRATION has to be calibrated with the field data. Traffic data collected from the study area on February 23rd, 1996, during 16:00 to 18:30 hrs was used for calibrating INTEGRATION. Some traffic variables, such as traffic flow and speed at the observation points, were selected as reference outputs. OD volumes were given as inputs of the program. Volume and speed from traffic detectors at nine observation points were used as references for INTEGRATION output checking. Several simulation runs were conducted by changing input parameters. For each run, we calculate the error between actual data and generated data. The criterion for output checking, or the error index, was selected to be the combination of volume error and speed error, which is the same as the normalized error, J, in the numerical experiments. As in the case of the calibration of the macroscopic model, all data points were used for the statistical evaluation.

Traffic demand pattern

In order to assess the effectiveness of probe data, a variety of traffic data conditions were considered by varying inflow volumes at the entrance of the road section, and the inflow/outflow at the on/off ramps:

Case A: low density with low main inflow.
Case B: low density, but widely ranged with a peak around the middle of simulation period.
Case C: low density with high inflow.
Case D: wide range with high inflow at the main line and an on-ramp.
Case E: wide range rising main inflow from moderate to high with moderate ramp inflows.
Case F: wide range with high inflow at the main line and on-ramps in the first half of simulation.

Cases A, B and C represent low-density and smooth traffic conditions, while cases D, E and F contain a wide range of traffic conditions from low- to high-density. From all the simulated data, case A has the smoothest conditions, while in cases B and C the variation in speed is slightly larger and more obvious. In case D, traffic speed continues to drop from the beginning of the simulation right through to its end. In case E, the speed drops around the middle of the simulation, then returns to low-density conditions at the end. In case F, around the middle of simulation there are speed drops in some segments, while there are some segments where speed remains high and some segments where speed remains low for the whole simulation period. It was assumed that, if available, probe data could be obtained from anywhere in the study network at any specific interval. Both probe data and fixed detector data can be transmitted to the traffic control center in real-time. To reduce fluctuation, the data was aggregated into 3-minute blocks for both the fixed detectors and probe vehicles.

NUMERICAL EXPERIMENTS

Identification of Model Parameters

Macroscopic traffic flow model parameters

Firstly the model parameters of the macroscopic model were identified using the real traffic data. τ and ν were constrained in the range from 0 to 9999, κ was from 0 to 200, and α was from 0 to 1. For NLT, the model parameters were estimated by changing the initial values of 50 sets randomly. In BCT, three sets of initial parameters were recommended by Cremer and Papageorgiou (1981), Papageorgiou *et al.* (1989), and Cremer and May (1986) – who were the first to apply them. Furthermore, different numbers of complex points, which yielded 50 sets of initial values, were also examined.

Experimental results indicated that, in NLT, different initial values often resulted in absolutely different solutions. In other words, the parameters strongly depend on the initial values. Furthermore, little improvement would be gained even if the program started with different initial values. The possible reason for this is that the objective function is nonlinear and has a lot of extreme values. Due to the nature of the NLT being based on derivatives, it is very difficult to escape from a local minimum once entrapped.

As shown in Figure 4, the BCT produced better estimates for the RMSE of volume and spot speed than the NLT in both cases. The difference between the results from the BCT and the NLT in Case 1 is not significant, as both techniques can work well in the case of smooth traffic. The initial values had little effect on the final solutions because the BCT has a mechanism that generates a number of random points automatically while avoiding a local minimum. Consequently, the method successfully yielded the parameters that were substantially different from the initial values. Moreover, the calculation process of the BCT is much simpler than the NLT because it does not require any derivative and matrix operations.

In Figure 4, the RMSE error is still large for Case 2, in which the traffic situation rapidly changed from an off-peak state to a morning peak state. To treat this phenomenon more precisely, the data sets of Case 2 were divided into two parts - before and after the abrupt change of speed. Then the parameters were identified separately for both time periods by the BCT. Figure 5 exhibits the RMSE of both flow and speed for each time period. Compared with those in Figure 4, the RMSE decreased in both time periods, especially in the second time period. The separation was effective in improving both RMSE's. Therefore, the model parameters should be identified according to traffic situation. For example, the value of τ or the relaxation coefficient in the first half of Case 2 is smaller than the one in the second half. This is because; the high-density case in the second half requires more time to compare the low-density case for traffic to turn into the equilibrium condition. ν/τ reflects the rate of change in speed due to the anticipated density ahead. The values of α that are close to 1 in the calibrating results indicate that the traffic volume is mainly contributed from the downstream traffic state. Table 1 summarizes the macroscopic model parameters calibrated here.

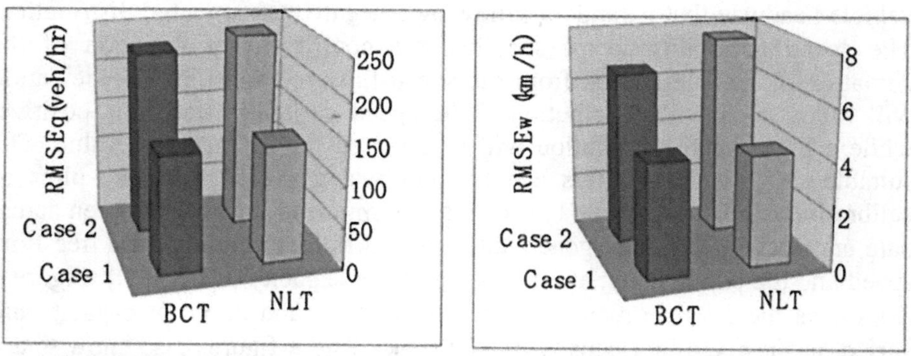

(1) RMSE of Traffic Volume. (2) RMSE of Speed.

Figure 4. Identification of the Parameters of the Macroscopic Model.

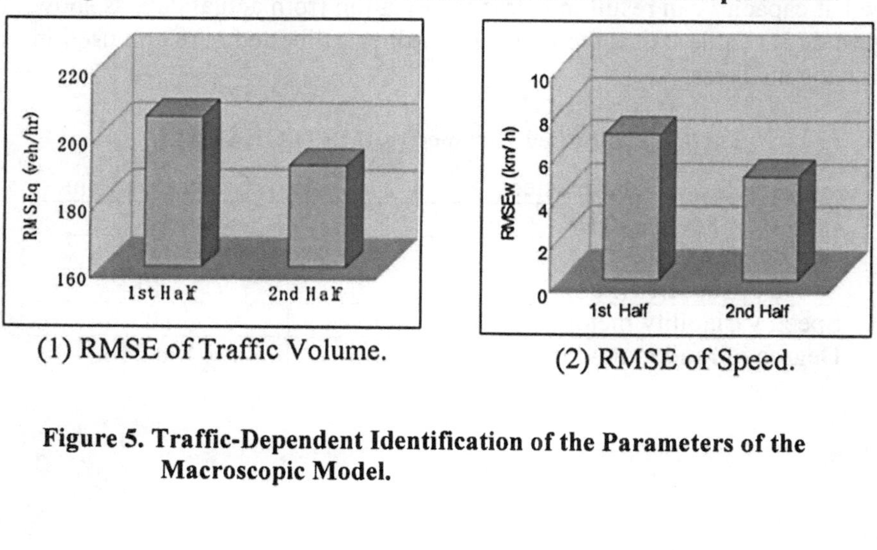

(1) RMSE of Traffic Volume. (2) RMSE of Speed.

Figure 5. Traffic-Dependent Identification of the Parameters of the Macroscopic Model.

Table 1. Calibrated Parameters of the Macroscopic Model.

Parameters	Case 1	Case 2	
		1st Half	2nd Half
τ (sec)	64.1	51.1	75.6
ν (km^2/hr)	64.8	35.5	100.0
κ (vpk)	40.0	40.0	40.0
α	0.92	1.0	1.0

Parameters of INTEGRATION

In the calibration of INTEGRATION, five parameters were determined including the free-flow speed, the jam density, the speed at capacity, the degree of randomness in departure headway, and the speed variability factor. The degree of randomness in departure headway indicates the fraction of vehicle headway that is random following a negative exponential distribution. The speed variability factor specifies the coefficient of variation of the deviation of vehicle speeds from the speed-flow relationship. The deviation will follow a normal distribution if the speed variability factor is positive, while a log-normal distribution will be adopted for the negative value. The suitable range of parameters can be simply determined from the plots of calibration results. Figure 6 (1) shows the normalized error, J, between actual data and INTEGRATION generated data at different values of the free-flow speed and the jam density, at a fixed speed at a capacity of 40 (kph). Figure 6 (2) shows the J values when the free-flow speed and speed at capacity are varied, at the fixed jam density of 120 (vpk). These figures also show that a large deviation from the actual data can occur if the parameters are not selected properly. For example, an improper pairing of free-flow speed and speed at capacity can result in a drastic deviation from actual data as shown in Figure 6 (2). Table 2 lists the parameter values calibrated here and used in the following analyses.

Table 2. Calibrated Parameters of INTEGRATION

Parameter	Calibrated Value
Free-flow speed (kph))	100
Jam density (vpk)	120
Speed at capacity (kph)	40
Speed variability factor	0
Degree of randomness in departure headway	0.2

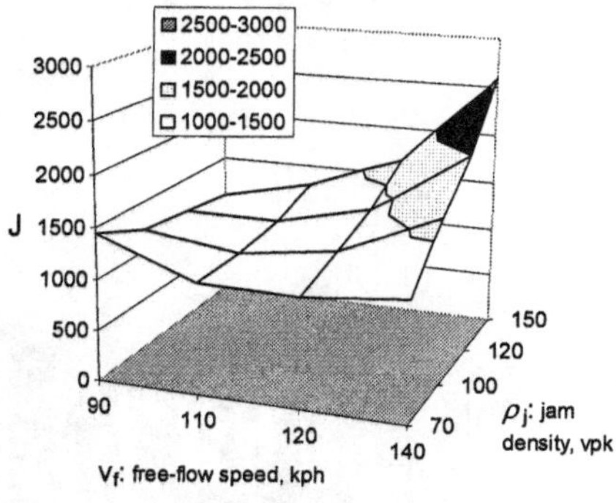

(1) J Vs. Free Flow Speed and Jam Density.

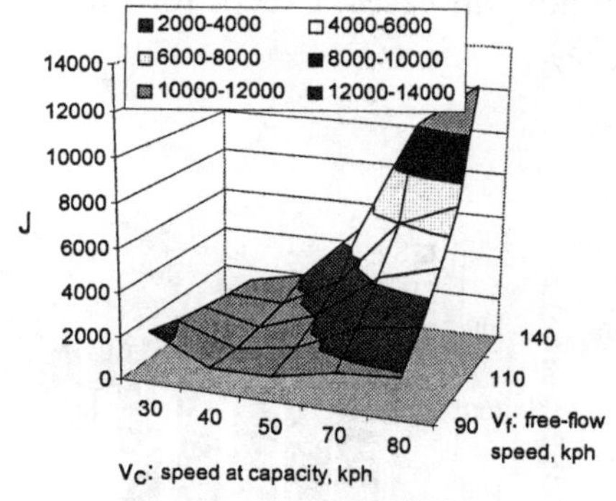

(2) J Vs. Free-flow Speed and Speed at Capacity

Figure 6. Sensitivity Analysis of INTEGRATION Model Parameters.

Efficiency of the Probe Data

Four estimation scenarios, differing in their estimation technique and the level of traffic information, were compared to investigate how the probe vehicle data works for each data case mentioned above.

S1: the macroscopic model without KFT.
S2: S1 + KFT using the supplementary fixed detector data.
S3: S1 + KFT using the probe vehicle data.
S4: S1 + KFT using both the probe vehicle data and the supplementary fixed detector data.

In this numerical experiment, it was assumed that the detector data at the entrance and exit of the road section were available for all scenarios. One supplementary detector was inserted at segment 5 for scenarios 2 and 4. In scenarios 3 and 4, 5% of probe vehicles, from the total amount of traffic in the network with a transmitting frequency of 1 report/second, were adopted. Consideration as to what is an appropriate probe fraction and frequency of probes can be referred to in the Sanwal and Walrand (1995), and Nanthawichit *et al.* (2003) report.

Traffic states

Figure 7 summarizes the overall estimation results. They show that the KFT contributes to the improvement of estimation accuracy. Compared to the scenario that does not use the KFT (S1), the estimation errors decrease in almost all cases, except for the *RMSEq* in case D of scenarios 2 (S2) and 3 (S3).

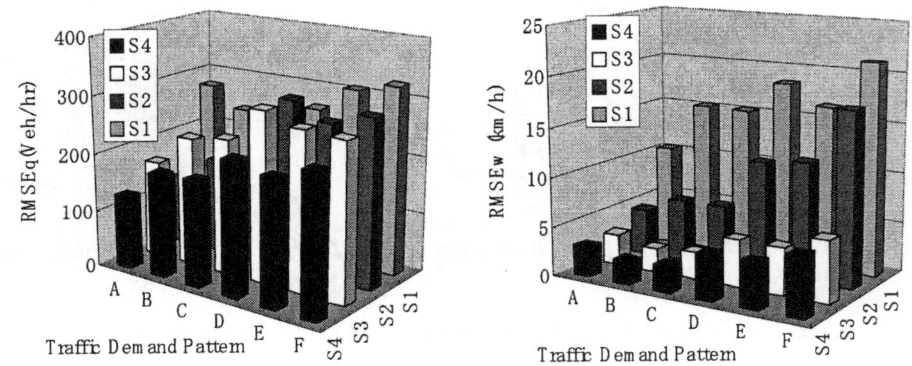

Figure 7. Effect of Detector Data and Probe Vehicle Data on the Estimation of Traffic States.

In case D for scenarios 2 and 3, the estimation errors in volume are larger than in scenario 1. The main reason for this, being that the deterioration from the output fluctuation outweighs the improvement from the KFT correcting process. However, if fixed detector and probe data were used in tandem, as in scenario 4, the estimation results improve.

Scenario 4 provides the best results for almost all cases, except in some *RMSEq* cases, where low-density conditions as in cases A to C, result in the *RMSEq* of scenario 4 being slightly poorer than that of scenario 2. The estimation results of scenario 4 provide a very small error for all indices: the errors in volume range from 128 to 185 vph in the low-density cases (A to C), and from 217 to 242 vph in the wide range density cases (D to F). Errors in speed are about 2 to 3 kph in low-density cases, and about 4-6 kph in wide range density cases.

The estimation method used in scenario 4 reduced the combination errors, *J*, by 81%-87% compared with those of scenario 1. The volume errors, *RMSEq*, and the speed errors, *RMSEv*, reduced by 12%-53% and 70%-85%, respectively. Thus, the usage of both detector and probe data was very effective in improving estimation accuracy. Nevertheless, cost-effective analysis is still required to determine if it is worthwhile to acquire more data in order to get a certain improvement in traffic state information, particularly in cases with similar results like scenario 3 and scenario 4. However, the authors believe that the results from scenario 3 and scenario 4 would be more deviating in a large and complex network case.

Figure 8 (1) shows the error in speed profiles of case C for scenarios 1 and 4, and Figure 8 (2) shows the same for case E. The profiles of the estimated speeds from scenario 4 (S4) accurately follow the actual ones, even where there is an abrupt change of regions as shown, for example, around the time step of 5000 to 6000 seconds in segments 5 and 6 in Figure 8 (2). On the other hand, the macroscopic model (S1) sometimes fails to capture the actual traffic variation. It is noticeable from Figure 8 that the error in speed estimation from scenario 4 is relatively low when comparing it to scenario 1. This suggests that a macroscopic model itself sometimes cannot capture uncertainty in a traffic flow, such as traffic incidents.

Scenario 2, which used only fixed detector data, did not yield good results, especially in speed under high-density and fluctuating conditions. One reason could be that the distance between adjacent detectors may have influenced the estimation errors of *RMSEv* and *J*. Here a dense installation of fixed detectors (2.8 km in this numerical experiment) is required to obtain the estimation as accurate as scenarios 3 and 4.

CONCLUSIONS

To conclude, this paper described a method for treating probe vehicle data together with fixed detector data in order to estimate the traffic state variables of traffic volume, space mean speed and density. This method used a macroscopic model along with the Kalman filtering technique (KFT). Prior to the development of this model, the parameters of traffic simulation models that were used to simulate the traffic states and to generate the hypothetical traffic data were estimated. This was done in order to minimize the discrepancy with the real data measured on the Yokohane Line of Metropolitan Expressway in Tokyo, Japan. In the newly developed method, the traffic states described by the macroscopic model are adjusted according to the KFT algorithm. The method was verified with several traffic data sets generated by the INTEGRATION simulation program. Four different scenarios according to the estimation method and available traffic data were examined for a single freeway section.

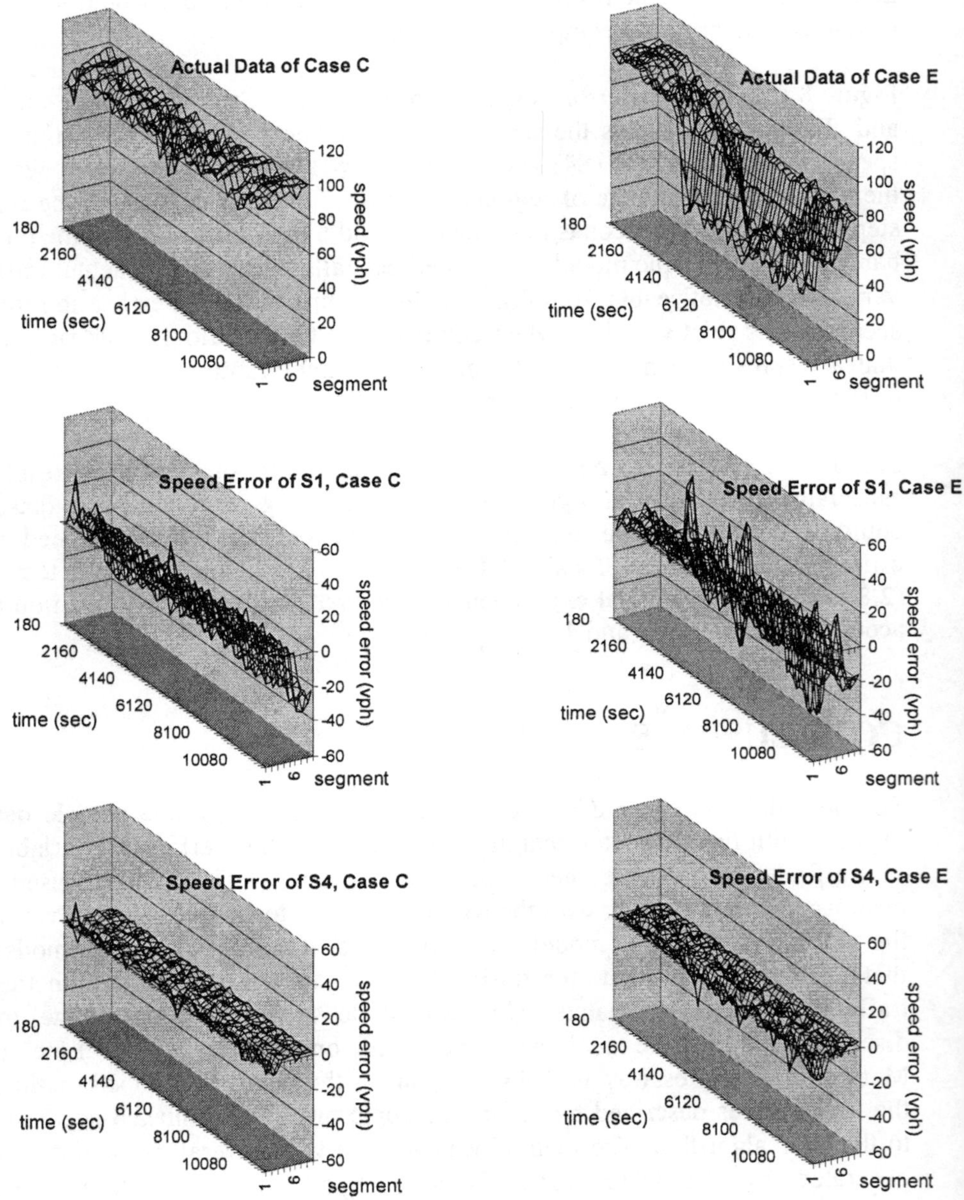

(1) Case C (Low Density) (2) Case E (Wide Range)

Figure 8. Comparison of Simulated Data and its Estimated Values for Different Scenarios.

Major findings are:

- The Box Complex technique (BCT), a sort of random searching method, was superior to the Nonlinear Least Square technique (NLT) technique in estimating the macroscopic model parameters.
- Some model parameters strongly depend on the traffic condition. Estimation precision can be improved if the model parameters are adjusted in accordance with the traffic condition.
- To use a commercial model such as INTEGRATION, careful calibration of model parameters is required. A large deviation from the actual data can occur if the parameters are not selected properly.
- The KFT contributes to improving the state estimation accuracy. For the road section studied, the normalized variance, J, was reduced by 30%-87% when compared to what was estimated using the macroscopic model only.
- In traffic state estimation, the method using both fixed detector and probe data provides the smallest errors compared to those using only probe data, and using only detector data.
- In the case of simple networks, like freeways in this study, the proposed method can capture the traffic variation properly without any incident information, even when the traffic conditions change abruptly.

Results from numerical experiments support the effectiveness of using the combined data from detectors and probe vehicles to estimate the traffic state. Introducing the use of data from other sources could also reduce the need to densely install the traffic detectors. Both the macroscopic model and the numerical technique adopted in this study were very simple. Thus, a highly complex model might not be necessary if available traffic data was able to cover most parts of a network. However, the findings are valid for only a small single road section. For further analysis it is recommended that a procedure to reduce fluctuation occurring, when the probe data is applied as observation variables, be developed. The practicality of the proposed method on the diverse road configurations, particularly on a large network, should be investigated. The applications of the method should be extended to the travel time prediction problem. Furthermore, consideration should be given to cost-effectiveness before bringing this method into practice.

ACKNOWLEDGMENTS

This research was supported by a grant from the Ministry of Education, Science, Sports and Culture, Government of Japan. Lastly, traffic data contributed by the Metropolitan Expressway was greatly appreciated.

REFERENCES

Brackstone, M., Fisher, G., and McDonald, M. (2001). The Use of Probe Vehicles on Motorways, Some Empirical Observations, *Proc. the 8th World Congress on ITS*, Sydney, Australia.

Box. M.J. (1965). A New Method of Constrained Optimization and a Comparison with Other Methods. *Computer J.*, **8**, pp42-52

Cathey, F.W., and Dailey, D.J. (2002). Transit Vehicles as Traffic Probe Sensors, T*RB 81st Annual Meeting*, Preprint **02-2228**, Washington, D.C.

Chen, M., and Chien, S. (2001). Dynamic Freeway Travel Time Prediction Using Probe Vehicle Data: Link-based vs. Path-based, *TRB 80th Annual Meeting*, Preprint **01-2887**, Washington, D.C.

Choi, K., and Chung, Y.S. (2001). Travel Time Estimation Algorithm Using GPS Probe and Loop Detector Data Fusion, *TRB 80th Annual Meeting*, Preprint **01-0374**, Washington, D.C.

Cremer, M. (1979). *Der Verkehrsfluss auf Schnellstrassen*, Springer Verlag, New York.

Cremer, M., and May, A.D. (1986). An Extended Traffic Flow Model For Inner Urban Freeways, *Proc. 5th IFAC/IFIP/IFORS Intern. Conf. On Control in Transportation Systems*, Vienna, Austria, pp383-388.

Cremer, M. and Papageorgiou, M. (1981). Parameter Identification for a Traffic Flow Model, *Automatica*, **17- 6**, pp837-843.

Daganzo, C (1995). Requiem for Second-order Fluid Approximations of Traffic Flow, *Transpn. Res.*, **35B**(4), pp277-286.

Hellinga, B., and Fu, L. (1999). Assessing the Expected Accuracy of Probe Vehicle Travel Time Reports, *J. Transportation Engineering*, **125-6**, ASCE, pp524-530.

Hellinga, B., and Gudapati, R. (2000). Estimation Link Travel Times for Advanced Traveller Information Systems, *Proc., CSCE 3rd Transpn. Specialty Conference,* London.

Hellinga, B., and Fu, L. (2002). Reducing Bias in Probe-based Arterial Link Travel Time Estimates, *Transpn. Res.,* **10C**, pp257-273.

Ivan, J.N., Schofer, J.L., Bhat, C.R., Liu, P., Koppelman, F.S., and Rodriguez, A. (1993). Arterial Street Incident Detection Using Multiple Data Sources, *Plans for ADVANCE.*

Kuester, J. L. ,and Mize, J. H. (1973). Box (Complex Algorithm) in *Optimization Techniques with Fortran*, McGraw-Hill, Inc., New York.

Lebacque, J.P., and Lesort, J. B. (1999). Macroscopic Traffic Flow Models: A Question of Order, *Proc., 14th Int. Symposium on Transportation and Traffic Theory,* Jerusalem, Israel, pp3-25.

Michalopoulos, P.G., Yi, P., and Lyrintzis, A. S. (1993). Continuum Modelling of Traffic Dynamics for Congested Freeways, *Transpn. Res.*, **27B**(4), pp315-332.

Nanthawichit, C., Nakatsuji, T., and Suzuki, H. (2003). Dynamic Estimation of Traffic States on a Freeway Using Probe Vehicle Data, *J. Infrastructure Plan. and Man.*, **730/IV-59**, JSCE, pp43-54.

Oberstein, K. (1997). Collection and use of floating car data experiences from Berlin, *Proc., 4th World Congress on ITS,* Berlin, Germany.

Papageorgiou, M., Blosseville, J.M., and Hadji-Salem, H. (1989). Macroscopic Modelling of Traffic Flow on the Boulevard Peripherique in Paris, *Transportation Research,* **13B**(1), pp29-47.

Payne, H. J. (1971) Models of Freeway Traffic and Control, in *Simulation Council Proc. Ser: Mathematical Models of Public Systems*, **1-1**, pp51-61.

Payne, H.J., Brown, D., and Cohen, S.L.(1987). Improved Techniques for Freeway Surveillance, *Transportation Research Record* **1112**, TRB, pp52-60.

Pourmoallem, N., Nakatsuji, T., and Kawamura, A.(1997). A Neural-Kalman Filtering Method for Estimating Traffic States on Freeways, *J. Infrastructure Plan. and Man.*, **569/IV-36**, JSCE, pp105-114.

Rakha, H. *INTEGRATION Release 2.30 for Windows: User's Guide-Volumes 1 and 2*, M. Van Aerde & Assoc., Ltd., 2001.

Sanwal, K. K., and Walrand, J.(1995). Vehicles as Probes, in California PATH Working Paper, Institue of Transportation Studies, University of California, Berkley.

Sen, A., Thakuriah, P.V., Zhu, X.Q., and Karr, A. (1997). Frequency of Probe Reports and Variance of Travel Time Estimates, *J. Transportation Engineering*, **123-4**, July/August, pp209-297.

Sethi, V., Bhandari, N., Koppelman, F.S., and Schofer, J.L. (1995). Arterial Incident Detection Using Fixed Detector and Probe Vehicle Data, *Transpn. Res,* **3C**(2), pp99-112.

Suzuki, H., and Nakatsuji, T. (2002). A New Approach to Estimate Freeway OD Travel Time Based on Dynamic Traffic States Estimation Using Feedback Estimator, *J. Infrastructure Plan. and Man.*, **695/IV-54**, JSCE, pp137-148.

CONSISTENCY OF TRAFFIC SIMULATION AND TRAVEL BEHAVIOUR CHOICE THEORY

Noboru HARATA
Institute of Environmental Studies, University of Tokyo, 7-3-1 Hongo Bunkyo-ku, Tokyo, JAPAN
nhara@ut.t.u-tokyo.ac.jp

ABSTRACT

In order to evaluate ITS options and dynamic TDM options, it has become important to understand how to integrate traffic simulations and dynamic travel behaviour choice models. Although Traffic simulation is well organised, visual and understandable, it looks weak in regards to representing user behaviour. From the standpoint of travel behaviour analysis, this paper points out the consistency problems and specific points in the integration of traffic simulation and dynamic route choice model or dynamic time-of-day choice model. It is recommended that fundamental concepts, such as choice set and user segmentation be incorporated appropriately into traffic simulations.

INTRODUCTION

Traffic simulation has the following advantages when compared with traffic assignment:

1) By tracing vehicle movements (passing links and their timing), it gives understandable output, such as travel speed, waiting queue length and congestion duration.
2) Results are more powerfully explained with visual presentation.

As a result it may then become common understanding for further discussion. For example, in order to reduce traffic impact in a development, visual presentations of the traffic simulation would have a great effect on a developer's decision.

Traffic simulation also has well-known disadvantages which are as follows:

1) The uniqueness of the solution is not guaranteed.
2) The fixed time-sliced (e.g. one hour or 15 minutes) OD table is not appropriate in many applications.
3) Generally, it assumes behavioural rules that are too simple and may violate the basic rules in the travel behaviour analysis, such as choice-set consideration and the decision maker segmentation.

Due to the uniqueness of these problems, it is usual to use average results of 30 or 50 times the simulations in order to evaluate the traffic management options. As a consequence, it is difficult to understand the meaning of the "with" and "without" comparison by the simulation models, when it is compared with the traffic assignment evaluation results with unique solutions.

On the other hand, travel behavior analysis has been developed in many ways. It is possible to develop route choice models and departure time choice models, in many ways, to evaluate ITS options and dynamic TDM options. It is common to use choice set and user segmentation in order to represent these behaviors. Also, many combinations of route choice models and traffic simulations have been developed. In particular, the SOUND model gives a basic structure with some consideration to captive and choice travelers.

However, I believe that there is a lot of room for improvement in order to make traffic simulation more behavioral. For example, if we try to evaluate dynamic pricing options or dynamic travel information options in area-wide transport studies, it is as difficult as traffic simulation itself and has a limited scope of analysis. We should consider both the vehicle movements and the choices of route and/or time of day, because these choices change drastically according to traffic congestion. Consequently, a possible powerful analytical tool for these evaluations is needed to discuss the consistency of traffic simulation and travel behaviour choice theory.

This paper discusses the following issues accordingly:

1) A basic structure for traffic simulation and travel behaviour choice model
2) A consistency problem with traffic simulation and travel behaviour choice theory
3) Case 1. Traffic simulation and route choice
4) Case 2. Traffic simulation and departure time choice
5) Summary

A BASIC STRUCTURE

As Yoshii (1999) has summarized, there has been many traffic simulation models that incorporate route choice behavior or route choice rule. Among them, I think that the SOUND model gives a basic structure between route choice model and traffic simulation.

In the SOUND, route choice model is applied at five time-intervals. Between the intervals, traffic simulation moves each vehicle according to the determined route in the latest route choice.

The SOUND structure of traffic simulation and route choice model is written as Figure 1. It is a loop of route choice routine and vehicle movement routine. It can describe route choice behaviour, dynamic traffic conditions and the relationship between them. However, choice set and user segmentation are not handled well. As a result, it has been named a "basic structure".

With an area-wide road network data, a simulation model with a basic structure can evaluate dynamic travel information and traffic management options more causally and accurately than those without a route choice routine.

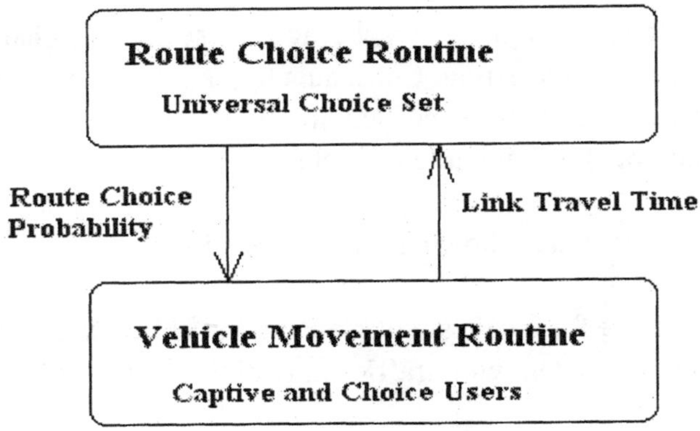

Figure 1. A Basic Structure of Simulation model.
Note: Modified on Figure 2.2.1 in Yoshii (1999).

Although this paper discusses the consistency of traffic simulation and the travel behaviour theory, it excludes behavioural elements in Vehicle movement routine, such as: variations of desired speed, acceleration and deceleration speed, and gap acceptance.

A CONSISTENCY PROBLEM

Modelling Assumptions

From the standpoint of travel behaviour analysis, it is natural to discuss how to incorporate basic concepts of travel behaviour analysis into traffic simulation consistently if any combination of them is required.

First of all, it is essential to consider choice set (that is available alternatives to travellers) in order to represent travel behaviour. However, traffic simulation as well as standard traffic assignment usually pays no attention to choice set. For example, traffic simulation assumes the universal choice set where all vehicles can use all possible routes on the road network.

Secondly, it is common wisdom to find effective user segmentation in order to explain travel behaviour well. Many travel behaviour studies have pointed out that responses to travel information or pricing will differ widely according to the traveller's attitude.

These basic concepts are explained fully in Moshe Ben-Akiva and Steven R. Lerman (1985). The Japanese' experience pertaining to travel behaviour models up until 1993 is summarised by Harata, Morikawa and Yai(1993).

For example, in order to design effective TDM measures to alleviate traffic congestion in some corridors, the following steps are typically recommended after finding a bottleneck (TDM study group (2002)):

1. Find captive users and choice users.
2. Do market segmentation of the choice users.
   For example, the travel-time-conscious traveller, the cost-conscious traveller, and the comfort conscious traveller are typical market segments.
3. Design TDM for specific segmentation.
4. Do field experiments

Because it is said that a 10 % reduction in peak traffic flows is enough to reduce traffic congestion, these steps are aimed at finding a tailor-made TDM for a specific market segment, such as a luxury bus service for the comfort-conscious segment. This in turn will result in a 10 % reduction in peak traffic flows.

Exchanging Input and Output

Traffic simulation simultaneously traces vehicle movements and constructs traffic conditions as a whole. On every second, it can calculate the travel time of any OD combination as a sum of link travel time on an available route. Theoretically, it is also possible to calculate the generalised cost or any LOS of any OD combination. As a result, they will be used as Input LOS in travel choice models to estimate choice probability.

The estimated choice probability will transform the actual route or time of day by using random numbers. Therefore, if my assumption is correct, the same choice probability will result in a different route or time of day according to an observed random number.

In this Input/Output circulation, the following consistencies should be kept:

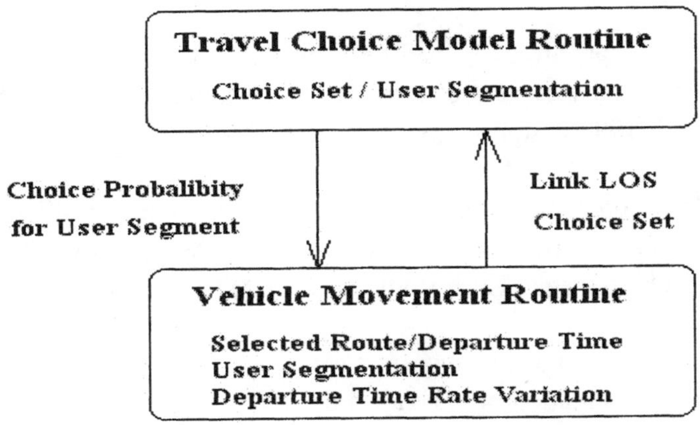

Figure 2. A Consistent Structure.

1) The Output LOS from traffic simulation should be consistent with Input to travel choice models.
2) The meaning of the variation resulting from the same set of choice probabilities should be explained. What kind of implication could be given to the simulation results?
3) Departure time distribution for the fixed time-sliced OD flows should be consistent with that of the distribution implicitly used in the departure choice model and in the calculation of delays at intersections.

CASE 1. TRAFFIC SIMULATION AND ROUTE CHOICE

Review

As Yoshii (1999) highlighted, there have been traffic simulation models with the Basic Structure. They are popular, used widely and are highly reputable. Here as Table 1 and Table 2, summary tables were written by extracting the specific points only relevant to the consistent structure.

Choice Set

The universal choice set is common amongst the route choice routine of traffic simulation models. They assume that any driver can use any route.

Table 1. Traffic Simulation with the Basic Structure; Part 1.

	SATURN	CONTRAM	INTEGRATION
Vehicle Movement	(Delay at Intersections)		
Model type	Hydro/Queue	Vehicle/Queue	Vehicle/Queue
Departure Rate	Flat	Flat	N.A.
Choice Route	Directional Flows	Minimum Path	Minimum Path
Segmentation	None	None	None
Route Choice			
Model type	User equilibrium	Incremental assignment	Incremental assignment
Choice Set	Universal	Universal	Universal
Segmentation	None	None	None
Network	Arterial link and node Two-level network	Arterial Link and node (Max=9999links)	Arterial Link and node

Table 2. Traffic Simulation with the Basic Structure; Part 2.

	FHWA	MITSIM	SOUND
Vehicle Movement	(Three level)	(Microscopic)	
Model type	Vehicle/Q-K, Following	Vehicle/Following	Vehicle/Following
Departure Rate	N.A.	N.A.	Flat
Choice Route	N.A.	Individual route	Individual route
Segmentation	None	Vehicle-type	Captive and Choice
Route Choice			
Model type	Incremental assignment	Time saving/ Threshold	Dial Assignment
Choice Set	Universal	Universal	Universal
Segmentation	None	Vehicle-type + Guided	Captive and Choice
Network	Arterial link and node with lane configuration	Arterial link and node With segments	Arterial link and node Aggregate one for Dial

Although the Dial assignment has a selection algorithm of relevant routes in order to avoid looped routes, and MITSIM has detailed link information of traffic controls for each vehicle-type, choice set is far from being available in route choice models.

On the other hand, route choice models use empirically found rules. Actual route data would give enough information about many criteria. This then determines the availability of various routes. Detour factors in distance and direction are typical ones. There may be a maximum number of right-turns. Or, there may be a maximum number of right turning vehicles. When the road network includes toll road and on-off ramp combinations, it can be selected as a sum of minimum generalised cost routes for drivers with different time values.

PHL Bovy & E.Stern (1990) explained a hierarchical series of choice sets and the "Label method" to determine the feasible choice set. The same method was used to identify the available route for the Kita-Kyushu expressway in 1979. With a total of 601 ODs, the "distances less than 2km" label reduced the number of ODs that can use the expressway to 55. Other labels, including "travel time differences greater than 10 minutes" and "travel time ratios greater than 2.0" reduced them to 331 each.

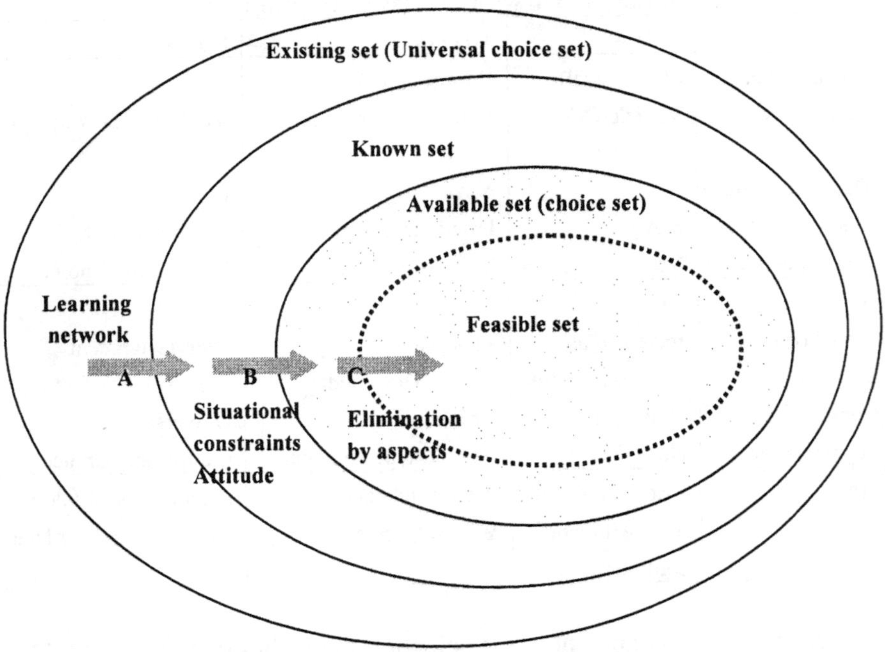

Figure3. A hierarchical series of choice sets.

Note: Modified on Figure 3.3 in "Route Choice: Wayfinding in Transport Networks, PHL Bovy & E.Stern(1990)"

User Segmentation

Travel time minimisation is the most commonly used criteria in traffic simulation. However, in reality, many different drivers perform route choice using different measures. For example, the Metropolitan Expressway survey revealed that car users put emphasis on travel time and small goods vehicle users put emphasis on travel distance.

The Japan Highway OD survey in 2000 showed that there were wide variations of on-off ramp combinations for the same OD. Those who did not use the nearest interchanges included: 18 % cars, 19% small goods vehicles, and 29% heavy goods vehicles. The reasons for HGVs were, travel time and so on. The variation can be approximately reproduced by considering a driver's value on time difference.

To evaluate the effects of travel time information, it is common for travel behaviour analysis to discriminate users according to whether they utilise pre-trip information and/or en-route information. More attention to user segmentation in simulation models would improve the rationality of them.

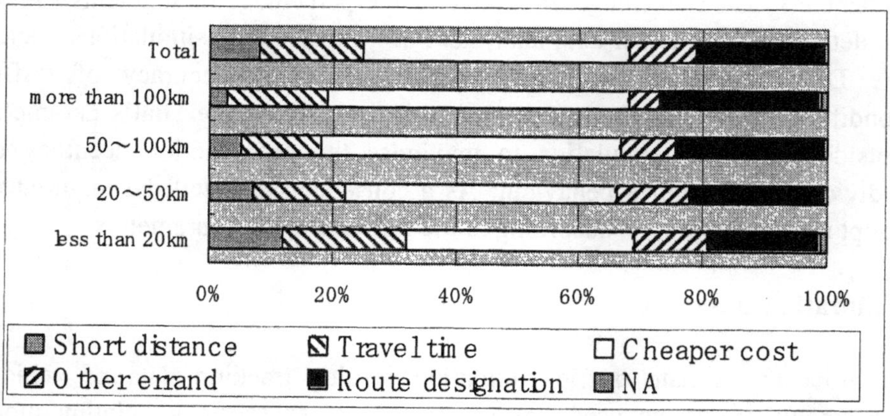

Figure 4. The HGVs' reasons why they do not use the nearest interchanges.

Exchanging Input and Output

In order to consider choice set in route choice routine, vehicle movement routine should have output choice sets for each OD with link LOS. Because vehicle movement routine can calculate various LOS and relative attractiveness of detour factors and "travel time difference/travel cost difference", it would be possible to create a sub-routine to output choice set on a pre-determined criteria.

With user segments, vehicle movement routine will output LOS of available routes, for each user segment. For example, with three user segments of distance-conscious users and two of generalised cost-conscious users - who belong to a different value of time class, LOS of available routes for three user segments will be needed as Input to travel choice model routine. Available routes for distance users may be fixed to the minimum distance path. For two generalised cost-conscious users, it will be possible to assume values of time distribution for the two groups. This will then give five or six representative values of time for the two user groups. With these representative values, available routes for one generalised cost user group may be defined as the sum of the best route for each representative value of time. If some of the best routes do not satisfy the pre-determined choice set criteria, they will be excluded from the choice set.

One additional point to be mentioned is, how to decide behavioural parameters. The percentage of captive or the percentage of segments should be determined on actual data analysis. Also, most traffic simulations decide the Dial's parameter with maximum reproduction accuracy of traffic conditions. However, there are some points to decide the Dial's parameter outside the traffic simulation to maximise the reproduction accuracy of individual route choice behaviour. As a consequence, it will be required to adopt user segmentation according to the different Dial's parameters.

Validation data

In order to validate traffic simulation results, tracking data of vehicle movements are to be used. Probe car data gives traffic simulation more reliable validation data than before. It becomes possible to understand a day-to-day variation of route choice behaviours and traffic conditions. GPS, PHS and intelligent tags can be used for this purpose.

CASE 2. TRAFFIC SIMULATION AND DEPARTURE TIME CHOICE

Review

Temporal redistributions of OD table will have great impacts on the level and duration of traffic congestion. For example, Oneyama (2001) estimated the impacts to Tokyo's 23-wards area with a pre-determined temporal redistribution pattern. However, how to set a realistic temporal redistribution is not as clear as how to set a spatial redistribution for route choice behaviour.

Travel behaviour analysis on temporal dimensions has been done mainly about commuting, because there is a clear restriction on arrival time. There are many discrete choice models of sliced time period alternatives and continuous choice models on departure time or arrival time. It is well known that IIA property of the logit-type model and discrete sliced time alternative produce an unrealistic response.

However, if a realistic choice set can be set from empirical data or activity diary data, it will be possible to model temporal choices of departure time within a relatively small time-span as found with commuting behaviour.

Also, even for shopping behaviour and social behaviour, if realistic choice set can be derived from activity-diary data, it will be possible to analyse these types of temporal choices by a discrete choice model. Realistic choice set may be derived from: diary data, a fixed activities schedule, and location and opening hours of destination facilities. Some activity/travel scheduling models have already developed choice set models of temporal choice. This methodology may be also powerful in identifying temporal choice set of commuting movements.

Choice Set

At a minimum, there are two-level temporal choices of trip timing. The first level is the temporal choice between space-time prisms. The second level is the exact trip timing choice within the specific space-time prism.

For the first-level of temporal choice, it seems quite logical to find the space-time prism by analysing activity and travel data analysis. (L.D.Burns (1979).

For example, regular workers have a fixed work schedule from 9 to 17 at a fixed place. They can do additional activities such as shopping and golf practices before 9am, after 5pm or during their lunch break.
If one is supposed to be back home by 7pm to have dinner with someone, he/she has a two-hour time window from 5pm to 7pm. How far he/she can go within the time window depends on the speed of the available mode of travel to him/her. The movable area within the specific time window is so-called the 'space-time prism'. It is also possible to find relevant space-time prisms before 9am or during the lunch break. If the lunch break is not long enough to take lunch and go shopping, then the space-time prism in the lunch break will disappear.

On the other hand, with the second-level temporal choice, the most common methods are to ask drivers directly about the possible alternative timing of the trip. However, it is natural to have some doubts about these responses because it is difficult to answer even with good information available about temporal variation of travel time. There is not enough data to determine the possible time range of trip timing and to construct a realistic and reliable model of trip timing at this detailed level.

More empirical data should be analysed in order to find some rule to determine the choice set of the second-level trip timing.

User Segmentation

Although there are limitations of required data, it seems better to discriminate drivers according to restrictions on arrival time and attitudes toward possible risks of delays. Many models demonstrate the difference of these segments, mainly in one simplified OD analysis. For a large network analysis, Yoshii (1999) used segmentation according to sensitivity to travel time information. Some combination of these segmentations would be useful to evaluate departure time responses to ITS options or dynamic TDM options.

Also, a day-to-day variation of traffic conditions is recognised and can recently be measured by various methods. The variation relates travel time reliability that must be an important factor in trip timing choice. It seems possible, to undertake many trials of traffic simulations, and to understand the possible variation of traffic conditions with the same choice probability and the same road network.

Exchanging Input and Output

With discrete choice models of sliced time alternatives exchanging Input and Output must exchange LOS and probabilities for each sliced time period. With given probabilities of sliced time alternatives, traffic simulation models will output LOS for all time periods. As a result, discrete choice models will up-date choice probabilities with the LOS of sliced time period

With continuous choice models on departure time or arrival time, simulation models must exchange LOS and probabilities for any time points. With the continuous distribution of departure time or arrival time, traffic simulation models will output LOS for any time points. With the revised LOS, continuous choice models on departure time or arrival time will update the continuous distribution.

For the simplicity of the exchange, it will be better to develop a continuous choice model that will give a continuous distribution of departure times directly.

Validation data

In order to validate traffic simulation results, a day-to-day variation of traffic conditions will be used. Again, current probe car data gives traffic simulations much better validation data than before.

If there is an opportunity to collect probe car panel data over many months like the Leicester Environmental Road Pricing Scheme, it will be a gift for trip timing model development.

A SUMMARY

In order to evaluate ITS options and dynamic TDM options, how to integrate traffic simulations and dynamic travel behaviour choice models becomes important. Although some researchers have tried to solve this problem, they have not dealt well with some important behavioural points. Although Traffic simulation is beautiful, visual, and understandable, it still has questionable points in representing user behaviour.

A consistent LOS problem may be solved technically when a large-scale comparative analysis on these issues will be done in near future. However, a consistency problem on modelling assumption appears problematical because no simulation model pays much attention to this fundamental issue.

From the standpoint of travel behaviour analysis, it is interesting that there have been no extensive discussions about choice set in traffic simulation development. Choice set is one of the most fundamental concepts in representing user behaviour. In addition, traffic simulation has not often adopted user segmentations. It is therefore strongly recommended to use segmentation by time value in evaluating pricing options and arrival time restrictions in evaluating ITS options.

More discussion is needed between travel behaviour analysts and traffic simulation developers on this issue. An integration of traffic simulation and dynamic route choice model will be the first prioritized area to discuss this issue empirically.

It is hoped that this paper gives some direction for full-scale integration of traffic simulations and dynamic travel behaviour choices.

ACKNOWLEDGEMENT

I express my special thanks to Prof. Kuwahara in allowing me to write on this topic, and to Dr. Yoshii and Dr.Horiguchi for their kind replies to my questions about the SOUND. Discussion also with my student Mr.Maruyama helped me to think carefully about departure time choice problems.

REFERENCES

Harata, N, Morikkawa,T, and Yai,T.(1993). Review and Perspective of Travel Behavior Analysis Focusing On Disaggregate Travel Demand Models, *Journal of Infrastructure Planning and Management*, **No.470, IV-20**, pp97-104 (in Japanese).

i-Transport Lab. (2001). Manual of SOUND/A-21 (in Japanese).

L.D.Burns. (1979). *Transportation, Temporal, and Spatial Components of Accessibility*, Lexington Books.

Maruyama, T., Harata, N. and Ohta, K. (2002). Application of combined network equilibrium model to mega metropolitan area, *Infrastructure Planning Review*, **Vol.19** (in Japanese).

Moshe.Ben-Akiva and Steven R.Lerman. (1985). *Discrete Choice Analysis*, The MIT Press.

Oneyama, H., IRYO, T.and Kuwahara, M. (2001). Simulation analysis of traffic congestion alleviation and environmental improvement by demand spreading over time in Tokyo Metropolitan area, *The Proceedings of the Infrastructure Planning Review Annual Meeting*, **24** (in Japanese).

P.H.L.Bovy., E.Stern.(1990). *Route Choice : Wayfindings in Transport Networks*, Kluwer Academic Publishers, pp309.

TDM study group. (2002). Successful Park and Ride, Japan Society of Traffic Engineers (In Japanese).

Yoshii, T. (1999). Development and application of dynamic assignment simulation model applicable to a large-scale network, p.178, Doctor of Engineering Thesis, University of Tokyo (in Japanese).

DRIVER'S ROUTE CHOICE BEHAVIOR AND ITS IMPLICATIONS ON NETWORK SIMULATION AND TRAFFIC ASSIGNMENT

Takayuki Morikawa and Tomio Miwa
Graduate School of Environmental Studies, Nagoya University, Chikusa-ku, Nagoya 464-8603, Japan
morikawa@civil.nagoya-u.ac.jp
miwa@trans.civil.nagoya-u.ac.jp

Shinya Kurauchi
Dept. of Geotechnical and Environmental Eng., Nagoya Univ., Chikusa-ku, Nagoya 464-8603, Japan
kurauchi@civil.nagoya-u.ac.jp

Toshiyuki Yamamoto
Department of Civil Engineering, Nagoya University, Chikusa-ku, Nagoya 464-8603, Japan
yamamoto@civil.nagoya-u.ac.jp

Kei Kobayashi
Chodai Co. Ltd, 20-4, 1-chome, Nihonbashi-Kakigaracho, Chuo-ku, Tokyo 103-0014, Japan
kobayashi-kei@chodai.co.jp

ABSTRACT

The principle of driver's route choice has long been the shortest path with a fixed time penalty of a toll. It is also understood that traffic is assigned on the road network based on user equilibrium with perfect information assumption.

This paper demonstrates two empirical studies that pose questions to these traditional assumptions, to better understand route choice behavior. Multi-class user equilibrium assignment with imperfectly informed drivers' classes is applied to a metropolitan area network first. Next, route choice behavior is directly observed and analyzed using probe' car data.

INTRODUCTION

Shortest-path routing has been used as the principle of driver's route choice in traditional traffic network assignment problems ever since Wardrop (1952) enunciated the user equilibrium. The generalized cost, which includes time and monetary cost, has also been adopted as a driver's criterion of route choice on the traffic network including toll routes. Especially in Japan, where toll expressways are quite common, operators of toll roads have often adopted the concept of generalized cost in traffic management. The basic premise of this is the assumption of a rational traveler who chooses the route that offers the least perceived costs (Willumsen, 2000). Nevertheless, many field observations on the drivers' route choice behaviors in the real world have suggested that drivers do not always select the route with shortest travel time or the least generalized cost (Chen, *et al.*, 2001).

The factors affecting the route choice, however, are not fully known to us yet. COMSIS Corporation (1995) showed in their literature review that several factors had been observed to influence the route choice in prior research, with the distinction between the pre-trip route selection and the en-route selection. The factors include: prior learning and experience, general locale such as the distinction between urban and rural road, credibility of information, perception of alternate routes, time savings, monetary cost, comfort and convenience, trip purpose, observation of traffic, reason for diversion, and timing of information. The last three are applicable only to en-route selection whereas the other factors are applicable to both pre-trip route selection and en-route selection. These various factors still need to be verified in the following studies.

The verification of the effects of these factors has not been fully accomplished yet, partly because of the lack of available data. Traditionally, the traffic counts on road-side and sample Origin-Destination (O-D) matrices obtained from road traffic census surveys or person trip surveys have been used for the traffic network analyses. The traffic counts and the O-D matrices, however, have little information on the route choice. We need more detailed information to examine the effects of various factors on route choice by both direct observations on the route choice behavior other than the traffic counts and the O-D matrices.

Abdel-Aty, *et al.* (1993) classified such direct observations into five categories; in the context of the route choice behavior influenced by Advanced Traveler Information System (ATIS): field experiments, route choice surveys, interactive computer simulation games, route choice simulation and modeling, and stated preference approach. Among these, field experiments are said to be the most accurate and representative method. However, field experiments have been labor-intensive and expensive into obtaining information. Therefore prior experiments were often limited to a relatively small sample size. Recently, Intelligent Transport Systems (ITS) technologies have enabled sample size enlargement in regards to monitored field experiments. The probe cars, which are equipped with a device to monitor their own location, can offer researchers rquiteelatively accurate information of the chosen route, even when the driver doesn't recognize himself where he is driving. The probe cars can also provide a great deal of additional information related to the route choice, such as: speed, acceleration, deceleration, braking, blinker use, etc. Thus, the probe cars are thought to be a promising source in providing detailed information on the route choice behavior.

The collection of the detailed information on driver's route choice behavior alone, however, does not assure to provide a better understanding of route choice behavior. We still need appropriate theories to model the route choice. Most of the prior studies have applied a random utility theory into the route choice model, which assumes that a driver knows all alternatives of routes and chooses the route with the highest utility. In real road networks, however, there can be a tremendous number of alternative routes, and it is doubtful that the driver knows all the alternative routes. Even if the driver knows a set of alternatives, it is unrealistic to assume that he calculates the utilities of all the alternatives by considering all the factors affecting the utility of the route. Especially, when the driver knows the real travel time only after he completes the trip, as such the travel time of each route is a probabilistic variable at the time he makes the choice. A normative theory of choice under risk, the expected utility theory, is often criticized by many empirical studies.

Researches of psychology or marketing have provided many behavioral theories other than the random utility theory. The objectives of this paper are to provide a framework of the route choice model from the viewpoint of the consumer behavior theory, and to propose an alternative to the random utility model. In addition, preliminary results of the data collection with probe cars in Japan will be presented, and the potential of the probe car data for examination of an alternative route choice model will be discussed. The remainder of the paper is structured as follows: the following section reviews driver's route choice behaviour from the view point of the consumer behavior theory – which provides the framework of the route choice model; section 3 provides a framework application – the traffic assignment model especially considers the imperfect knowledge of travel time on routes; section 4 presents the preliminary results of the data collection of the probe cars; and section 5 is a brief summary and conclusion.

REVIEW ON DRIVER'S ROUTE CHOICE BEHAVIOR

For conventional network simulation and traffic assignment, multinomial logit models are widely used to represent the driver's route choice behavior. Recently, various types of discrete choice models with flexible structures have been developed (see Bhat, 2000 and Koppelman *et al.*, 2000 for recent comprehensive review) and applied to the route choice contexts (e.g., Vovsha *et al.*, 1998).

The decision makers of discrete choice models implicitly assumed that 1) they have perfect information on all the available alternatives and their attributes; 2) they also have perfect information on all the possible outcomes; and 3) they choose the alternative with highest utility (expected utility under risks). Under these assumptions, the decision maker will reach the unique and best alternative instrumentally, and thus models with such assumptions can be called an "instrumental rationality" model (Heap *et al.*, 1992).

The above assumptions may be appropriate for relatively simple choice contexts such as travel mode choice. It is, however, unrealistic for drivers to choose the route in a perfectly rational way where numerous alternative routes with uncertain travel times exist. Many researchers - mainly in the area consumer behavior and psychology - have been suggesting that actual decisions are far from the one postulated in instrumental rationality models (e.g., Wright, 1975). They also propose that alternative approaches called "bounded rationality" (e.g., Simon, 1987) relax the assumption of instrumental rationality by considering the limitation of a human beings' cognitive resources. Although most of these suggestions are derived from the observations of actual market choices of commodities and consumer durables and of hypothetical choices in experimental situations, the ideas can be extended to the driver's route choice behavior.

Route choice decisions are characterized by the following aspects: 1) decision-making under uncertainty, especially in travel time and network configuration, 2) decision-making among numerous alternatives under severe time pressure, and 3) repeated choice where factors such as past experience, learning, and habit strongly influence each decision. The remainder of the section briefly reviews the existing models and the empirical finding related to the above aspects, and then discusses how we could model the route choice behavior in network simulation and traffic assignment.

Specification of Choice Set

Route choice decisions nominally avail to each driver numerous alternative routes. It is however, unrealistic to expect the driver to compare the many attributes of all available routes before making a choice. In addition, it is theoretically and empirically verified that the misspecification of choice sets yields biased parameter estimates of choice models and misleads demand forecasting (e.g. Williams *et al.*, 1982). Therefore, it is very important for the analysts to precisely specify the set of routes considered by the driver.

Figure 1 illustrates the taxonomy of the specification approach for choice set originally proposed by Cascetta *et al.* (2002). In most of the network simulations and traffic assignment models, all loop-less routes are enumerated, thus implicitly assuming that all the drivers have perfect information of network configuration. On the other hand, the algorithm proposed by Dial (1971), which is widely applied in stochastic user equilibrium model, can be interpreted as assuming the imperfect information of drivers implicitly in the sense that it enumerates only a subset of routes that are likely to be chosen using the concept of "link likelihood".

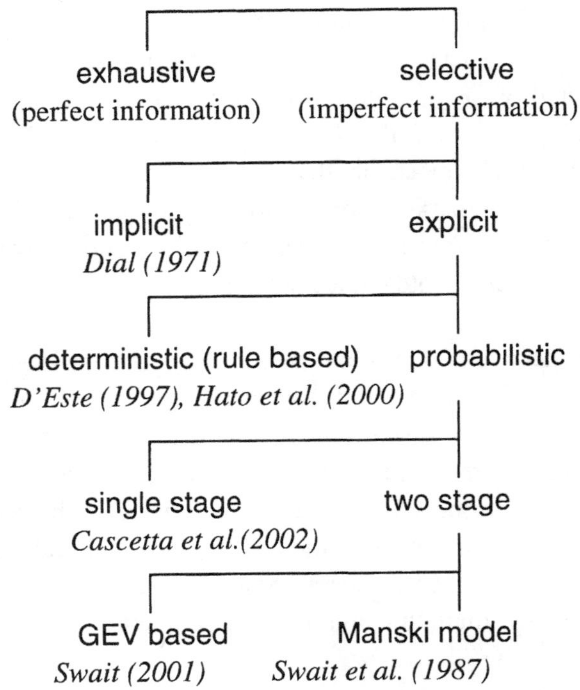

Figure 1. Specification Approaches for Choice Set.
Adapted and modified from Cascetta *et al.* (2002).

When we investigate the effects of information provision systems such as ATIS, driver's knowledge of available routes as well as congestion levels should be taken into account in the model. Generation process of choice set, therefore, is a key factor of route choice models for simulation tools or network assignment.

It is widely agreed that a decision maker is likely to make a two-stage choice: 1) a screening stage in which alternatives are eliminated using the simplified rules in order to be manageable; and 2) a choice stage in which the most preferred alternative is chosen through an elaborate process (e.g., Payne *et al.*, 1993). The model proposed by Manski (1977) is the prototype of this two-stage representation in probabilistic form. Various types of choice set generation models have been developed and incorporated into the Manski's formulation (e.g., Swait *et al.* 1987). Another probabilistic model which adheres to this process was proposed by Swait (2001), where both stages are incorporated into a single model through the specification of GEV generating function.

The above two approaches are rigorously formulated and have been successfully applied in mode choice context. It is, however, difficult to apply them to a route choice context since they have to evaluate all the possible subsets of alternatives, which may not be computationally tractable. Thus, it is realistic to adopt the deterministic or rule based model to represent the screening process in network simulation and traffic assignment. D'Este (1997) and Hato *et al.* (1999) developed the algorithm for network assignment in which Tversky's EBA concept (Tversky, 1972) is embedded to create choice sets of acceptable routes.

Another approach that seems applicable for the real transport network was proposed by Cascetta *et al.* (2002). In this approach, the availability of each route is represented by binary choice model, and then incorporated into the utility function in logarithmic form as one of the explanatory variables. Since this approach reduces the two-stage consideration into a single-stage, models such as multinomial logit are straightforwardly applicable.

Representation of Perceived Travel Time

The main factors which cause uncertainty in a driver's route choice situation are, network configuration and travel time. In classical network simulation and traffic assignment models, the deterministic and objective values are used for travel time, where all the drivers are implicitly assumed to have perfect information regarding traffic conditions. This assumption is apparently unrealistic, and models with such assumptions can not capture the effect of the information provision of travel time. Furthermore, factors other than travel time, e.g., perceived uncertainty of travel time, may be of significance in route choice.

Figure 2 illustrates the taxonomy of the representation approach to driver's perceived travel time. Random utility models, which usually include the objective value of travel time in the utility functions, can be interpreted as assuming the imperfect information of drivers in the sense that the difference between objective and perceived travel times and the heterogeneity of perceptions are included in the probabilistic component.

However, it would not be enough to investigate the effect of information provision since it would reduce the cognitive efforts caused by uncertainty even if he/she does not change the route. Two approaches can be taken to explicitly

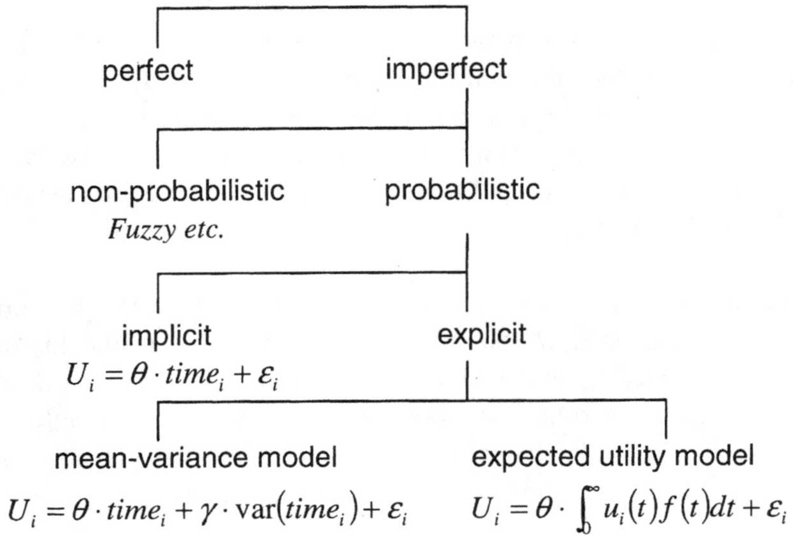

Figure 2. Representation of Perceived Travel Time.

consider the variation of travel time. One is called a mean-variance model (e.g., Jackson *et al.*, 1982) in which both mean travel time and its variance are used as the explanatory variables in the utility function. The other approach is based on the expected utility theory, where the probabilistic distribution of travel time is directly incorporated. Although these two approaches are relatively cumbersome, since the former requires asking or observing the variance of travel time and the latter is the open-form model, they would be useful in analyzing the effects of information provision at an individual level.

Recently, different types of approaches, which do not adopt the probabilistic representation of travel time, have been proposed. Akiyama (1993), for example, utilizes the fuzzy set theory to represent the driver's perception of travel time; and Fujii *et al.* (2001) developed a model of a driver's departure time decision where the drivers are postulated to perceive the travel time as an interval of possible resultant value. These approaches seem to represent the actual perception closely, and show great promise to understand the route choice decisions.

Decision Making Rules

Discrete choice models such as multinomial logit generally use the linear-in-attributes utility functions. Such models are called "compensatory model" in the sense that a low score with one attribute (e.g., high travel cost) can be compensated by a high score with another (e.g., short travel time). The linear-in-attributes models have been successfully applied due to their computational operation ability for both estimation and prediction.

The decision makers in such models are explicitly or implicitly modeled as "they evaluate all the attributes of all the available alternatives simultaneously, and then select the alternative with the highest utility considering trade-offs from amongst all the attributes". This assumption, however, may be inadequate under the route choice situation since it requires a large amount of cognitive effort and time while the drivers must make decisions from amongst numerous alternatives under severe time pressure.

On the other hand, there exists a wide variety of decision-making rules that have been proposed in the literature on consumer behavior and psychology (see Payne *et al.*, 1993 for recent comprehensive review). They have suggested that individuals employ more simplified decision rules, such as the satisficing concept of Simon (1955) and non-compensatory rules, in order to avoid the search costs associated with the gathering of necessary information in the above situations. Particularly, hierarchical or sequential decision rules, such as lexicographic and elimination by aspects (Tversky, 1972), where alternatives are processed attribute-by-attribute, are considered to be closer to the actual choice process and supported by empirical findings in a number of studies (e.g. Wright, 1975). These heuristics hold the accuracy of decisions but require low cognitive effort compared to compensatory models even if the number of alternatives increases and the time pressure becomes severe (Payne *et al.*, 1993).

Although some of the non-compensatory rules have been applied to the screening process as mentioned above, much more work should be done to address this research topic.

Dynamics of Route Choice Decisions

Most traffic assignment techniques assume the network equilibrium. This implies that the decision makers change their behavior instantaneously according to the traffic conditions, their preferences are invariant over the time period, and decisions are independent of past decisions. On the other hand, route choice decisions are repeated day by day and thus strongly influenced by past experiences. Therefore, it is necessary to incorporate the dynamic aspects of route choice behavior into the simulation tools to capture the effect of dynamic traffic management such as route guidance information.

Repeated choices may result in: 1) an adjustment of perceived travel time through short-term learning, 2) a change in decision-making rules through long-term learning, in order to reduce the cognitive effort, and 3) a construction of habit which finally no longer requires the decision-making process. Among the variety of theoretical and empirical findings in past research, the following studies are particularly worth being addressed.

Two different views can be made on the effects of repeated choice on drivers' perceived travel time. One is "rational expectation" as proposed by Muth (1961), where drivers' subjective expectation for travel time coincides with the objective value through the long-term learning. Kobayashi (1994) theoretically verifies that system converges to an equilibrium state under the assumption of rational expectation. On the other hand, Horowitz (1984) and Iida *et al.* (1992) suggest that drivers tend to make adaptive expectations since the usage of previous travel experiences varies from driver to driver. In such cases, traffic condition does not always converge to the equilibrium state, and it sometimes converges to other stable states.

Similarly, two different views exist on drivers' decision-making rules. Kobayashi *et al.* (2001) demonstrated that any decision-making rules are reduced to the minimax rule through the long-term learning under the assumption of bounded rationality. On the other hand, Nakayama *et al.* (1999) empirically verified that several decision-making rules still exist in converged "deluded equilibrium" using the micro-simulation model in which the drivers change the rule according to the experience.

Repeated choice may also affect the timing of decision-making. Since the cognitive resources of human beings' are limited, people tend to construct their habits through the iterative choices to save the cognitive efforts, which finally no longer requires the decision-making process. This behavioral aspect is salient especially in a route choice situation where the drivers must make decisions from among numerous alternative routes under severe time pressure. In such situations, decision-making may occur only if the driver perceives the difference in transportation environment from the ordinary state. In contrast, conventional models assume that drivers make decisions at every specific time period and/or when they reach each node. Such models may overestimate the route changing behavior and mislead the effect of information provision.

Modeling Approach

As briefly described in this section, there are a variety of approaches and behavioral insights. This causes a new problem of what approach or behavioral aspects we should adopt in a network simulation and traffic assignment. The answer to this question is found in the adaptive use of approaches according to the purpose of the studies faced by analysts. Stochastic user equilibrium model, for example, may even play an important role in analyzing average traffic conditions of large networks, while dynamic traffic simulation models are essential to investigate the effect of daily traffic management such as signal control and route guidance information.

Another problem will arise when incorporating the various behavioral aspects into the models. Since the choice set, perceived travel time, past experiences, and decision making rules vary across drivers, the heterogeneity among drivers should be explicitly considered. It is, however, difficult to directly observe the decision making process of each driver. One possible way to introduce the heterogeneity is as follows:

Step 1) develop the model including various types of decision makers which is typically observed through empirical research,

Step 2) conduct several traffic simulations by changing the distribution of those groups, and

Step 3) select the one which best fits the observed traffic state.

In the next section, we will show an empirical study of traffic assignment models where several groups of drivers with imperfect information are assumed using this approach.

APPLICATION OF NETWORK ASSIGNMENT WITH IMPERFECT INFORMATION

Ordinary traffic assignment models are based on Wardrop's first principle (1952) which goes "under equilibrium conditions traffic arranges itself in congested networks such that all routes between any origin-destination pair have equal and minimum costs while all unused routes have greater or equal costs." Under this condition no driver can reduce his/her costs by switching routes. Real traffic, however, has a dynamic nature and drivers are cruising on the road network with imperfect information on the future travel time and even on the network configuration. This situation can partly be observed by the probe car data presented in the next section where quite a few supposedly well-informed taxi drivers take the wrong routes.

On the other hand, in Japan, on-board car navigation equipments have sold more than ten million sets. Most of the recent models have the function of VICS (Vehicle Information & Communication System), one of the ATIS in Japan, that provides real-time congestion level to the navigation system. ATIS is believed to have the traffic flow approach the equilibrium state by rerouting some drivers to faster routes. In other words, no drivers are better off with ATIS if the prevailing traffic is under the equilibrium with perfectly informed drivers. In this section, traffic assignment models with imperfectly informed drivers are proposed to better represent the real traffic flow and evaluate the benefit of ATIS.

The Models

A traditional approach in traffic assignment that incorporates uncertainty in travel time (or generalized costs) is the stochastic user equilibrium (SUE) model (Daganzo and Sheffi, 1977). The SUE model assumes that every driver minimizes the perceived cost that is distributed with some probability density function. The dispersion parameter of the density function represents the magnitude of the variability in perceived costs. Yang (1998) proposed a multi-class user equilibrium model where each class has its own dispersion parameter to characterize different uncertainty levels among drivers.

According to our preliminary application of SUE models to the city street network in a large metropolitan area, however, SUE and non-stochastic assignments show little difference in reproducing the observed link flows. Therefore, imperfect information is represented in more direct ways in the traffic assignment models proposed in this section. We assume that imperfectness in information has two aspects: 1) imperfect travel time and 2) imperfect network configuration. Multi-class user equilibrium models are also employed where drivers in one class are perfectly informed and those in the other class have only imperfect information.

Route choice behavior of drivers in the class of imperfect travel time is modeled as follows. Although those drivers may have knowledge of the network configuration, they are not informed of the real-time travel time and consequently may not be able to choose the shortest route. In the proposed model, they are assumed to choose the route according to the travel time at no traffic condition. As a result, the traffic of this class is assigned by the "all-or-nothing" technique. We will denote this class "ImPerfect Information on Travel time Class" or "IPIT Class".

The drivers in the class of imperfect network configuration, on the other hand, might not be aware of some of the routes themselves. If the route that is not included in the driver's choice set is actually the shortest one, he/she cannot take the shortest route. In the model, the choice set of those drivers consists of only routes by arterial roads (three or more lanes and some connecting two lane roads). Figure 3 shows the actual network and the perceived network for the drivers in this class. We will denote this class "ImPerfect Information on Network configuration Class" or "IPIN Class". The traffic of the IPIN Class is assigned by the deterministic user equilibrium to the perceived network.

The traffic of the drivers' class of perfect information, or PI Class, is assigned by means of the ordinary deterministic user equilibrium technique to the full network. Two models are used in assigning all the traffic:

Model 1: All the drivers are either in PI Class or IPIT Class.

Model 2: All the drivers are either in PI Class or IPIN Class.

The model with the three classes could be used but, as stated below, IPIT Class is buried with PI Class and would be of no significance in fitting the observed traffic flow.

With Model 1, since the traffic of IPIT Class is assigned to the shortest path in the flow independent way, the traffic of PI Class is assigned by the user equilibrium technique after the IPIT traffic is assigned. The ratio of the two classes is changed by trial-and-error to best fit the observed traffic flow.

With Model 2, the following modified Frank-Wolfe algorithm is applied.

Step1: Initial feasible solution
Set iteration number n = 1 and obtain the initial feasible link flow $\{x_a^{(n)}\}$ as follows:
1) Load IPIN Class traffic
Find the shortest path on the perceived network for each O-D pair at the free flow level and load the all the IPIN Class traffic on the shortest path.
2) Load PI Class traffic
Find the shortest path on the full network for each O-D pair at the free flow level and load the all the PI Class traffic on the shortest path.
3) Calculate the total traffic
Add the two class traffic volume to obtain the total traffic $\{x_a^{(n)}\}$.

Step2: Link cost update
Update the link travel time $\{ta(x_a^{(n)})\}$ for the current flow $\{x_a^{(n)}\}$.

Step3: Trial solution
Find the shortest path and load all the traffic.
1) Load IPIN Class traffic
Find the shortest path on the perceived network for each O-D pair at the current flow level and load the all the IPIN Class traffic on the shortest path.
2) Load PI Class traffic
Find the shortest path on the full network for each O-D pair at the current flow level and load the all the PI Class traffic on the shortest path.
3) Calculate the total traffic
Add the two class traffic volume to obtain the total traffic $\{y_a\}$.

Step4: Search for descendent vector
Set $x_a^{(n+1)}=x_a^{(n)}+ \alpha (y_a-x_a^{(n)})$ and find the optimum step-size α that minimizes the objective function to obtain traffic volume $x_a^{(n+1)}$.

Step5: Convergence Check
If the convergence criterion is satisfied, stop. If not, set n=n+1 and go back to Step2.

Empirical Analysis

The case study area, Metropolitan Nagoya, has a diameter of approximately 50 km with 279 traffic zones (the same number of centroids). The road network has 1304 nodes and 4303 links and includes tolled expressways. Daytime 12 hour (7am – 7pm) traffic is assigned on the network.

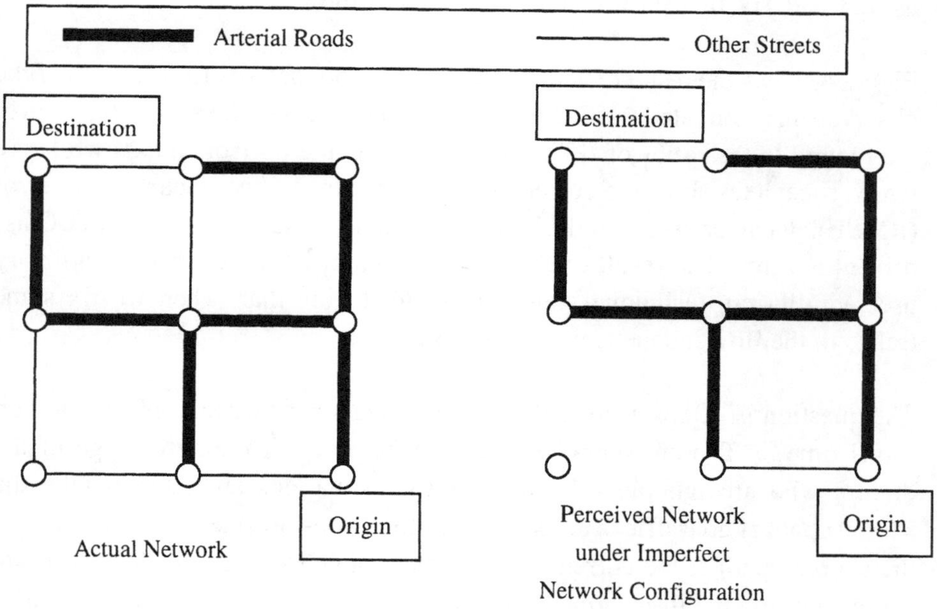

Figure 3. Perceived Network under Imperfect Information on Network Configuration

The link-cost function used is a BPR type with the following form:

$$t = t_0 \left\{ 1 + \alpha \left(\frac{q}{c} \right)^{\beta} \right\} \tag{1a}$$

$$t_0 = \gamma_0 + \gamma_1 X_1 + \gamma_2 X_2 \tag{1b}$$

t = link travel time
t_0 = free flow travel time

q = traffic volume
c = volume capacity
X_1 = density of signalized intersections
X_2 = inverse of speed limit
$\alpha, \beta, \gamma_0, \gamma_1, \gamma_2$ = unknown parameters

A set of unknown parameters are estimated for each of the six road categories using Road Traffic Census Data of this area as shown in Table 1.

Figure 4 shows the result of Model 1 where the goodness-of-fit measures of the observed and estimated traffic volumes on links. The goodness-of-fit measures are plotted to the ratios of IPIT Class drivers on the horizontal axis while the two vertical axes show the correlation coefficients and root mean square error (RMSE). It can be seen that the best fit is obtained when the ratio of IPIT Class drivers is zero. The result of this traffic assignment shows that the ordinary user equilibrium technique yields are better fitting than when we fix some traffic to the flow independent shortest paths.

The question is "Does it mean that all the drivers have perfect information on travel time?" The answer would be "Maybe not." If the percentage of the drivers who are ignorant of current traffic situations and always take the shortest path at no traffic is not large, the other well-informed drivers who take the shortest path at the current traffic may be able to cover the "naïve" route choice done by the ill-informed drivers. If this is the case, the resulting traffic flow is the same as the ordinary user equilibrium that is obtained under the assumption that all the drivers have perfect information.

If the ratio of IPIT drivers were large enough, their naïve route choice would not be covered by PI drivers and the better fit to the observed flow would be obtained at a certain ratio of IPIT drivers in Figure 4. The obtained result tells us that there is no evidence that too many drivers take the flow's independent shortest paths to distort the user equilibrium pattern.

Model 2 performs in a significantly different way as shown in Figure 5. The best goodness-of-fit value is obtained at the point where 20% of drivers are categorized in IPIN Class. In the next step, in the application of Model 2 we stratified all the trips by trip purposes and then changed the ratios of IPIN Class drivers for each segment. When we categorized 80% of private trips and 10% of business trips into IPIN Class drivers, the assigned traffic best fit to the observed traffic as shown in Figure 6. In that case, the total IPIN Class drivers composed of 26% of total drivers. We obtained roughly the same goodness-of-fit values when we assumed that 100% of the private trips, which comprised 30% of total trips, made by IPIN Class drivers.

The findings from the above analysis can be summarized as follows:

1) We could reject the assumption that the majority of drivers take the length-wise shortest route due to ignorance of current travel time and that consequently the disequilibrium traffic state is realized.
2) It is found that too many drivers use only arterial roads to keep the user equilibrium state. This may be attributed to the limited choice set of routes for some drivers and/or the preference bias toward arterial roads.
3) We obtained the best fit to the observed traffic when we set 20 – 30% of drivers to have a limited choice set which is composed of arterial roads. Most of those naïve drivers are found in private trips rather than commuting or business trips.

Table 1. Parameter Estimates of Link-Cost Function (t-statistics in parentheses).

Road Type	α	β	γ_0	γ_1	γ_2	# of samples
Intercity Expressway	0.362 (3.3)	4.53 (5.3)	0.0097 (21.7)		0.143 (4.4)	220
Urban Expressway	0.408 (7.3)	3.02 (6.2)	0.0138 (41.6)		-	342
Arterial Roads with 5+ Lanes	1.57 (2.7)	1.95 (3.5)	0.0164 (3.4)	0.0037 (12.4)	0.290 (1.2)	1233
Arterial Roads with 3-4 Lanes	1.35 (1.8)	2.99 (8.0)	0.0124 (6.2)	0.0024 (17.1)	0.740 (7.4)	2858
Streets with 2 Lanes	0.351 (12.3)	1.32 (2.5)	0.0006 (0.4)	0.0037 (23.5)	1.05 (18.9)	8061
Streets with 1 Lane	0.154 (8.3)	1.82 (2.1)	0.0231 (18.9)	-0.0002 (-0.3)	0.0539 (1.2)	308

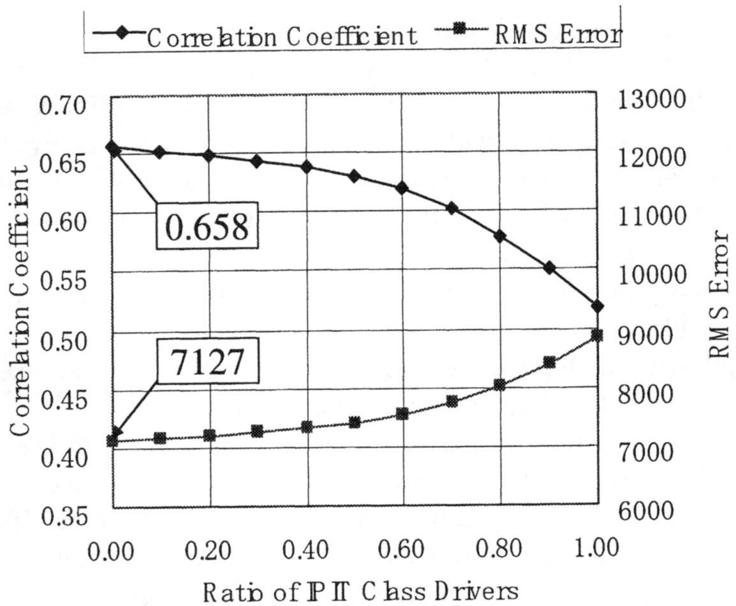

Figure 4. Fit to Observed Flows with respect to Ratio of IPIT Drivers.

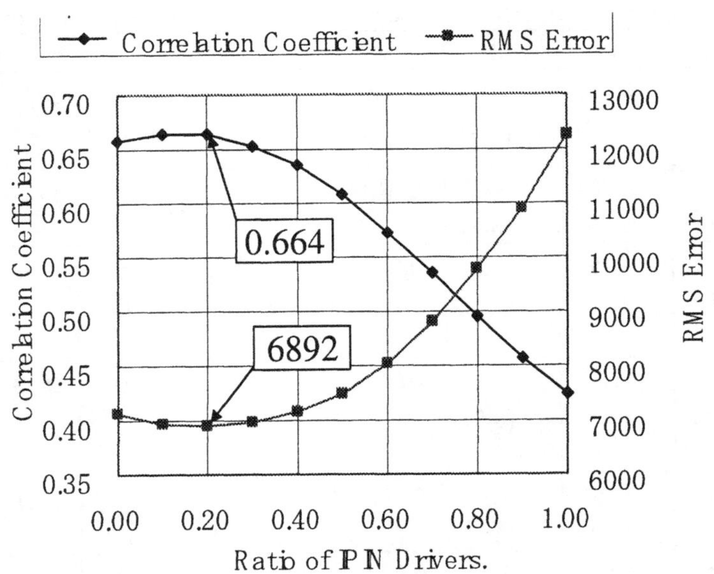

Figure 5. Fit to Observed Flows with respect to Ratio of IPIN Drivers.

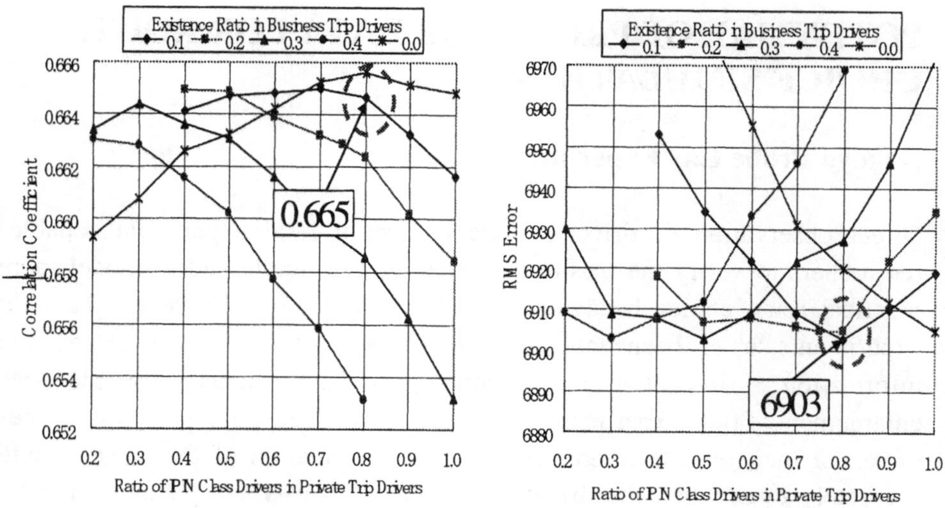

Figure 6. Fit to Observed Flows with respect to Ratio of IPIN Drivers in Stratified Drivers.

Benefit Evaluation of Information Provision

When the traffic is apart from the user equilibrium state due to imperfect information, user benefit accrues by providing information and approaching the traffic to the more efficient user equilibrium. The total travel time was compared between the two states.

We assumed that all the private trip drivers currently have imperfect information on network configuration as seen in the previous analysis. By providing information on the full network, for instance, by means of on-board car navigation equipment, the prevailing traffic will approach the user equilibrium state and consequently the total travel time reduces by 4.6 million minutes per day which is approximately 3.6% of total travel time. The monetary value of this travel time reduction is equivalent to 97 billion JPY or 800 million USD per year by applying the standard value of time of 57.5 JPY per minute.

According to the assignment results, 94% of travel time savings accrues to the previously imperfectly informed drivers. The average monetary benefit to such a driver is 70,900 JPY/vehicle/year or 600 USD. A car navigation unit of 150,000 JPY pays itself off in approximately two years.

POTENTIAL OF PROBE CAR DATA FOR ROUTE CHOICE SIMULATION

Nagoya Probe Car Experiment

Direct observation of driver's route choice in field experiments enables researchers to verify the traditional behavioral principle and to model more practical route choice behavior for traffic simulation. However, the field experiments have been labor-intensive and expensive in obtaining the information. Recently, ITS technologies have enabled sample size enlargement of field experiments. Probe cars, which are equipped with devices to sensor their own locations, can provide information of the chosen route accurately even when the driver doesn't recognize where he is driving by himself. In this chapter, the preliminary result of the data collection with probe cars in Japan is presented, and the potential of the probe car data for examining route choice model is discussed.

The probe car data of this study was collected in the Nagoya Metropolitan Area over two months (Jan 2002 – Mar 2002). This experiment was conducted by *Internet ITS Project*, which is sponsored by the Ministry of Economy, Trade, and Industry and implemented by *Internet ITS Research Group* which included Keio University, Toyota, Denso, and NEC. In this experiment, 1,500 taxies in Nagoya served as probe cars, in cooperation with 32 Nagoya Taxi Association member companies. Probe information such as location, velocity, acceleration and windshield wiper operation was provided via the Internet. Data transmission was event-based, and the frequency of events observed during the two month experiment period is shown in Table 2. Events of "Mileage", "Start" and "Stop" consist of 30-35% respectively and they cover more than 96% of all events.

This study analyses the routes taken by taxis with passengers since taxis without passengers often just cruise arterial roads to find passengers. With regards to the route choice decisions made by taxis with passengers, it is assumed that taxi drivers choose the route and that passengers decide whether they use a tolled expressway or not. It is generally also considered that taxi drivers have a wealth of knowledge of the road network and respond to traffic conditions appropriately. Hence, the results of route choice analysis in this study may not be directly applied to the general drivers.

Table 2. Events when the information is transmitted to the operation center.

Event	Definition	Percent
Mileage	300 meter run without any other events	31.3%
Start	Vehicle starts	35.1%
Stop	Vehicle stops	29.8%
Passenger load/unload	Passenger's load and unload	
Time	550 seconds passes without any other events	
Engine start and stop	Engine starts and stops	3.8%
Hazardous movement	Exceeding speed limit, excessive acceleration and deceleration	

Observations of Route Choice Behavior

Trips between Nagoya Airport and Nagoya CBD which include Nagoya Station, the Government Center, and the CBD Core are extracted for the route choice analysis because taxis are very heavily used in those O-D pairs. Figure 7 illustrates the O-D pairs and the rough network. Their distances are approximately 10-15 km. Although there are numerous routes between these O-D pairs, they can be integrated into six routes characterized by six bridges over the *Shonai River*. Since these bridges create bottlenecks, drivers' route choice behavior can be characterized by bridge choice or bottleneck choice

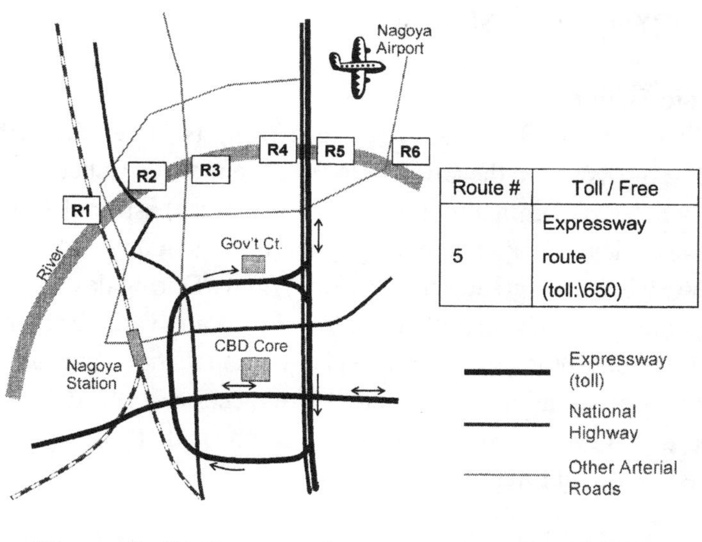

Figure 7. Study area of route choice observations.

behavior. Among the six, one route uses a tolled expressway where, the toll is 650 JPY (5.50 USD), and the ring expressway in the CBD is clockwise and one way.

Tables 3-5 show number of trips, mean travel time, standard deviation of travel time and mean travel distance on each route during four periods of time (morning peak: 7:00-9:00, day-time off peak: 9:00-17:00, evening peak: 17:00-19:00, and night-time off peak: 19:00-7:00). We can see that the majority of trips between Nagoya Airport and CBD use Route 4 (free) and Route 5 (toll). Route 5 is especially the most popular for most of the time periods and O-D pairs. We can calculate the mean value of travel time (VOT) under the assumption of the symmetrical distribution of VOT from the ratio of free and tolled route users.

(1) Station to Airport
During two off-peak periods Route 4 is used more heavily than Route 5 (toll). During morning peak Route 3 is significantly faster than Route 4 but few use this faster route. The mean VOT varies 50-100 JPY/min (0.4-0.8 USD/min) according to time periods. During hours when free routes are congested (during morning peak, day-time off-peak, and evening peak), it is about 60 JPY/min (0.5 USD/min), which is very close to the standard VOT used in Japan. When free routes are not congested, i.e., during night-time off-peak, it is about 90 JPY/min (0.8 USD/min).

(2) Airport to Station
For this direction, the distance of Route 5 (toll) is larger than free routes because the expressway in the CBD is clockwise one way. Hence, time savings of using Route 5 are marginal (not more than 5 minutes) especially during peak hours. Nevertheless, Route 5 is most popular through a day and consequently the mean VOT becomes greater than 100 JPY/min. During day-time off-peak, although Route 2 is better, in both mean travel time and variance than Route 4, the number of Route 4 users is four times larger than that of Route 2. We could say from these observations that Route 4 is unnecessarily heavily used when we assume that drivers have perfect information or that drivers' decision is purely based on travel time.

(3) Airport to/from the Government Center and the CBD Core
Between the Government Center and Nagoya Airport, time-savings of Route 5

are not more than 5 minutes because of the short usage of the expressway. During the evening peak 36% trips use Route 5 in the direction from Nagoya Airport to the Government Center, although the free route is four minutes faster. Furthermore, from Nagoya Airport to the CBD Core during the morning peak, the free route is five minutes shorter but 44% of trips use Route 5. Thus, we could observe from probe cars that a substantial number of drivers, even taxi drivers, make seemingly irrational decisions.

Table 3. Route choice observations between Nagoya Airport and Nagoya Station.

Nagoya Station → Airport

Morning Peak (2hrs)

Rt.	# of trip	Mean Time (min)	SD	Distance (km)
1	0	--	--	--
2	0	--	--	--
3	2	25.9	1.1	12.2
4	66	34.3	7.7	13.0
5	122	23.7	5.4	13.6
6	0	--	--	--

Day-time Off-Peak (8hrs)

Rt.	# of trip	Mean Time (min)	SD	Distance (km)
1	1	35.3	0.0	13.2
2	0	--	--	--
3	6	33.1	3.3	13.0
4	181	34.4	4.4	13.3
5	160	23.1	5.5	13.6
6	1	50.2	0.0	13.4

Evening Peak (2hrs)

Rt.	# of trip	Mean Time (min)	SD	Distance (km)
1	0	--	--	--
2	0	--	--	--
3	1	42.7	0.0	13.5
4	10	33.7	6.7	11.4
5	34	22.5	4.7	13.5
6	1	43.3	0.0	13.8

Airport → Nagoya Station

Morning Peak (2hrs)

Rt.	# of trips	Mean Time (min)	SD	Distance (km)
1	0	--	--	--
2	4	29.2	5.2	14.3
3	0	--	--	--
4	12	29.5	4.2	13.4
5	22	29.7	6.9	15.9
6	0	--	--	--

Day-time Off-Peak (8hrs)

Rt.	# of trips	Mean Time (min)	SD	Distance (km)
1	1	32.1	0.0	13.5
2	63	32.6	4.2	14.2
3	3	35.1	4.0	13.6
4	243	34.7	5.0	13.5
5	304	28.9	5.8	16.5
6	0	--	--	--

Evening Peak (2hrs)

Rt.	# of trips	Mean Time (min)	SD	Distance (km)
1	0	--	--	--
2	18	37.0	4.8	14.5
3	0	--	--	--
4	71	32.9	6.1	14.2
5	96	31.4	6.4	16.8
6	0	--	--	--

Night-time Off-Peak (8hrs)

Rt.	# of trip	Mean Time (min)	SD	Distance (km)
1	1	31.7	0.0	15.8
2	1	26.5	0.0	14.2
3	1	31.7	0.0	13.6
4	52	29.7	4.1	12.7
5	32	23.1	5.1	13.7
6	0	--	--	--

Night-time Off-Peak (8hrs)

Rt.	# of trips	Mean Time (min)	SD	Distance (km)
1	0	--	--	--
2	100	30.7	4.1	14.3
3	2	24.6	1.5	12.9
4	282	31.6	4.6	13.7
5	371	26.2	4.8	16.8
6	2	43.0	9.8	15.8

Table 4. Route choice observations between Nagoya Airport and Government Center.

Government Center→ Airport

Morning Peak (2hrs)

Rt.	# of trips	Mean Time (min)	SD	Distance (km)
4	14	23.0	3.6	9.8
5	8	22.1	3.2	10.4

Day-time Off-Peak (8hrs)

Rt.	# of trips	Mean Time (min)	SD	Distance (km)
4	31	24.4	4.5	10.2
5	19	20.2	4.8	10.1

Evening Peak (2hrs)

Rt.	# of trips	Mean Time (min)	SD	Distance (km)
4	6	24.9	2.6	10.4
5	10	18.2	4.5	11.1

Night-time Off-Peak (8hrs)

Rt.	# of trips	Mean Time (min)	SD	Distance (km)
4	5	22.4	5.5	9.6
5	1	20.5	0.0	10.6

Airport → Government Center

Morning Peak (2hrs)

Rt.	# of trips	Mean Time (min)	SD	Distance (km)
4	0	--	--	--
5	3	23.0	6.3	9.4

Day-time Off-Peak (8hrs)

Rt.	# of trips	Mean Time (min)	SD	Distance (km)
4	43	28.5	5.0	10.5
5	22	23.9	4.9	10.1

Evening Peak (2hrs)

Rt.	# of trips	Mean Time (min)	SD	Distance (km)
4	16	26.6	4.3	10.0
5	9	30.9	4.1	11.5

Night-time Off-Peak (8hrs)

Rt.	# of trips	Mean Time (min)	SD	Distance (km)
4	16	24.4	4.9	9.2
5	8	21.0	5.7	11.8

Table 5 Route choice observations between Nagoya Airport and CBD Core.

CBD Core→ Airport

Morning Peak (2hrs)

Rt.	# of trips	Mean Time (min)	SD	Distance (km)
4	35	27.1	4.5	11.5
5	72	21.6	3.9	13.0

Day-time Off-Peak (8hrs)

Rt.	# of trips	Mean Time (min)	SD	Distance (km)
4	137	30.1	5.1	12.0
5	227	23.7	4.1	13.5

Evening Peak (2hrs)

Rt.	# of trips	Mean Time (min)	SD	Distance (km)
4	6	33.7	4.8	12.1
5	19	22.3	3.8	12.4

Night-time Off-Peak (8hrs)

Rt.	# of trips	Mean Time (min)	SD	Distance (km)
4	63	24.6	4.4	11.7
5	175	20.6	2.7	13.6

Airport → CBD Core

Morning Peak (2hrs)

Rt.	# of trips	Mean Time (min)	SD	Distance (km)
4	5	32.5	3.8	12.4
5	4	37.8	6.7	12.4

Day-time Off-Peak (8hrs)

Rt.	# of trips	Mean Time (min)	SD	Distance (km)
4	346	34.2	5.8	12.1
5	178	29.9	7.0	12.9

Evening Peak (2hrs)

Rt.	# of trips	Mean Time (min)	SD	Distance (km)
4	44	33.5	7.0	12.3
5	49	27.1	6.4	11.5

Night-time Off-Peak (8hrs)

Rt.	# of trips	Mean Time (min)	SD	Distance (km)
4	214	31.0	4.8	12.3
5	132	27.4	5.7	12.9

CONCLUSIONS

This paper focused on driver's route choice behavior. It especially examined the appropriateness of the user equilibrium assumption on travel time. The assumption of driver's having perfect information on traffic state is unrealistic particularly on congested urban road networks. Even the information on network configuration may well be imperfect except for commuter's regular corridors.

This paper, firstly, the traditional user equilibrium traffic assignment model was applied to a metropolitan area. By applying the multi-class user equilibrium model, we obtained the best fit to the observed traffic flows when

20–30% of drivers, mainly private trip drivers, chose only arterial roads. It was found this causes unnecessary congestion on arterial roads. By providing information of the full network to make the equilibrium state, about a 70,900 JPY (600 USD) reduction in travel time accrues to such drivers. We cannot however conclude, whether this concentration on arterial roads is attributed to the imperfect information on network configuration or simply due to the preference bias to arterials.

Secondly, probe car data is analyzed to directly observe route choice behavior. We investigated a heavily traveled O-D pair where a toll route and several free routes could be used. It was found that 30–70% travelers used the toll route, implying that VOT distributes. We observed that multiple free routes are used among which travel times differed 10–20%. It was also observed that the toll route is often slower but many drivers chose it. Thus, a substantial number of drivers make seemingly irrational decisions. Imperfect information and other factors apart from time and cost may partly explain these observations.

A route choice model is a basic component of traffic simulators. Factors affecting route choice and the decision making mechanism must be discovered to build a plausible route choice models. Observing route choice behavior had been very costly until ITS technologies, such as probe cars, emerged. Therefore, normative models such as user equilibrium have been widely used to represent traffic flows. We found that user equilibrium traffic assignment models perform fairly well especially when some modifications, such as multi-class users, to represent daily traffic are made. Route choice observation made through probe car data, however, revealed seemingly irrational behavior and disequilibrium states. Such assignment models, therefore, cannot be reproduced because of the dynamic nature of traffic.

Due to the evolving ITS-driven transportation policies such as dynamic route guidance, a need for reliable traffic simulators will increase rapidly. Accordingly, intensive studies on route choice behavior are called for.

REFERENCES

Abdel-Aty, M. A., Vaughn, K. M., Kitamura, R. and Jovanis, P. P. (1993). Impact of ATIS on drivers' decisions and route choice: A literature review, Research report of California PATH program of the University of California, UCB-ITS-PRR-93-11.

Akiyama, T. (1993). Transport network analysis using fuzzy set theory, *Infrastructure Planning Review*, **11**, pp13-27 (in Japanese).

Bhat, C.R. (2000). Flexible model structures for discrete choice analysis, In *Handbook of Transport Modelling* (D.A. Hensher and K.J. Button), pp71-90.

Cassetta, E., Russo, F., Viola, F. A. and Vitetta, A. (2002). A model of route perception in urban road networks, *Transportation Research Part B*, **36**, pp577-592.

Chen, T.-Y., Chang, H.-L. and Tzeng, G.-H. (2001). Using a weight-assessing model to identify route choice criteria and information effects, *Transportation Research Part A*, **Vol. 35**, pp197-224.

COMSIS Corporation (1995). Analysis of travelers' preferences for routing: literature review, Report submitted to Federal Highway Administration, U.S. Department of Transportation, DTFH61-95-C-00017.

D'Este, G. (1997). Hybrid route choice procedures in a transport network context, *Journal of the Eastern Asia Society for Transportation Studies*, **2(3)**, pp737-752.

Daganzo, C.F. and Sheffi, Y. (1977). On Stochastic Models of Traffic Assignment, *Transportation Science*, **Vol. 14**, pp42-54.

Dial, R.B. (1971). Probabilistic multipath traffic assignment which obviates path enumeration, *Transportation Research*, **5**, pp33-111.

Fujii, S. and Kitamura, R. (2001). Decision frame for departure time choice under uncertainty, *Infrastructure Planning Review*, **18(3)**, pp491-495 (in Japanese).

Hato, E. and Asakura, Y. (2000). Incorporating bounded rationality concept into route choice model for transport network analysis, *European Transport Conference 2000 in Cambridge, Behavioural Modelling*, **441**, pp1-12.

Heap, S. H., Hollis, M., Lyons, B., Sugden, R. and Weale, A. (1992). *The Theory of Choice*, Blackwell.

Horowitz, J.L. (1984). The stability of stochastic equilibrium in a two-link transportation network, *Transportation Research Part B*, **18**, pp13-28.

Iida, Y., Akiyama, T. and Uchida, T. (1992). Experimental analysis of dynamic route choice behavior, *Transportation Research Part B*, **26**, pp17-32.

Jackson, W.B. and Jucker, J. V. (1982). An empirical study of travel time variability and travel choice behavior, *Transportation Science*, **16**, pp460-475.

Kobayashi, K. (1994). Information, rational expectation and network equilibria – an analytical perspective for route guidance systems, *The Annals of Regional Science*, **28**, pp369-393.

Kobayashi, K. and Matsushima, K. (2001). Bounded rationality and traffic behavior modeling: an analytical perspective, *Journal of Infrastructure Planning and Management*, **688/ IV-53**, pp5-17 (in Japanese).

Koppelman, F.S. and Sethi, V. (2000). Closed-form discrete-choice models, In *Handbook of Transport Modelling* (D.A. Hensher and K.J. Button), pp211-227.

Manski, C. (1977). The structure of random utility models, *Theory and Decision*, **8**, pp229-254.

Muth, J.F. (1961) .Rational expectations and the theory of price movements, *Econometrica*, **29**, pp315-335.

Nakayama, S., Fujii, S. and Kitamura, R. (1999). Drivers' learning and road transportation system behavior: a dynamic study toward the complex-systems analysis of Road transportation, *Infrastructure Planning Review*, **16**, pp753-761 (in Japanese).

Payne, J., Bettman, J. and Johnson, E. (1993). *The Adaptive Decision Maker*, Cambridge University Press, New York.

Simon, H. A. (1987). Bounded rationality, In *The New Pargrave: Utility and Probability* (J. Eatewell et al.), W. W. Norton & Company.

Simon, H. A. (1955). A behavioral model of rational choice, *Quarterly Journal of Economics*, **69**, pp99-118.

Swait, J. (2001). Choice set generation within the generalized extreme value family of discrete choice models, *Transportation Research Part B*, **35**, pp643-666.

Swait, J. and Ben-Akiva, M. (1987). Incorporating random constraints in discrete models of choice set generation, *Transportation Research Part B*, **21**, pp91-102.

Tversky, A. (1972). Elimination by aspects: a theory of choice, *Psychological Review*, **79**, pp281-299.

Vovsha, P. and Bekhor, S. (1998). Link-nested logit model of route choice: overcoming route overlapping problem, *Transportation Research Record*, **1645**, pp133-142.

Wardrop, J. (1952). Some theoretical aspects of road traffic research, *Proceedings of the Institution of Civil Engineers, Part II*, pp325-362.

Williams, H. C. W. L. and Ortuzar, J. D. (1982). Behavioural theories of dispersion and the mis-specification of travel demand models, *Transportation Research Part B*, **16**, pp91-102.

Willumsen, L.G. (2000). Travel networks, In *Handbook of Transport Modelling* (D.A. Hensher and K.J. Button), pp165-180.

Wright, P. (1975). Consumer choice strategies: Simplifying vs optimizing, *Journal of Marketing Research,* **12**, pp60-67.

Yang, H. (1998). Multiple equilibrium behaviors and advanced traveler information systems with endogenous market penetration, *Transportation Research.-B*, **Vol.32, No.3**, pp205-218.

An Overview of PCATS/DEBNetS Micro-Simulation System: Its Development, Extension, and Application to Demand Forecasting

Ryuichi Kitamura
Department of Urban Management, Kyoto University, Sakyo-ku, Kyoto 606-8501, JAPAN
Department of Civil and Environmental Engineering, University of California, Davis, USA
rkitamura@termws.kuciv.kyoto-u.ac.jp

Akira Kikuchi
Department of Urban Management, Kyoto University, Sakyo-ku, Kyoto 606-8501, JAPAN
kikuchi@term.kuciv.kyoto-u.ac.jp

Satoshi Fujii
Department of Civil Engineering, Tokyo Institute of Technology, JAPAN
Toshiyuki Yamamoto
Department of Geotechnical and Environmental Engineering, Nagoya University, JAPAN

ABSTRACT

The micro-simulator of individuals' daily travel, PCATS, and a dynamic network simulator, DEBNetS, are integrated to form a simulation system for urban passenger travel. The components of the simulation system are briefly described, and three areas of on-going system improvement are described, i.e., (i) introduction of stochastic frontier models of prism vertex location, (ii) adoption of a fine grid system for quasi-continuous representation of space, and (iii) use of MCMC algorithms to handle colossal choice sets. Application case studies demonstrate that micro-simulation is a practical approach for demand forecasting and policy analysis, especially in the area of demand management.

INTRODUCTION

Traffic flow on road networks is dynamic. The flow rate on a roadway section varies substantially in the course of the day; the direction of the dominant traffic flow changes over time; and queues at bottlenecks grow, then dissipate. Simulating traffic flow on networks along a time axis is an effective approach to represent network dynamics. In fact a number of network simulators have been developed and their effectiveness in traffic control and transportation planning is being evaluated.

Inputs to dynamic network simulation may take on the form of a set of origin-to-destination (O-D) trip matrices by time of day, each of which represents trip interchange over a certain time interval within a day. Since such intervals are artificial and real traffic flow varies continuously over time irrespective of the set of time intervals adopted by the researcher, it would be more desirable to adopt shorter intervals for improved accuracy. At the limit where the length of the interval diminishes, inputs are represented by individual trips to be loaded onto the network along a continuous time axis.

The conventional, trip-based passenger travel demand forecasting methods, such as the four-step procedures, have been unable to provide inputs to dynamic network simulation. This is simply because these forecasting methods have not incorporated a time axis into their analytical frameworks. The spatial dimension has been captured in these methods by subdividing the study area into traffic zones. The time dimension, on the other hand, has never been represented properly in any analytical manner. This is quite curious in light of the fact that congestion, the long-time preoccupation of transportation planners, is a consequence of the concentration of demand in time, as well as in space. The O-D trip matrices by time interval that many transportation planning agencies have developed are by and large based on daily O-D traffic; heuristic procedures and rules of thumb are applied to the daily traffic to generate trip matrices by time interval.

One approach to generating inputs to dynamic network assignment is to simulate individuals' travel behavior and generate trips over a time axis; car trips generated in the simulation can be loaded onto the road network as they are generated. This approach has many advantages over the use of trip matrices by time of day. First, the latter method is incapable of capturing the effect of various planning measures, e.g., congestion pricing or parking surcharges by time of day. Second, the latter method is valid only when demand can be viewed as inelastic. The foundation of the former approach, on the other hand, is the recognition that travel demand is elastic. It is also based on the recognition that individuals may modify their behaviors when faced with changes in their travel environment. Consequently this approach is capable of accounting for behavioral changes caused by planning measures.

An increase in travel time due to congestion may cause the individual to change the route, departure time, destination, travel mode, or any combination of these. It may also be the case that the traveler chose not to make the trip altogether. These, and possibly other reactions of individuals, can be represented while simulating individuals' behaviors when reactions to changes in the travel environment are properly captured in the simulation procedure. Furthermore, when the simulation of individuals' behaviors and the simulation of network flow are interactively performed repeatedly, it may become possible to simulate individuals' learning and adaptation to changes in the travel environment, thus simulating the process in which individuals' behaviors collectively approach "equilibrium."

Dynamic network simulation also has many advantages. It has many applications in real-time traffic control as well as the prediction of network performance. More precise estimation of the magnitude and duration of congestion will become possible by dynamic network simulation. This will in turn make the prediction of the cost of congestion more precise. The analysis of transient phenomena, such as non-recurrent congestion due to a traffic accident, can be performed more appropriately. In addition, the environmental impacts of automotive traffic can be assessed more accurately because of the explicit inclusion of the time dimension. For example, the CO_2 concentration at an intersection can be estimated more adequately using simulation outputs. Roles of automotive traffic in the "heat island" phenomenon now found in large metropolitan areas can be better examined. The formation of photochemical smog, which depends on the clock time when NO_x and other contributing substances are emitted, can also be estimated more adequately.

Returning to the simulation of individuals' behaviors discussed above, it is worthy to note in this context that it offers additional merit of being capable of addressing how various planning measures may influence the patterns of trip chaining and the frequency of cold starts. Conventional methods of demand forecasting, which are trip-based, do not adequately capture how patterns of trip chaining may be influenced by policy measures, and how these changes may in turn affect traffic flow and pollutant emissions, as well as the frequency of cold starts. The combination of the simulation of individuals' travel behaviors and dynamic simulation of network flow can thus be a very powerful tool for traffic management and planning.

As a more exact and versatile tool for traffic management, demand forecasting and policy analysis, a micro-simulation model system of individuals' daily travel behaviors and network flow has been developed at Kyoto University. The main body of the system comprises two simulators, PCATS and DEBNetS. PCATS stands for Prism Constrained Activity-Travel Simulator, and, as its name indicates, simulates an individual's daily activity and travel while explicitly incorporating Hägerstrand's space-time prism (Hägerstrand, 1970). DEBNetS, which stands for Dynamic Event-Based Network Simulator, is a micro-meso scale simulator of network flow. These two simulators are integrated to perform, in the nomenclature of the conventional four-step procedures, trip generation, trip distribution, modal split and network assignment. The critical difference from the conventional procedures, however, is the inclusion of a time axis; trips are generated and network flow is simulated along time in the PCATS/DEBNetS system. Another important difference is that PCATS simulates not individual trips but the series of trips generated by an individual over a course of the day, while maintaining the spatial continuity of the trips and coherently representing various constraints on trip making.

In this paper, the PCATS/DEBNetS system is first outlined, with emphasis on PCATS. Also discussed is a synthetic household generator and demographic simulator called HAGS. HAGS is used to generate data for long-term forecasting with PCATS/DEBNetS. Ongoing efforts to extend the capabilities of the simulation system are then described. These are:

- incorporating models of prism vertex locations,
- adoption of a fine (10m × 10m) grid system instead of a system of large traffic zones, and
- use of a Markov Chain Monte Carlo (MCMC) method to handle colossal choice sets.

Following this, results of empirical applications are summarized. Problems that can be addressed with these enhancements are then discussed.

OUTLINE OF THE PCATS/DEBNETS SYSTEM

Three simulators, PCATS, DEBNetS and HAGS, are outlined in this section, with more emphasis placed on PCATS because it has several unique features that are noteworthy. HAGS stands for Household Attribute Generating System and generates disaggregate data of household and person attributes.

There are several important differences between the conventional travel demand forecasting procedures and the proposed micro-simulation system that are expected to facilitate more authentic representation of travel demand. They are:

- Clock time is the key dimension of the simulation system; activities and trips are generated for each person, then properly expanded and loaded on the network along a continuous time axis;
- The series of activities and trips made by an individual are simulated for the period of one day; thus spatial and temporal continuity of activities and travel is maintained and interdependencies among trip attributes are represented; and
- Hägerstrand's prisms are evaluated for each individual in the simulation along with coupling constraints associated with private modes of travel and operating hours of public transit; activities and trips are generated within the confines of these constraints (Kitamura, Fujii et al., 2000).

Incorporating space-time prisms and other constraints into PCATS implies that the travel environment is depicted more realistically for each individual. No individual and household attributes are aggregated by zone to eliminate loss of information and resulting statistical inefficiency and prediction error; in fact, as discussed later in this paper, parts of on-going effort are dedicated to the use of a fine grid system instead of traffic zones. The PCATS-DEBNetS system thus takes full advantage of the advent of fast computing and data management capabilities that are now available and eliminate many of the constraints under which the four-step procedures were constructed in the 50's and 60's.

PCATS[1]

Behind the development of PCATS lies the recognition that various constraints imposed on individuals' activity and travel are not well

[1] The discussions of this section are based on Kitamura, Fujii et al. (2000).

represented in conventional models of travel behavior.[2] Emphasized in PCATS, therefore, are the constraints imposed on the individual's movement in geographical space along time. Because the speed of travel is finite while the time available for travel and activity is limited, the individual's trajectory in time and space is necessarily confined within a certain region. This region is called "Hägerstrand's prism." PCATS first identifies the set of prisms that govern an individual's behavior, then generates activities and trips within each prism while observing coupling constraints involving private travel modes and operating hours of public transit.[3]

PCATS first determines for each individual the periods in which he is committed to engage in a certain activity, or a bundle of activities, at a predetermined location.[4] These periods are called "blocked periods." The complement of the set of blocked periods for an individual comprises the set of "open periods." A Hägerstrand's prism is established for each open period, i.e., given the mode of travel being used, it is determined for each zone whether the zone can be reached within the open period and, if so, how much time can be spent in the zone.

Blocked periods for workers are typically determined by work schedules. As in Damm (1982), then, a worker's day may be assumed to include three prisms: one before work, one during the lunch break, and one after work. The beginning time of the first prism before work and the ending point of the last prism after work are not observed. As discussed later in this paper, stochastic frontier models have been developed to determine unobserved locations of prism vertices (Kitamura, Yamamoto et al., 2000; Pendyala et al., 2002; Yamamoto et al., 2002), and are being incorporated into PCATS.[5]

In PCATS, the probability associated with a daily activity-travel pattern is decomposed into a series of conditional probabilities, each associated with an activity bundle and the trip to reach the location where it will be pursued,

---

[2] For details, see Fujii et al. (1997) and Kitamura & Fujii (1998). The latter reference contains validation results of PCATS.

[3] In the current version of PCATS a zone system is used to construct a prism, which defines the universe of alternatives that exist for each activity-travel decision (activity type, duration, location and travel mode).

[4] In this study the set of activities pursued at a location is called an activity bundle, and the activity bundle is treated as the unit of analysis. Therefore there always is a trip between two successive activity bundles. It is in general unknown from standard travel survey data if an activity bundle is fixed in time and location. In the exercises presented in this paper, work, work-related business and school activities are treated as fixed activities.

[5] The results reported in this paper are based on versions of PCATS that had not incorporated the stochastic frontier models.

given current conditions and the past history of activity engagement. In this sense PCATS has a sequential structure. Models of activity engagement are now being developed to adopt a two-tier structure in which the decision of activity engagement in the respective prisms is first made, then attributes of activities and trips are determined (Kitamura et al., 2001), and will be incorporated into PCATS in the future.

The conditional probability of pursuing an activity bundle is further decomposed to yield the following three sets of model components: 1) activity type choice models, 2) destination and mode choice models, and 3) activity duration models. The activity type choice models are two-tier nested logit models. The upper tier comprises three categories of activity bundles: (A) in-home activities, (B) activities at or near the location of the next fixed activity, and (C) general out-of-home activities. Nested under the first category are two lower-level alternatives: (A-1) engage in out-of-home activities subsequently in the prism, and (A-2) do not engage in out-of-home activities within the prism. The alternatives nested under (C) include the following activity types: meal, social, grocery shopping, comparison shopping, hobbies and entertainment, and sports and recreation. These are defined in terms of the trip purposes as adopted in the travel diary data used for the implementation of the PCATS components. The destination and mode choice models are also nested logit models. Alternative destinations constitute the upper-level alternatives, and available travel modes are nested under each destination alternative.

The type of the first activity bundle in an open period is determined using an activity type choice model. The models are specified such that the probability that a given activity type will be selected decreases as the time available in the prism becomes shorter relative to the distribution of activity durations for that activity type. Given the activity type, a destination-mode pair is next determined using a destination and mode choice model.

As noted earlier, geographical zones are used in the versions of PCATS that are used to produce results reported in this paper. The extension of a prism is evaluated for each travel mode, and destination-mode pairs are excluded from the choice set if they do not fall in the prism for the mode. Again, the amount of time available at the destination is one of the determinants of the choice probability along with the attributes of the destination zone and the trips by respective travel modes to the destination. In the application examples presented later in this paper, travel modes are classified into auto, public transit, walk, and bicycle in the case study for Kyoto, and into auto, public transit, and "others" in the study for Osaka.

Following these, the duration of the activity at the destination is determined using the activity duration model corresponding to the activity type. PCATS initially adopted standard hazard-based duration models with Weibull distributions (Fujii et al., 1997; Kitamura, Fujii et al., 1998, 2000). Standard hazard models, however, do not appear adequate when accounting for the presence of prism constraints because the hazard is presumably influenced by the presence of constraints. For example, it is unlikely that an individual would terminate an activity just a few minutes before the time when a constraint so dictates. Instead, he would extend the activity by a few minutes so that there will be no window of time which is too short to engage in another activity and therefore will be wasted. This situation, then, can be best represented by the distribution of activity durations that has a spike (probability mass) at the time point when the constraint dictates that the activity be terminated. It was suspected that this is one of the reasons why activities tended to be over-generated in an earlier version of PCATS (Iida et al., 2000). To resolve this problem, split population survival models (Schmidt & Witte, 1989) are adopted in PCATS. The split population survival model assumes a two-stage structure. In the first stage, it is determined whether the entire time available in the prism is allocated to the activity (i.e., the activity is terminated when the constraint dictates so). Given that the entire amount of available time is not allocated to the activity, then the time allocated to the activity is determined in the second stage. The version of PCATS applied in Osaka (see "APPLICATION CASES") incorporates split survival models.

Once the attributes of an activity bundle are all determined, the procedure is repeated for the next activity bundle in the same prism. Activity and travel in each open period is thus simulated by recursively applying these model components, while considering the history of past activity engagement. The procedure is repeated until each open period is filled with activities. Note that activity starting and ending times are determined based on the simulated activity durations and, in case of auto trips, travel times obtained from DEBNetS.

DEBNetS[6]

DEBNetS takes as its input auto trips produced by PCATS and replicates the dynamics in network traffic flow along the time axis. DEBNetS determines each vehicle's speed based on simplifying rules, adopts event scanning, and thereby reduces computational requirements. The behaviors of individual vehicles are represented by applying speed (u)-density (k) relationships to

[6] This section is based on Kitamura, Fujii et al. (1998). The development of DEBNetS is presented in Fujii et al. (1998).

roadway segments, with the assumption that each segment is internally homogeneous (the u-k relation proposed by the Bureau of Public Roads is used in the simulation). A network link is therefore divided into a number of homogeneous segments. As a vehicle enters a segment in the simulation, the travel speed of this vehicle is determined based on the number of vehicles in the segment at the time of its entrance, using the u-k relationship specified for the segment. The time the vehicle exits the segment is determined based on the travel speed thus evaluated.

Because of this representation of vehicular movement on a network, computation is required of each vehicle only when a vehicle exits a segment to enter a new segment. If the vehicle is located at a branching point, then a segment to enter is selected based on shortest path information (updated every 15 minutes in the application cases reported later in this paper), and the travel time to traverse that segment is determined. If the vehicle is not at a branching point, then only the travel time on the new segment needs to be calculated. Computation is thus required infrequently for each vehicle, only as often as there are segments on the route taken from the origin to the destination. Because of this, event scanning is adopted in DEBNetS. Trips produced by PCATS are loaded onto the network exactly at the clock time when they are supposed to start.

In the case studies, the CO_2 emissions as a result of a vehicle trip, say trip i, is evaluated as $C_i = \sum_{j \in \Omega_i} L_j f_c(u_{ij}, \delta_i)$, where C_i denotes the total CO_2 emissions by vehicle trip i in the study area, Ω_i the set of link segments on the route of trip i, L_j the length of segment j, u_{ij} the mean travel speed to traverse segment j in trip i, δ_i the type of the vehicle used for trip i, and $f_c(u_{ij}, \delta_i)$ the emissions factor per unit distance for vehicle type δ_i at speed u_{ij}. Vehicles are classified into: small vehicles, which include passenger cars and light duty trucks, and others. The emissions factor, f_c, was developed based on the mean fuel consumption rate by vehicle type and speed class available from the City of Kyoto, and the mean automotive CO_2 emissions per liter of fuel available from the Japanese Ministry of Environment. The emissions by individual trips thus estimated are aggregated in the simulation to produce total emissions in the study area by time of day.

Many decisions simulated in PCATS, e.g., travel mode choice and destination choice, are in part based on auto travel time information, which is output from DEBNetS. Auto trips that are fed to DEBNetS, on the other hand, are output from PCATS. It is in general the case that travel times used in PCATS are not identical to those in DEBNetS. For this reason, these two simulators are applied iteratively until convergence in travel times is obtained.

DEBNetS Enhancement and Validation[7]

Earlier applications of the PCATS-DEBNetS system have indicated that the upstream propagation of congestion must be appropriately represented, and that computational time must be reduced for practical application of the simulation package. For the former, procedures are introduced to regulate the rate of the outflow from the last (most downstream) segment of a link to better reflect the capacity of the traffic lights as well as congestion in the downstream link.

Because of the way traffic flow is simulated in DEBNetS, computational requirements can be effectively reduced by event scanning. Event scanning, however, calls for the task of sorting events according to the time of their occurrences. The time required for this sorting increases as the number of vehicles in simulation increases, and can be a substantial portion of the total computation time. A scanning scheme, which is a combination of event scanning and periodic scanning, is therefore adopted to reduce this sorting time; traffic flow is simulated by event scanning, while events are sorted and vehicle data are updated by periodic scanning. This reduced computational time quite substantially to about one-seventh.[8]

Using the empirical morning peak benchmark traffic data made public for the purpose of validating simulation models by the Road Traffic Simulation Systems Clearing House,[9] a validation run was performed with DEBNetS. With a value of 0.95, the correlation coefficient between simulated hourly link traffic volumes and observed volumes indicates good fit (Fig. 1). Correlation coefficients of 10-minute traffic volumes have an average of 0.830 between 8:00 A.M. and 10:00 A.M (Fig. 2).[10] Link traffic speeds observed with

---

[7] This section is based on Kikuchi et al. (2002)

[8] The application to the City of Osaka (Kikuchi, Fujii et al., 2000; Kitamura et al., 2001) involved a network with 2,994 links, 1,050 nodes, 292 centroids and a total of 990,575 trips. A DEBNetS run for a 24 hour period took 17 hours and 31 minutes using a personal computer with two Pentium III, 800 MHz CPUs and Sun OS 5.8, with shortest path trees updated every 15 simulation minutes. With the improved scanning scheme, the same run took 2 hours and 25 minutes with shortest path trees updated very 5 simulation minutes.

[9] http://trans1.ce.it-chiba.ac.jp/ClearingHouse/

[10] Correlation coefficients between 7:50 A.M. and 7:59 A.M. are excluded as they were heavily influenced by the initial condition of the simulation, which started with an empty network. Correlation coefficients have an average of 0.859 between 8:00 A.M. and 8:29 A.M., 0.842 between 8:30 A.M. and 8:59 A.M., 0.822 between 9:00 A.M. and 9:29 A.M., and 0.798 between 9:30 A.M. and 9:59 A.M.

10-min. intervals, on the other hand, are not well represented with an overall correlation coefficient of 0.32, suggesting needs for further improvement of the simulator.

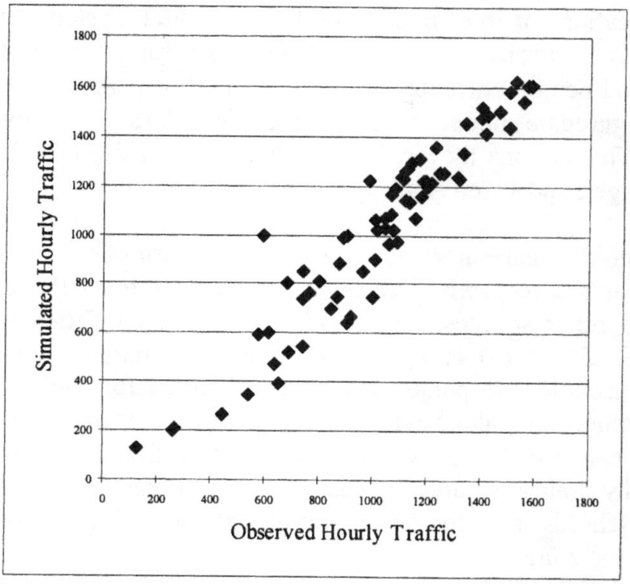

Figure 1. Correlation between Observed and Simulated Hourly Link Traffic Volumes.

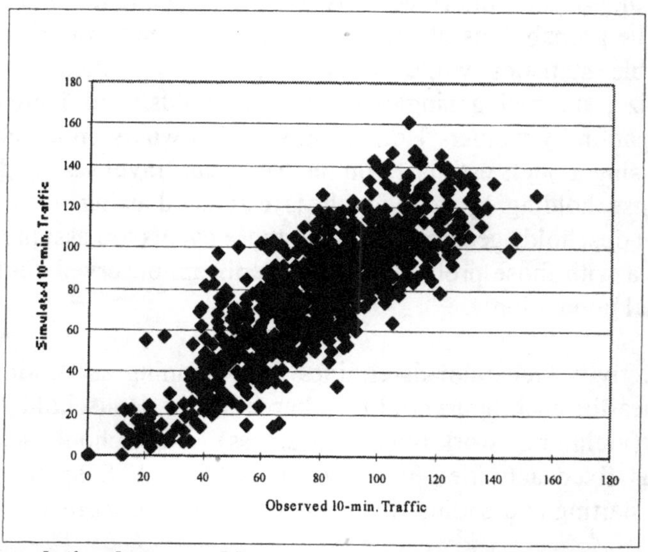

Figure 2. Correlation between Observed and Simulated 10-min. Link Traffic Volumes.

HAGS[11]

The case study for Osaka described in "APPLICATION CASES" of this paper is an application of the micro-simulation model system to long-range forecasting with a horizon year of 2020. This calls for the generation of future household and person attributes at the household or person level by zone, so that the disaggregate model components of the micro-simulation system can be applied. This is done by HAGS, which comprises: Household Distributor, Household Ager, and Fixed Activity Generator.[12]

The Household Distributor determines the distribution of attributes of households in the respective zones based on data from the census, travel surveys, and other sources. An iterative proportional fitting (IPF) method (Beckman et al., 1996) is applied to base-year marginal distributions of pertinent household and person attributes, obtained for each zone from the census and other data, along with their area-wide joint distributions obtained from base-year travel survey data, to yield a frequency distribution of households by their attributes for each zone. Each zone is then populated by cloning households from the travel survey data according to the distribution obtained for the zone.

The Household Ager simulates the aging of the base-year households generated by the Distributor through the horizon year. Simulated events include: birth, death, marriage, divorce, employment, and residential relocation. The probabilities of these events are obtained from the census and other available statistics while accounting for the trends of decreasing household size, increasing single-person households, and increasing labor force participation by women. Household vehicle ownership is endogenously determined using models estimated on the base-year travel survey data. Future driver's license holding by individuals is estimated using a simple cohort model. New households generated through these events are assumed to stay in the study area with those probabilities that replicate observed social changes in the regional population.

The Fixed Activity Generator determines the beginning and ending times of fixed activities for each household member from the Household Ager. Since only work (including work-related business) and school activities are considered as fixed activities in the versions of PCATS so far developed, work/school starting and ending times are generated probabilistically for each

[11] This section is based on Kitamura, Fujii et al. (1998).
[12] Details are reported in Nishida et al. (2000)

worker or student based on their distributions observed in the base-year travel survey data. The household and person attributes thus generated are input to PCATS in the micro-simulation.

EXTENSION

This section presents some of the improvements being made to the PCATS-DEBNetS system. These improvements are expected to make the simulation system more realistic and versatile.

Incorporating Models of Prism Vertex Locations

One of the important features of PCATS is its representation of various constraints imposed on individuals' activity and travel. In particular, Hägerstrand's prisms are evaluated for each individual to represent the spatial and temporal constraints imposed on him. The vertex of a prism (its beginning or ending point), however, is not always observable.[13] In earlier versions of PCATS, a fixed time of day is arbitrarily selected for unobserved prism vertex. This, of course, is not a desirable approach because there is no theoretical or empirical basis for the selection of the particular time of day, and also because there is no reason to believe that every individual's prism vertex is located at the same time point.

Stochastic frontier models have been developed to estimate the location on a time axis of an unobserved prism vertex (Kitamura, Yamamoto et al., 2000; Pendyala et al., 2002; Yamamoto et al., 2002). A stochastic frontier model depicts the value of an unobserved variable based on the relation that the unobserved variable is greater (or smaller) than some observed variable. In this case, the relations hold that the location on a time axis of an unobserved origin vertex is earlier (smaller) than the beginning time of the first trip observed in the prism, and that the location of an unobserved terminal vertex is later (greater) than the ending time of the last trip in the prism.

Results of model estimation have indicated that an hour of commute pushes back the terminal vertex of a worker's evening prism by 52.4 minutes; a female member of a household with children has a terminal vertex that is about 100 minutes earlier than that of her male counterpart; and the terminal

[13] The terminal vertex of a worker's morning prism before work, is set in PCATS at the work beginning time, while its origin vertex is unobservable. For a worker's evening prism after work, on the other hand, the origin vertex can be set at the work ending time, while the terminal vertex is not observable.

vertex of a younger single person tends to be located later. It is also shown that a worker's terminal vertex is located 96 minutes earlier on a day when he is not working. It has also been shown there exists a mean difference of 1.46 hours between the observed trip ending time and the estimated location of the terminal vertex of a worker's evening prism (Kitamura, Yamamoto et al., 2000).[14]

In the studies referenced above, the location of a prism vertex was estimated by the stochastic frontier models using person and household attributes and commute characteristics as explanatory variables. The causal relation assumed in these stochastic frontier models, where commute characteristics affect the location of the prism vertex, is reversal of that assumed in PCATS; in PCATS, prism constraints influence commute characteristics. Models are therefore re-estimated for PCATS implementation without commute characteristics as explanatory variables, except for commute distance which is treated as exogenous.

Earlier versions of PCATS produced too many trips in later hours of the day, which can be attributed at least in part to the arbitrary value of 3:00 A.M. used as the location of the terminal vertex of an evening prism. This problem can be resolved by introducing the modified stochastic frontier models into PCATS and adopting more realistic values for prism vertices.

Adoption of a Fine Grid System: From Zones toward Coordinates

As earlier discussions of this paper have alluded to, the primary reason for the use of traffic zones in conventional demand forecasting models, including the four-step procedures, is the computational limitations that existed in the 50's

---

[14] There remains the question of whether the vertex location as depicted by the stochastic frontier models in fact represents the prism constraint in the strict sense of Hägerstrand. One could argue that what the models depict may represent a threshold which an individual subjectively holds as the earliest possible starting time or the latest possible ending time for a trip, which may not necessarily coincide with actual constraints that are governing. For example, a commuter may believe that he cannot possibly leave home before a certain time in the morning, but he may do so for a business trip. Models of prism vertices are estimated in this study with empirical data without any information on the individual's beliefs or perceptions of prism constraints. Yet, observed travel behavior is governed by subjective beliefs and perceptions, e.g., "I must return home by midnight" or "I cannot possibly leave home before 6:30 A.M." Thus some ambiguity does exist about the nature of what the models depict. It is nonetheless considered reasonable to assume here that they offer a useful measure for the practical purpose of determining the earliest possible departure time or latest possible arrival time for a trip (Kitamura, Yamamoto et al., 2000).

and 60's. Information collected at household and individual levels through travel surveys has been aggregated to zonal values (e.g., means or medians) and some models are estimated using zonal statistics. It has been pointed out that this leads to gross statistical inefficiency because most information is lost in the process of aggregation.

Less attention has been directed to the problem of error and low resolution in spatial representation that result from the use of traffic zones. A traffic zone typically covers quite a large area, introducing inaccuracies in representing level-of-service (LOS) attributes of a trip and creating difficulties in handling short trips typically made on foot or by bicycle.

The most typical case for the former may be found in the representation of walking in a transit trip. Since all LOS attributes are zone-based, one representative value is assigned to the walking distance to a transit stop for all transit trips generating from a zone. Walking distance, however, varies greatly from location to location within a zone. Using a representative value may lead to serious errors in analysis. This applies to non-linear models, including discrete choice models, even when the representative value is an accurate zonal average.

Short trips are difficult to represent with a zone system because they tend to be internal to a zone, or, "intra-zonal." Again, the same, "representative," zonal values are assigned to all intra-zonal trips. All intra-zonal trips, then, would have the same probability of being made by auto or on foot. This would make the analysis unrealistic, especially when zones are large as is the case in the fringe of a metropolitan area. In fact mode choice models used to exclude non-motorized modes of travel in most planning regions. Presumably this is at least in part due to the difficulty in dealing with short trips within the framework of conventional zone systems. One approach to overcome these shortcomings and extend the applicability of the model, is to adopt finer zones, or, a coordinates system to represent spatial location.

Attempts have been made to develop a location reference system and to construct models on it as components of PCATS. In Kikuchi, Kobata et al. (2000), a fine grid system is defined and the study area is subdivided into 10 m × 10 m parcels, and the location of each parcel is referenced using a coordinates system. The study area is a rectangular area (13 km east to west, 11 km south to north) that centers around the central business district of Kyoto, Japan. The grid system produced about 140 million 10 m × 10 m parcels. After eliminating parcels on which no opportunities for activities exist (e.g., river water, forests, railroad tracks, roadways), approximately 74 million parcels qualify as potential trip destinations (Kikuchi, Fujii et al., 2001).

The location of railroad stations, bus stops and other transportation facilities are input to the database using geographical information system (GIS) software.[15] Parcel-to-parcel LOS attributes are determined first by simulating auto and bus traffic on a network of major roadways for each hour of the day.[16] This simulation produced the travel time, number of transfers, and transit fare between each pair of nodes on the network. LOS information between 10 m × 10 m parcels is obtained by systematically inter/extrapolating the LOS data obtained from the simulation. The duration of a walk or bicycle trip is evaluated by applying a constant to the parcel-to-parcel straight-line distance obtained by the GIS software.

The land use data developed in the study is based on information compiled for 3,635 neighborhood units in the City of Kyoto. Each neighborhood unit typically comprises of housing units on the two block faces that share a street segment. Since no information is available on how land uses are distributed within each neighborhood unit, they are uniformly distributed to parcels that lie within a neighborhood unit. Obviously this is an approximation. Ideally the land use database should be developed based on information on each plot of land, as is done in Portland, Oregon.

Models of destination and mode choice are developed based on the grid system and the land use database thus developed (Kikuchi, Kobata et al., 2000), and applied to evaluate selected TDM measures (Kikuchi, Fujii et al., 2001). Such applications call for the development of methodologies to efficiently handle the huge number of alternatives that are involved in choice models defined on the grid system. The following subsection is concerned with such methodologies.

Application of Markov Chain Monte Carlo (MCMC) Methods to Handle Colossal Choice Sets

Even with the advent of fast computers, simulating discrete choices requires a substantial amount of time when the choice set is large, because evaluating choice probabilities for all alternatives in the choice set requires a substantial amount of computation time. For example, consider the multinomial logit model, where the probability that alternative j will be chosen by individual i from the choice set, Ω_i, is given as $PD_i(j) = \exp(\beta' X_{ij}) \big/ \sum_{l \in \Omega_i} \exp(\beta' X_{il})$. To simulate a choice according this choice model, one must evaluate $PD_i(j)$

---

[15] The GIS software used in this study is SIS (Spatial Information System) V5.2, Informatix, Inc.

[16] At this point, this simulation is independent of the network traffic simulation of DEBNetS.

for all alternatives in Ω_i, which requires that the denominator of $PD_i(j)$ be evaluated. This, however, involves computing J_i exponential functions, where J_i is the number of alternatives in Ω_i. Although this wouldn't be a problem when zones are used as the alternatives of destination choice, it imposes serious computational problems when J_i is as large as hundreds of thousands.

When destination choice is formulated as a multinomial logit model, the Markov Chain Monte Carlo (MCMC) algorithm applies quite well to simulate choices. The algorithm is shown schematically in Fig. 3. In the figure, r_i refers to an alternative. Once the procedure is repeated large enough a number of times and the influence of initial condition has diminished, r_i's can be drawn with large intervals to form a sample of alternatives that are drawn according to the choice probabilities as indicated by the model.[17]

Most critical in this application is the fact that the algorithm only requires the ratio of two choice probabilities, but not choice probabilities themselves. The ratio,

$$\gamma = PD_i(r_j)\big/PD_i(r_k) = \exp(\beta' X_{ir_j})\big/\exp(\beta' X_{ir_k}),$$

does not involve the denominator of the logit choice probability, and thus can be evaluated very easily. This substantially reduces the computational requirements for choice simulation. The accuracy of the algorithm was tested by simulating destination choice in an abstract uniform circular city where the distribution of destination locations can be theoretically determined. The result indicated that the MCMC algorithm produced a distribution of destination locations that is statistically not different from the theoretical distribution (Kikuchi, Yamamoto et al., 2001).

An example of simulation results is presented in Fig. 4. The figure shows a set of destination locations that are chosen under the existing condition, and another set of locations that are chosen when the service level of public transit is improved. In this simulation, the individual who is located at "S," whose next fixed activity must take place at "N," is choosing a destination for a discretionary activity. It can be seen that the distribution of destination locations expands with the improvement in service level. In fact the sum of travel times from the current location (S) to the destination, then to the next

[17] Previous applications of MCMC algorithms in the transportation field can be found in Hazelton et al. (1996) and Yamamoto et al. (2001). Also see Hajivassiliou et al. (1996) and Chiang et al. (1999).

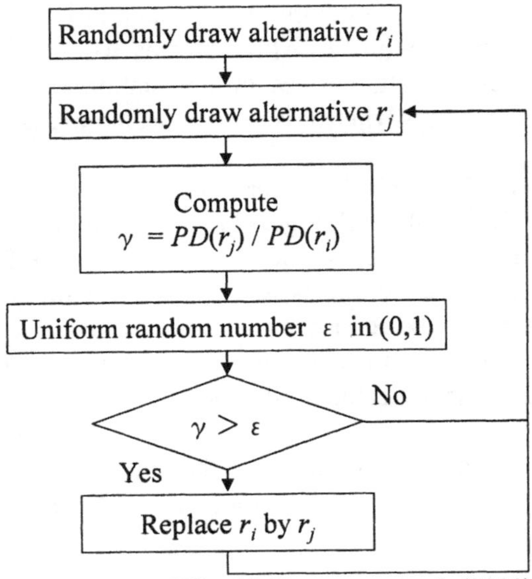

Source: Kikuchi, Yamamoto et al. (2001)

Figure 3. MCMC Algorithm.

fixed activity (N) increases from an average of 4,396 m to an average of 4,533 m with the improvement. The variance in travel distance increases from 2.1×10^6 to 3.6×10^6.

Another improvement being made is concerned with destination choice. The destination choice models of PCATS are being modified to better reflect recent findings on destination choice behavior. Using data from the Los Angeles metropolitan area, Kitamura, Chen et al. (1998) show that the travel time from a potential destination to the home base is as influential a factor of destination choice as the travel time from the origin to the potential destination. It has also been shown that closer locations tend to be chosen toward the end of the day, and distance to a destination is positively correlated with the time spent there. The study, however, does not incorporate the space-time prism into its analytical framework. The destination choice models in PCATS are now being reformulated to reflect these findings and improve their predictive capability. In addition, as noted earlier, the two-tier model of activity engagement and activity attributes is being developed for implementation in PCATS.

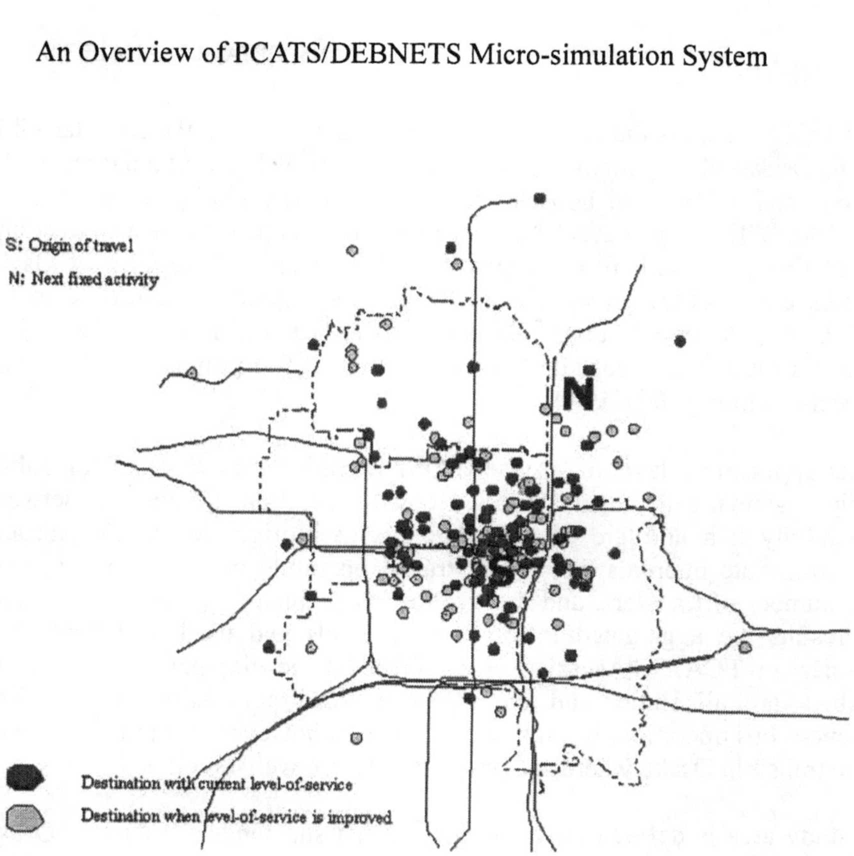

Source: Kikuchi, Yamamoto et al. (2001)
Figure 4. Distribution of Simulated Destination Locations.

APPLICATION CASES

The PCATS-DEBNetS system was first applied to evaluate the effectiveness of several transportation planning measures in reducing CO_2 emissions in the City of Kyoto, Japan (Fujii et al., 2000; Kitamura, Fujii et al., 1998, 2000). It was then applied in Osaka (Iida et al., 2000; Kawata et al, 1999; Kitamura, Fujii et al., 2000), Toyota (Kikuchi et al., 1999) and Ashiya, Japan.[18] It is currently being implemented in Tampa, Florida. The Osaka application deployed the demographic simulator, HAGS, to facilitate long-term forecasting. Drawing from Kawata et al. (1999), Iida et al. (2000) and Kitamura, Fujii et al. (2000), results from the Osaka application are presented in this section. This case study is based on a zone system.

[18] Also see Arentze et al. (2001)

The City of Osaka is the largest of the three major cities in the Kei-Han-Shin (Kyoto-Osaka- Kobe) metropolitan area of Japan. It has a population of 2.6 millions and 1.1 million households (as of October, 1995), and an area of 221.27 km^2. The city is served by ten lines of the Japan Rail's networks and 15 rail lines operated by other private rail companies. The City of Osaka operates eight subway lines and a people mover system. According to the 1990 household travel survey, the share of auto trips is about 17%, while public transit accounts for about 34% of all trips. The population of the city has been declining since 1990.

In this application, level-of-service (LOS) variables are evaluated for public transit systematically using public transit operation schedules, network connectivity data, and fare schedule data. For each origin-destination pair and with 10-minute intervals throughout transit operating hours, the total travel time, number of transfers, and transit fare are evaluated for the fastest route. The results are aggregated into 2-hour intervals and used as transit LOS variables in PCATS. At this stage of model development, transit LOS variables are all static, and the effect of road congestion, which does influences bus operation, is not represented (bus, however, is a minor mode of public transit in Osaka where subway networks are well developed.)

The study area is defined by the ring road that surrounds the City of Osaka and areas in its periphery. Residents in the study area, and those who reside outside the area but worked or studied in the area, are included in the simulation. The records of a total of 103,462 individuals were drawn from the results of the 1990 household travel survey. The home and work locations, other person and household attributes, and the fixed (work and school) activities of these individuals are retained, but all other activities and trips are deleted from the records. The resulting records, supplemented with land use data and network data, constitute the base of the simulation. A PCATS run on this database took approximately 6 minutes on a Pentium II (300 MHz) Linux machine. Outputs of PCATS are quite similar to trip records in a household travel survey data set, and indicate the purpose, origin, destination, mode, beginning and ending times of each trip, from which the location, beginning time and duration of each out-of-home activity can be inferred.

PCATS was first run to examine how well it replicates the 1990 data. The resulting mean number of trips per person per day is 2.87 for workers and 2.49 for non-workers. These compare with the average trip rates of 2.75 and 2.63, respectively, obtained from the survey data. The error in the PCATS prediction is 4.7% for workers and –5.3% for non-workers. Although these results still represent over- or under-prediction of quite a few trips, they at the same time represent a substantial improvement in the model system's accuracy compared with the earlier results from the Kyoto application.

The number of trips generated is shown by time of day in Figs. 5 and 6 for workers and non-workers, respectively. Overall trip generation is well replicated along the time axis by PCATS. Comparing the two figures indicates that trip generation by workers is better replicated than that by non-workers. This is presumably because non-workers tend to have larger unblocked periods and more degrees of freedom in their activities, making prediction more difficult.

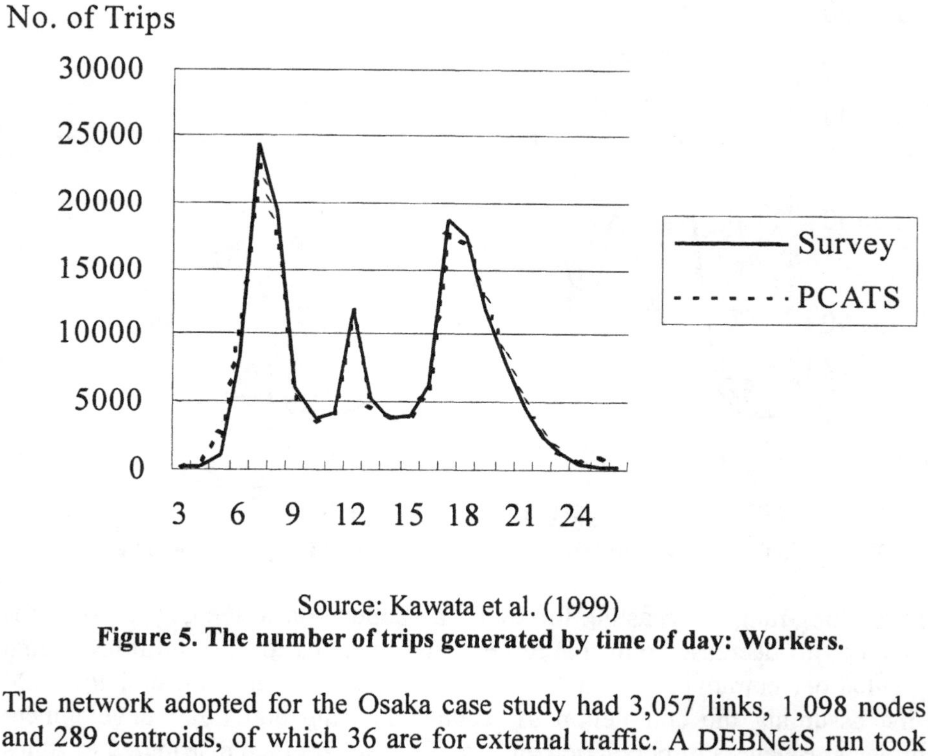

Source: Kawata et al. (1999)
Figure 5. The number of trips generated by time of day: Workers.

The network adopted for the Osaka case study had 3,057 links, 1,098 nodes and 289 centroids, of which 36 are for external traffic. A DEBNetS run took approximately 30 minutes using one processing unit of Fujitsu VPP-500.[19] A preliminary comparison of travel times indicated that DEBNetS overestimated travel speeds on toll roads (an average of 36.8 km/h was estimated while the observed 1994 average was 28.5 km/h), while reasonable estimates were obtained for surface roads (19.7 vs. 18.8 km/h). Presumably this was because the same assignment method as adopted by the regional planning agencies was used in this study, and the expressway tolls were converted to equivalent

[19] VPP-500 consists of 15 processing units, each having a capability of 1.6G flops.

travel times and added to the actual link travel times in the assignment. As a short-term solution to reduce the discrepancies, link constants were calibrated using a simple algorithm, and then added to the link travel time. This yielded an average estimated expressway travel speed of 29.7 km/h (Kikuchi, Fujii et al., 2000). In the long run, more behavioral route choice models will be incorporated into DEBNetS. The average absolute prediction error for daily traffic volume was 11.4% on 7 major links on the network, and a prediction error of 8.7% was obtained for screen-line traffic on 8 major screen lines.

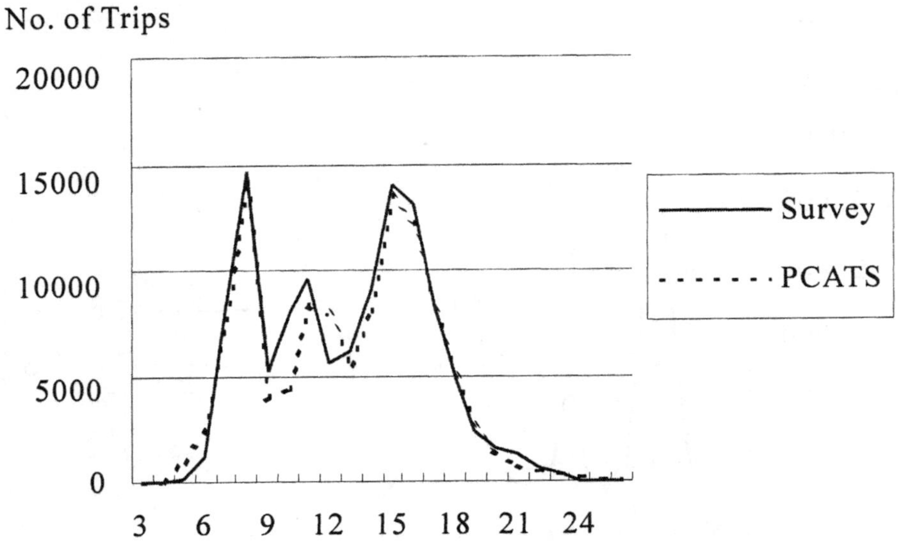

Source: Kawata et al. (1999)

Figure 6. The number of trips generated by time of day: Non-workers.

In the long-range forecasting for 2020, the population of the City of Osaka is assumed to decrease from the current 2.60 millions to 2.51 millions. Zonal population, employment and land use characteristics are adjusted for 2020 first assuming the completion of planned housing and other development projects. Balances of population and employment are then distributed to zones. Weights for individuals are determined based on the population age distributions by municipality for 2020, which were obtained using a cohort method. These weights are applied to household members aged by HAGS. As noted earlier, HAGS reflects trends towards smaller households, later marriage, and increasing labor-force participation by women in Japan.

The following policy packages are examined in the study:

Package 0. Do nothing
Package 1. Execute planned infrastructure development projects
Package 2. Package 1 plus circumferential and radial arterials and other facilities to disperse through traffic
Package 3. Package 1 plus new rail lines, LRT, and other measures to reduce auto use
Package 4. Package 1 plus minimal infrastructure projects (a circumferential roadway and LRT) and introduction of congestion pricing and transit malls in the central city.

Despite the decline in the total population in 2020, the number of auto trips increases substantially (Fig. 7). This is due to increases in driver's license holding among women and older individuals. Another factor is the large-scale residential development projects planned in the waterfront area, which will lead to increases in population in areas where public transit service levels are low. Vehicle-kilometers traveled within the City of Osaka, however, do not increase proportionally with the number of auto trips (Fig. 8). In fact, only Package 2, which is auto-oriented, yields a vehicle-kilometer total that is larger than that of Package 0 (do-nothing alternative). Reasons for this are difficult to pinpoint with the results so far tabulated, but it may be the case that dispersed residential locations tend to produce either shorter trips, or more trips that are made outside the city boundaries and therefore are not included in the tabulation here. It is also conceivable that the various measures implemented in the respective packages tend to shorten the length of auto trips.

Fig. 9 presents estimated CO_2 emissions within the City of Osaka for the respective policy packages. From Figs. 8 and 9, it can be seen that vehicle travel can be reduced by implementing the planned infrastructure projects (Package 1) and by developing public transit and adopting measures to suppress auto use (Packages 3 and 4). Only auto-oriented Package 2 produces more CO_2 emissions than does Package 0. The results make evident that investing in road facilities promotes more auto use. Interesting is the result that Package 4, which involves the development of a circumferential road, has more vehicle-kilometers traveled than Package 3, which involves only transit development; yet the former package results in less CO_2 emissions than the latter because of the implementation of congestion pricing. The effects of auto restriction measures in Package 4 can also be seen in Fig. 10 which shows vehicle-kilometers traveled within the CBD area; Package 4 has CBD vehicle-kilometers that are about 20% less than those of Package 0, and about 14% less than those of Package 3.

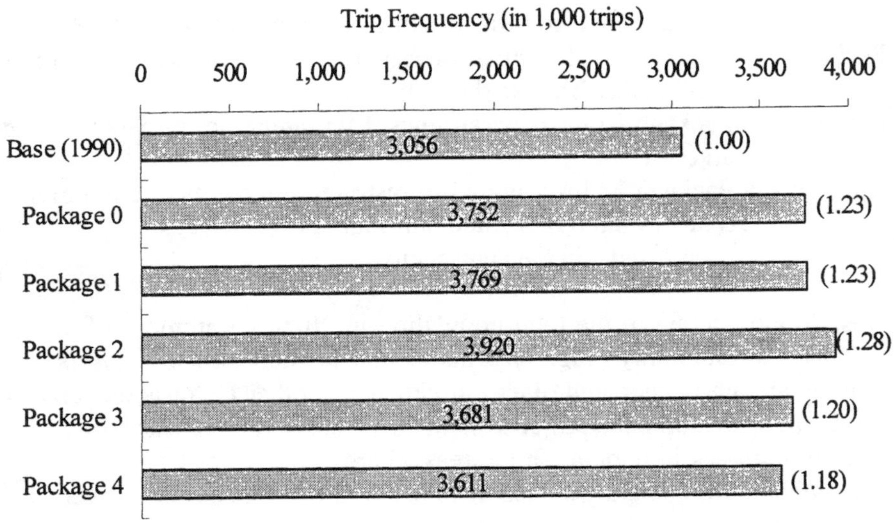

Source: Kawata et al. (1999)

Figure 7. Auto Trips Generated in the City of Osaka.

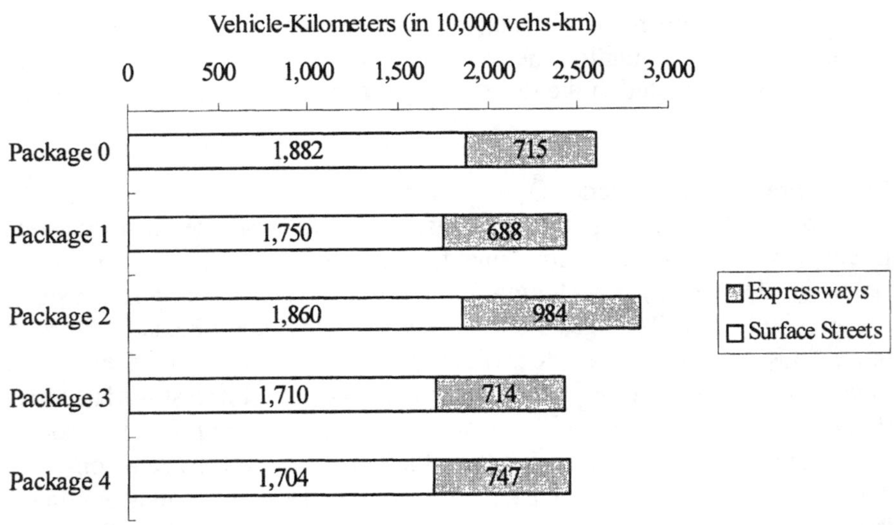

Source: Kawata et al. (1999)

Figure 8. Vehicle-Kilometers Traveled within the City of Osaka.

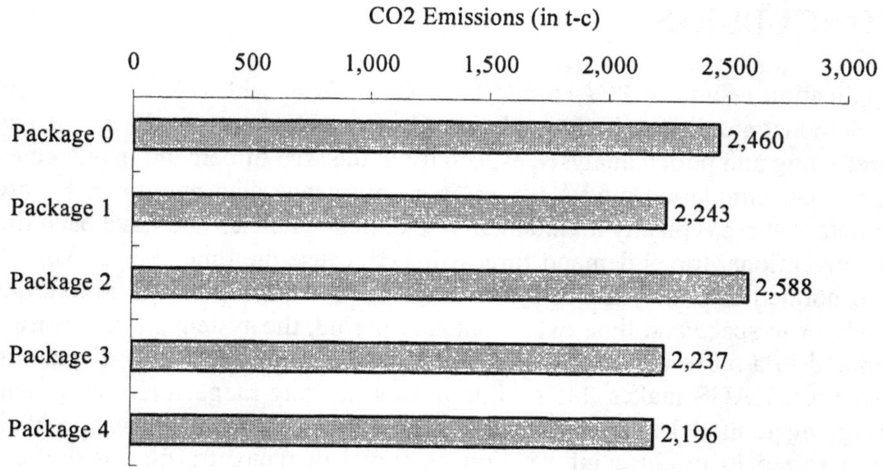

Source: Kawata et al. (1999)

Figure 9. CO_2 Emissions by Vehicles within the City of Osaka.

The infrastructure investment and transportation control measures in these packages tend to improve traffic flow. The mean auto travel speed increases by about 11% with Package 1 and Package 3, and by 21% with Package 4 which includes auto restriction measures in the CBD. The most auto-oriented Package 2 produces the smallest improvement in travel speed of 9%.

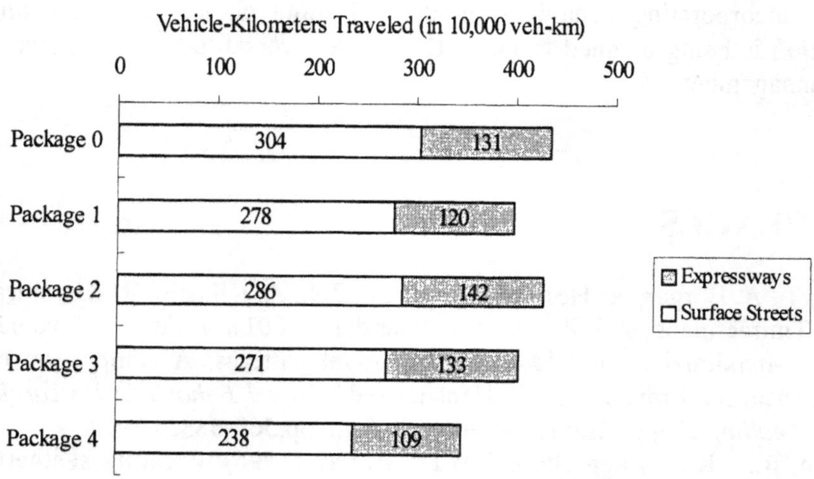

Source: Kawata et al. (1999)

Figure 10. Vehicle-Kilometers Traveled in Osaka CBD.

CONCLUSION

Application results of PCATS and DEBNetS, along with HAGS, are starting to demonstrate that micro-simulation is a practical approach for demand forecasting and policy analysis, especially in the area of demand management. The micro-simulation model system is by no means data hungry; it is based on data that are typically maintained by planning agencies and have been used in conventional travel demand forecasting. Because the time axis is explicitly incorporated into the model system and because it represents individuals' behavior in space and time over a one-day period, the system simulates travel demand in a more coherent manner and applies to a wide range of planning measures. HAGS makes this tool applicable to long-range forecasting using disaggregate models. The rich data the simulation system offers as output can be analyzed to produce information on travel demand, traffic condition, or pollutant emissions.

The application examples of the micro-simulation system presented in this paper are one of the earliest attempts in which all steps of the four-step procedures are performed by micro-simulation. As such, the model system is yet to be refined. This paper has described some of the refinements being undertaken to improve the PCATS/DEBNetS system. It is hoped these efforts will make the micro-simulation system a more practical, accurate and versatile tool for urban travel demand forecasting and policy analysis. In addition, incorporating behavioral models of route choice with real-time information is being planned to make DEBNetS a versatile tool for network traffic management.

REFERENCES

Arentze, T., A. Borgers, F. Hofman, S. Fujii, C. Joh, A. Kikuchi, R. Kitamura, H. Timmermans and P. van der Waerden (2001). Rule-based versus utility-maximizing models of activity-travel patterns: A comparison of empirical performance, In D. Hensher (ed.) *Travel Behaviour Research: The Leading Edge*, Elsevier Science, Oxford, pp.569-583.

Beckman, R.J., K.A. Baggerly and M.D. McKay (1996). Creating synthetic baseline populations, *Transportation Research A*, **30A**, 415-429.

Chiang, J., S. Chib and C. Narasimhan (1999). Markov Chain Monte Carlo and models of consideration set and parameter heterogeneity, *Journal of Econometrics*, **89**, 223-248.

Damm. D. (1982). Parameters of activity behavior for use in travel analysis, *Transportation Research*, **16A**(2), 135-148.

Hägerstrand, T. (1970). What about people in regional science?, *Papers of the Regional Science Association*, **24**, 7-21.

Hajivassiliou, V., D. McFadden and P. Ruud (1996). Simulation of multivariate normal rectangle probabilities and their derivatives: theoretical and computational results, *Journal of Econometrics*, **72**, 85-134.

Hazelton, M.L., S. Lee and J.W. Polak (1996). Stationary states in stochastic process models of traffic assignment: a Markov Chain Monte Carlo approach, In J.-B. Lesort (ed.) *Proceedings of the 13th International Symposium on Transportation and Traffic Theory*, Pergamon, Oxford, pp. 341-357.

Fujii, S., A. Kikuchi and R. Kitamura (2000). A micro-simulation analysis of the effects of transportation control measures to reduce CO_2 emissions: a case study in Kyoto City, *Traffic Engineering*, **35**(4), 11-18 (in Japanese).

Fujii, S., M. Okushima, A. Kikuchi and R. Kitamura (1998). Development of a network flow simulator and evaluation of travel time, *In the Proceedings of the 17th Annual Meeting of the Japanese Society of Traffic Engineers*, pp. 694-695 (in Japanese).

Fujii, S., Y. Otsuka, R. Kitamura and T. Monma (1997). A micro-simulation model system of individuals' daily activity behavior that incorporates spatial, temporal and coupling constraints, *Infrastructure Planning Review*, **14**, 643-652 (in Japanese).

Iida, Y., M. Iwabe, A. Kikuchi, R. Kitamura, K. Sakai, Y. Shiromizu, D. Nakagawa, M. Hatoko, S. Fujii, T. Morikawa and T. Yamamoto (2000). Micro-simulation based travel demand forecasting system for urban transportation planning, *Infrastructure Planning Review*, **17**, 841-847 (in Japanese).

Kawata, H., Y. Iida and Y. Shiromizu (1999). Case study of evaluation for comprehensive transportation policy, *The Proceedings of the Infrastructure Planning Review Annual Meeting*, **22**(1), 511-514 (in Japanese).

Kikuchi, A., S. Fujii and R. Kitamura (2001). Evaluation of transportation policies by micro-simulation of individuals' behaviors on a coordinates system, *City Planning Review*, **36**, 577-582 (in Japanese).

Kikuchi, A., S. Fujii, Y. Shiromizu and R. Kitamura (2000). Calibration of DEBNetS on a large-scale network, In *the proceedings of the 20th Annual Meeting of the Japanese Society of Traffic Engineers*, Tokyo, pp. 49-52 (in Japanese).

Kikuchi, A, Y. Kato, T. Macuchi, S. Fujii and R. Kitamura (2002).

Improvement and verification of dynamic traffic flow simulator "DEBNetS", *Infrastructure Planning Review*, **19** (in press, in Japanese).

Kikuchi, A., A. Kobata, S. Fujii and R. Kitamura (2000). A mode and destination choice model on a GIS database: from zone-based toward coordinates-based methodologies of spatial representation, *Infrastructure Planning Review*, **17**, 841-847 (in Japanese).

Kikuchi, A., T. Yamamoto, K. Ashikawa and R. Kitamura (2001). Computation of destination choice probabilities under huge choice sets: application of Markov Chain Monte Carlo algorithms, *Infrastructure Planning Review*, **18**(4), 503-508 (in Japanese).

Kikuchi, A., R. Kitamura, S. Kurauchi, K. Sasaki, T. Hanai, T. Morikawa,S. Fujii and T. Yamamoto (1999). Effect Analysis of Transportation Policies using Micro-Simulation Method - A Case Study of Toyota City -, *The Proceedings of the Infrastructure Planning Review Annual Meeting*, **22**(1), 817-820 (in Japanese).

Kitamura, R., C. Chen and R. Narayanan (1998). The effects of time of day, activity duration and home location on travelers' destination choice behavior, *Transportation Research Record*, **1645**, 76-81.

Kitamura, R. and S. Fujii (1998). Two computational process models of activity-travel behavior, In T. Gärling, T. Laitila and K. Westin (eds.) *Theoretical Foundations of Travel Choice Modelling*, Pergamon Press, Oxford, pp. 251-279.

Kitamura, R., S. Fujii, A. Kikuchi and T. Yamamoto (1998). Can TDM make urban transportation "sustainable"?: A micro-simulation study, Paper presented at *International Symposium on Travel Demand Management*, Newcastle, UK.

Kitamura, R., S. Fujii, T. Yamamoto and A. Kikuchi (2000). Application of PCATS/DEBNetS to regional planning and policy analysis: Micro-simulation studies for the Cities of Osaka and Kyoto, Japan, In *the Proceedings of Seminar F, European Transport Conference 2000*, pp. 199-210.

Kitamura, R., T. Yamamoto, K. Kishizawa and R.M. Pendyala (2000). Stochastic frontier models of prism vertices, *Transportation Research Record*, **1718**, 18-26.

Kitamura, R., T. Yamamoto, K. Kishizawa and R.M. Pendyala (2001). Prism-based accessibility measures and activity engagement, Paper presented at the *80th Annual Meeting of the Transportation Research Board*, Washington, D.C., January.

Nishida, S., T. Yamamoto, S. Fujii and R. Kitamura (2000). A household attributes generation system for long-range travel demand forecasting with disaggregate models, *Infrastructure Planning Review*, **17**, 779-787 (in Japanese).

Pendyala, R.M., T. Yamamoto and R. Kitamura (2002). On the formation of time-space prisms to model constraints on personal activity-travel engagement, *Transportation*, **29**(1), 73-94.

Schmidt, P. and A. Witte (1989). Predicting criminal recidivism using split population survival time models, *Journal of Econometrics*, **40**, 141-159.

Yamamoto, Y., R. Kitamura and R.M. Pendyala (2002). Comparative analysis of time-space prism vertices for out-of-home activity engagement on working days and non-working days, Submitted to *Geographical Analysis*.

Yamamoto, T., R. Kitamura and K. Kishizawa (2001). Sampling alternatives from a colossal choice set: an application of the MCMC algorithm, *Transportation Research Record*, **1752**, 53-61.

FSC
www.fsc.org
MIX
Papier | Fördert
gute Waldnutzung
FSC® C083411

Kolibri 360 GmbH
Ferdinand-Jühlke-Straße 7
99095 Erfurt, Deutschland
produktsicherheit@kolibri360.de